Current Developments in Mathematical Sciences

(Volume 3)

Fractional Calculus: New Applications in Understanding Nonlinear Phenomena

Edited by

Mehmet Yavuz

*Department of Mathematics and Computer Sciences,
Necmettin Erbakan University,
Konya 42090,
Turkey*

&

Necati Özdemir

*Department of Mathematics,
Balıkesir University,
Çağiş Campus, 10145 Balıkesir,
Turkey*

Current Developments in Mathematical Sciences

Volume # 3

Fractional Calculus: New Applications in Understanding Nonlinear Phenomena

Editors: Mehmet Yavuz & Necati Özdemir

ISSN (Online): 2589-272X

ISSN (Print): 2589-2711

ISBN (Online): 978-981-5051-93-3

ISBN (Print): 978-981-5051-94-0

ISBN (Paperback): 978-981-5051-95-7

need for a court order if at any point you breach any terms of this License Agreement. In no event will any delay or failure by Bentham Science Publishers in enforcing your compliance with this License Agreement constitute a waiver of any of its rights.

3. You acknowledge that you have read this License Agreement, and agree to be bound by its terms and conditions. To the extent that any other terms and conditions presented on any website of Bentham Science Publishers conflict with, or are inconsistent with, the terms and conditions set out in this License Agreement, you acknowledge that the terms and conditions set out in this License Agreement shall prevail.

Bentham Science Publishers Pte. Ltd.
80 Robinson Road #02-00
Singapore 068898
Singapore
Email: subscriptions@benthamscience.net

BENTHAM SCIENCE

CONTENTS

FOREWORD

In the past few decades, fractional derivatives and integrals have been recognized as powerful modelling and simulation tools for engineering, physics, economy and other application areas. Many physical laws are expressed more accurately in terms of differential equations of arbitrary order. The fractional derivatives and integrals and their potential uses have gained a great importance, mainly since they have become powerful instruments with more accurate, efficient and successful results in mathematical modelling of several complex phenomena in numerous seemingly diverse and widespread fields of science, especially engineering, finance and biology. As the fractional dynamical systems grow, mature and develop, it is very prominent to focus on the most promising novel directions that were worked out based on the novel methods and schemes handed over recently in the field.

The key objective of this book is to focus on recent advancements and future challenges on the basic foundation and applications of the fractional derivatives and integrals in dynamical systems.

This edited book received a number of submissions, out of which 10 high-quality chapters were accepted. The chapters of this book have a large variety of interesting and relevant subjects, namely, fractional partial differential equations, chaotic systems and control, heat conduction, numerical algorithms, complexity and fractional calculus with power law, exponential decay law and Mittag-Leffler non-singular kernel.

We congratulate the Editors, Dr. Mehmet Yavuz and Dr. Necati Özdemir, who were able to collect a variety of topics of relevance to the reader and we are sure that this book will be helpful to scientists doing research in different fields of fractional calculus.

Dumitru Baleanu
Department of Mathematics, Cankaya University,
06530 Balgat, Ankara, Turkey
Institute of Space Sciences, R76900 Magurele-Bucharest, Romania
Emails: dumitru.baleanu@gmail.com; dumitru@cankaya.edu.tr

Jordan Hristov
Department of Chemical Engineering,
University of Chemical Technology and Metallurgy,
8 Kliment Ohridsky Blvd., 1756 Sofia, Bulgaria.
Email: jordan.hristov@mail.bg

PREFACE

The investigation of fractional integrals and fractional derivatives has a long history, and they have many real-world applications because of their properties of interpolation between operators of integer order. This field has covered the classical fractional operators such as Riemann–Liouville, Weyl, Caputo, Grunwald–Letnikov, and so on. Also, especially in the last two decades, many new fractional operators have appeared, often defined using integrals with special functions in the kernel as well as their extended or multivariable forms. These have been intensively studied because they can also be useful in modelling and analysing real-world processes, because of their different properties and behaviours, which are comparable to those of the classical operators.

This book contains ten chapters in three sections. The first section, **Chaotic Systems and Control**, contains three chapters. In Chapter 1, Sene proposed a numerical procedure and its applications to a fractional-order chaotic system represented with the Caputo fractional derivative. In Chapter 2, Okundalaye *et al.* gave a new multistage optimal homotopy asymptotic method for solutions to a couple of fractional optimal control problems. In Chapter 3, Farman *et al.* studied a complex chaotic fractional-order financial system in price exponent with control and modelling.

The second part of the book, **Heat Conduction**, contains two chapters. In Chapter 4, Hristov proposed an attempt to demonstrate that the Duhamel theorem applicable for time-dependent boundary conditions (or time-dependent source terms) of heat conduction in a finite domain and the use of the Fourier method of separation of variable (superposition version) naturally leads to appearance of the Caputo–Fabrizio operators in the solution. In Chapter 5, Avcı and İskender Eroğlu considered the oscillatory heat transfer due to the Cattaneo–Hristov model on the real line modelled by a fractional-order derivative with a non-singular kernel.

The third section of the book, **Computational Methods and Their Illustrative Applications**, contains five chapters related to different types of real-life problems. In Chapter 6, Ghoreishi *et al.* applied the optimal homotopy analysis method for a nonlinear fractional-order model to HTLV-1 infection of CD4$^+$ T-cells. In Chapter 7, Durur *et al.* investigated the behavior analysis and asymptotic stability of the traveling wave solution of the Kaup-Kupershmidt equation with the conformable operator. In Chapter 8, Baishya *et al.* took into account the Caputo fractional order derivative in the mathematical analysis of a rumor-spreading model and presented interesting numerical results. In Chapter 9,

Veeresha *et al.* studied a unified approach for the fractional system of equations arising in the biochemical reaction without a singular kernel. In Chapter 10, Bora *et al.* investigated the hydro-morphodynamic effects induced by a non-powered floating object navigating in an approach channel using the CFD (Computational Fluid Dynamics) process.

We are very much thankful to all the contributors to this book for their valuable and productive works. The foreword for this book has been written by Prof. Dumitru Baleanu and Prof. Jordan Hristov. We would like to express our sincere gratitude for their guidance and support.

We are extremely grateful to Mrs. Humaira Hashmi (Editorial Manager Publications) and Mrs. Fariya Zulfiqar (Manager Publications) of Bentham Science Publishers who helped us in the publication process. We are also extending our thanks to Bentham Science Publishers for publishing this book.

We wish that this book will be especially useful to scientists doing research in the field of fractional calculus and to researchers at graduate level in this field.

Mehmet Yavuz
Department of Mathematics and Computer Sciences
Necmettin Erbakan University
Konya 42090
Turkey

&

Necati Özdemir
Department of Mathematics
Balıkesir University
Çağiş Campus, 10145 Balıkesir
Turkey

List of Contributors

Ali Akgul	Department of Mathematics, Art and Science Faculty, Siirt University, 56100 Siirt, Turkey
Aqeel Ahmad	Department of Mathematics and Statistics, University of Lahore, Lahore 54590, Pakistan
Asıf Yokuş	Department of Mathematics, Faculty of Science, Firat University, Elazig 23100, Turkey
Beyza Billur İskender Eroğlu	Department of Mathematics, Faculty of Arts and Sciences, Balıkesir University, Balıkesir, Turkey
Chandrali Baishya	Department of Studies and Research in Mathematics, Tumkur University, Tumkur 572103, India
Derya Avci	Department of Mathematics, Faculty of Arts and Sciences, Balıkesir University, Balıkesir, Turkey
Emel İrtem	Department of Civil Engineering, Doğuş University, Istanbul, Turkey
Hülya Durur	Department of Computer Engineering, Faculty of Engineering, Ardahan University, Ardahan 75000, Turkey
Jordan Hristov	Department of Chemical Engineering, University of Chemical Technology and Metallurgy, Sofia, Bulgaria
L. Akinyemi	Department of Mathematics, Lafayette College, Easton, Pennsylvania, USA
M. Sedat Kabdaşlı	Faculty of Civil Engineering, Istanbul Technical University, Istanbul, Turkey
M.S. Kiran	Research Centre, Department of Chemistry, GM Institute of Technology, Davangere 577006, India
Mehmet Yavuz	Department of Mathematics and Computer Sciences, Necmettin Erbakan University, Konya 42090, Turkey
Mohammad Ghoreishi	School of Mathematical Sciences, Universiti Sains Malaysia, 11800 Gelugor, Penang, Malaysia
Muhammad Farman	Department of Mathematics and Statistics, University of Lahore, Lahore 54590, Pakistan
Muhammad Umer Saleem	Department of Mathematics, Division of Science and Technology, University of Education, Lahore, Punjab 54770, Pakistan
Ndolane Sene	Section Mathematics and Statistics, Institut des Politiques Publiques, Cheikh Anta Diop University, Dakar Fann, Senegal
Necati Özdemir	Department of Mathematics, Faculty of Science and Arts, Balikesir University, Balikesir, Turkey
Oluwaseun O. Okundalaye	Department of Mathematical Sciences, Faculty of Science, Adekunle Ajasin University, Akungba-Akoko, Nigeria
Onur Bora	Department of Civil Engineering, Balıkesir University, Balıkesir, Turkey
P. Veeresha	Center for Mathematical Needs, Department of Mathematics, CHRIST (Deemed to be University), Bengaluru 560029, India

Parvaiz Ahmad Naik   School of Mathematics and Statistics, Xi'an Jiaotong University, Xi'an, Shaanxi 710049, P.R. China

Sindhu J. Achar   Department of Studies and Research in Mathematics, Tumkur University, Tumkur 572103, India

Wan A.M. Othman   Institute of Mathematical Sciences, University of Malaya, 50603 Kuala Lumpur, Malaysia

Numerical Procedure and its Applications to the Fractional-Order Chaotic System Represented with the Caputo Derivative

Ndolane Sene[1*]

[1]*Section Mathematics and Statistics, Institut des Politiques Publiques, Cheikh Anta Diop University, Dakar Fann, Senegal*

Abstract: This chapter focuses on a numerical procedure and its application to a fractional-order chaotic system. The numerical scheme will discuss the Lyapunov exponents for the considered model and characterize the chaos's nature. We will also use the numerical scheme to depict the phase portraits of the proposed fractional-order chaotic system and the bifurcation maps. Note that the bifurcation maps are used to characterize the influence of the different parameters of our considered fractional model. The impact of the initial conditions and the coexisting attractors will also be analyzed. With the coexistence, the new types of attractors will be discovered for our considered model. To confirm the investigations in this chapter, the proposed model will be applied to the electrical modeling. Therefore, the circuit schematic of the considered fractional model will be implemented in real-world problems. And we notice good agreement between the theoretical results and the results obtained after Multisim simulations. The stability of the equilibrium points of the presented model will also be focused on details and will permit us to delimit the chaotic region in general.

Keywords: Attractors, Bifurcation maps, Chaotic systems, Lyapunov exponents, Stability analysis.

INTRODUCTION

This chapter focuses on chaos theory in the context of fractional calculus. There exist many real-world applications of chaos theory in modeling electrical circuits [1, 2], engineering sciences, modeling electronics phenomena and others [4, 5]. Many differential equations admitting chaotic behaviors and hyperchaotic behaviors exist in two dimensions, three dimensions, four dimensions, five

---

*Corresponding author Ndolane Sene: Section Mathematics and Statistics, Institut des Politiques Publiques, Cheikh Anta Diop University, Dakar Fann, Senegal; E-mail: ndolanesene@yahoo.fr

Mehmet Yavuz & Necati Özdemir (Eds.)

dimensions, and others dimensions. Many of them are described by the integer-order derivative. This chapter will introduce the fractional operators in the modeling of a class of chaotic systems. The fractional calculus continues its expansion, with many discussions about these applications of the derivative in real-life problems. Many address the interpretations of the fractional operators, but the consensus is not unanimous. Many derivatives exist as the old operators; we have the Caputo derivative [6, 7] and the Riemann-Liouville derivative [6, 7], and many others. We also have recent fractional operators with new types of kernels as the Atangana-Baleanu derivative, the Caputo-Fabrizio derivative, conformable derivative, and others [11-17]. For recent developments of fractional calculus and fractional operators applications, the readers can look at the following papers for examples[11, 18-24].

LITERATURE REVIEW IN CHAOS

Modeling chaotic and hyperchaotic systems with fractional operators in fractional calculus was first proposed by Petras in studies [25, 26]. The literature of chaotic and hyperchaotic systems with integer derivatives and fractional derivatives is vast. In this part, we recall some of them. Petras studies various classes of fractional-order chaotic systems described by the Caputo derivative, the stability of the equilibrium points of the chaotic systems is also presented in its book [25]. An algorithm to obtain the Lyapunov exponents in the context of fractional calculus by Matlab software has been presented by Danca and his co-author [27]. Ren and co-authors presented a new chaotic system flow with the presence of a hidden attractor; this new system is known to belong to the jerk systems with no equilibrium points [28]. Rajagopal *et al.* presented the so-called chameleon fractional chaotic system [29]. Vaidyanathan and the co-authors presented a hyperchaotic system with five dimensions [2]. The authors also presented the circuit schematic of their model, and the results are represented in oscilloscopes. Pham *et al.* presented the coexistence attractors of a hidden chaotic system with no-equilibrium points [30]. The authors presented a new class of chaotic systems and developed some properties related to their presented novel model [31]. The authors developed control and synchronization of the fractional chaotic system using an active controller [32]. Diouf *et al.* proposed the phase portraits and bifurcation maps of the three-dimensional financial chaotic differential equation using a numerical scheme in the context of fractional calculus [33]. Sene used Caputo derivative to model financial model in four-dimensional space [34]. The system studied is hyperchaotic but sometimes with one positive Lyapunov exponent and sometimes with two positive Lyapunov exponents. The properties related to the Lyapunov exponents in fractional context are well detailed in this work. Sene *et al.* proposed an investigation related to chaotic and hyperchaotic systems described by the Caputo derivative [35]. They studied the proposed system the qualitative properties using the Lyapunov exponents and phase portraits. Sene analyzed the class of fractional-order chaotic system described by the Caputo derivative using bifurcation and Lyapunov exponents [36]. See also the study [37] for more investigations.

MOTIVATIONS AND NOVELTIES

Modeling fractional-order chaotic systems will be the main innovation of this chapter. The chaotic system will be represented using the Caputo derivative. To obtain the phase portrait, discretization, including the discretization of the Riemann-Liouville integral and the analytical solution, will be used. The fractional-order will generate different types of attractors, and the Lyapunov exponents will be used to classify them. The Lyapunov exponents in the context of the fractional derivatives, as proposed by Danca [27], will be illustrated. Note that one positive Lyapunov exponent means the existence of chaotic behaviors in general. The impact of the initial conditions, the coexisting attractors will be analyzed, and the variation of the proposed model's parameters will be illustrated using the bifurcation maps. The stability analysis will delimit the interval under which the chaotic behaviors exist when we use the Caputo derivative. Many other qualitative properties of the dynamic under investigations will be presented, illustrated, and discussed as possible in this chapter.

FRACTIONAL OPERATORS AND DERIVATIVES

Many operators exist in fractional calculus with and without singular kernel. In this chapter, we try to recall two of them which will be of great interest to us for our investigation. Caputo, Riemann, and Liouville propose the fractional derivatives most used in fractional calculus for decades. We defined the fractional integral before these derivatives, know as the Riemann-Liouville derivative, as the following definition.

The representation of the Riemann-Liouville integral of the function z is described in the following formula

$$(I^\alpha z)(t) = \frac{1}{\Gamma(\alpha)} \int_0^t (t-s)^{\alpha-1} z(s) ds,$$

where $\Gamma(\dots)$ defines the Gamma Euler function and and the order α is imposed to respect the condition $\alpha > 0$ [6, 7]. The fractional integral operator plays an essential role in discretization. Its discretized form is well known in the literature and can be found in Garrapa's investigations related to the proposition of the numerical schemes in the context of fractional derivatives.

The representation of the Riemann-Liouville derivative of the function z is described in the following form

$$D^\alpha z(t) = \frac{1}{\Gamma(1-\alpha)} \frac{d}{dt} \int_0^t z(s)(t-s)^{-\alpha} \, ds,$$

where $\Gamma(\dots)$ defines the Gamma Euler function and and the order α is imposed to respect the condition $\alpha \in (0,1)$ [6, 7]. This derivative received many investigations but is not used nowadays. The problem is in the initial condition, which includes the integral form. We knew that the initial states in integral forms are not physic. This representation is not realistic because the initial condition should be constant at the initial time. The inconveniences in the initial conditions and the fact that the derivative of the constant function does not give zero have motivated the introduction of the Caputo-Liouville derivative in the literature.

The representation of the Caputo-Liouville derivative of the function z is described in the following form

$$D^\alpha z(t) = \frac{1}{\Gamma(1-\alpha)} \int_0^t \frac{dz}{ds}(t-s)^{-\alpha} \, ds,$$

Where $\Gamma(\dots)$ defines the Gamma Euler function and and the order α is imposed to respect the condition $\alpha \in (0,1)$ [6, 7]. This derivative received an increasing number of papers and innovative applications in real-world problems. The initial condition explains the successes of this derivative.

Before closing this section, we recall some properties, including the previous derivatives and fractional integral. The first one is when the operators' orders converge to zero and one; then we recover respectively in both definitions the classical integral and the classical integer order derivative. In other words, we have the following relationships

$$\left(I^1 z\right)(t) = \int_0^t z(s)ds, \qquad D^1 z(t) = \frac{dz}{dt}.$$

The compositions rule of two Riemann-Liouville integrals and between fractional integral and Caputo derivative give the following relationships

$$I^\beta \left(I^\alpha z \right)(t) = \left(I^{\beta+\alpha} z \right)(t), \qquad D^\alpha \left(I^\alpha z \right)(t) = z(t).$$

FRACTIONAL-ORDER SYSTEM WITH CAPUTO DERIVATIVE

This section illustrates the used model in our investigation. The fractional differential system which we consider illustrating the chaotic system is the following system

$$
\begin{aligned}
D_c^\alpha x &= y, \\
D_c^\alpha y &= z, \\
D_c^\alpha z &= -ax - 3.5y - z - x^3,
\end{aligned}
\tag{1}
$$

the considered initial conditions for the model previously defined are given by

$$x(0) = x_0 = -0.2, \qquad y(0) = y_0 = 0.5, \qquad z(0) = z_0 = 0.2. \tag{2}$$

For the parameter of the previous system, we choose it as follows $a = -5.5$. The model Eqs. (1) is first proposed in Petras works [25]. In this chapter, we will give the numerical procedure of the proposed system. The proposed numerical scheme will be used in depicting the phase portraits in different planes. It will help us to represent the bifurcation diagram, too. Note that the bifurcation maps are utilized to analyze the topological changes in the model's behaviors when the different parameters of the model change. The characterization of the attractor will be investigated using the Lyapunov exponents. The Lyapunov exponents' concept is simple; it uses the Jacobian matrix and the numerical scheme and informs us of the chaotic attractor or hyperchaotic attractor. The number of positive Lyapunov exponents plays an important role in the nature of the dynamics' behaviors. The initial conditions have a significant impact on the nature of the chaos. Note that the chaotic systems are system sensitive to the initial conditions. The changes in the initial conditions and the model's parameters can create the coexistence attractors, which will also be addressed in this chapter. As initial conditions are important, we justify using the Caputo derivative, which considers the initial physical condition. Contrary to the Riemann-Liouville derivative, which initial conditions are in integral form with no physical sense. The introduction of the fractional derivative in modeling chaotic systems is to see the fractional integrator's influence in the circuit realization and take into account the memory effect. Another novelty in our analysis is to generate new types of coexisting attractors in the same and different planes as possible.

PROCEDURE OF THE SOLUTION OF MODEL

In this section, we describe the predictor-corrector procedure to solve the fractional differential equations presented in Eqs. (1). Numerical discretization plays an important role in fractional calculus. There exist many differential equations, complex and non-linear fractional equations where the Laplace transform, and the domain decomposition methods can not be adequate and are not trivial. For example, the convergence and the stability of the second cited method is not trivial in many contexts. The alternative is the use of numerical discretization. The convergence of the solutions can be proved easily with Newton's method and the stability with the Lipschitz criteria of the used drift functions. In our context, the utilization of the analytical solution to do the discretization has many advantages because the discretization of the fractional integral will be included. The solutions of the differential equation in term of the Riemann-Liouville integral are assigned in the following equations

$$
\begin{aligned}
x(t) &= x(0) + I^{\alpha} f(t, x_1), \\
y(t) &= y(0) + I^{\alpha} g(t, x_1), \\
z(t) &= z(0) + I^{\alpha} k(t, x_1).
\end{aligned}
\tag{3}
$$

We suppose that the function f, g, and k are obtained from the fractional model defined by Eqs. (1) under initial conditions (2)

$$
\begin{aligned}
f(t, x_1) &= y, \\
g(t, x_1) &= z, \\
k(t, x_1) &= -ax - 3.5y - z - x^3.
\end{aligned}
$$

The representations of the solutions (3) at the point of discretization t_n are assigned in the following equations

$$
\begin{aligned}
x(t_n) &= x(0) + I^{\alpha} f(t_n, x_1), \\
y(t_n) &= y(0) + I^{\alpha} g(t_n, x_1), \\
z(t_n) &= z(0) + I^{\alpha} k(t_n, x_1),
\end{aligned}
\tag{4}
$$

where $x_1 = (x, y, z)$. Using the predictor-corrector scheme in fractional context proposed by Garrapa [38] and setting $t_n = t_0 + nh$ where h represents the step size and the initial time supposed to be zero in this chapter. The Riemann-Liouville integral present in Eqs. (4) can be rewritten as the forms

$$x(t_n) = x(0) + h^\alpha \left[\bar{\kappa}_n^{(\alpha)} f(0) + \sum_{j=1}^{n-1} \kappa_{n-j}^{(\alpha)} f(t_j, x_{1j}) + \kappa_0^{(\alpha)} f(t, x_1^P) \right],$$

$$y(t_n) = x(0) + h^\alpha \left[\bar{\kappa}_n^{(\alpha)} g(0) + \sum_{j=1}^{n-1} \kappa_{n-j}^{(\alpha)} g(t_j, x_{1j}) + \kappa_0^{(\alpha)} g(t, x_1^P) \right],$$

$$z(t_n) = x(0) + h^\alpha \left[\bar{\kappa}_n^{(\alpha)} k(0) + \sum_{j=1}^{n-1} \kappa_{n-j}^{(\alpha)} k(t_j, x_{1j}) + \kappa_0^{(\alpha)} k(t, x_1^P) \right].$$

We express the predictor as the form given by the following

$$x^P(t_n) = x(0) + h^\alpha \sum_{j=1}^{n-1} \kappa_{n-j-1}^{(\alpha)} f(t_j, x_{1j},),$$

$$y^P(t_n) = y(0) + h^\alpha \sum_{j=1}^{n-1} \kappa_{n-j-1}^{(\alpha)} g(t_j, x_{1j}),$$

$$z^P(t_n) = z(0) + h^\alpha \sum_{j=1}^{n-1} \kappa_{n-j-1}^{(\alpha)} k(t_j, x_{1j}).$$

The parameter into the previous equations are expressed as the following forms

$$\bar{\kappa}_n^{(\alpha)} = \frac{(n-1)^\alpha - n^\alpha (n - \alpha - 1)}{\Gamma(2+\alpha)},$$

for n describing the set $1, 2, \ldots$, the parameters are expressed as the following form

$$\kappa_0^{(\alpha)} = \frac{1}{\Gamma(2+\alpha)} \text{ and } \kappa_n^{(\alpha)} = \frac{(n-1)^{\alpha+1} - 2n^{\alpha+1} + (n+1)^{\alpha+1}}{\Gamma(2+\alpha)}.$$

Without losing generality, the numerical discretizations of the f, g, and k are represented as the following

$$f(t_j, x_{1j}) = y_j,$$

$$g(t_j, x_{1j}) = z_j,$$

$$k\left(t_j, x_{1j}\right) \;=\; -ax_j - 3.5y_j - z_j - x_j^3.$$

Let the following relations $x(t_n)$, $y(t_n)$ and $z(t_n)$ be the approximate solutions of the considered fractional attractor system (1) and x_n, y_n and z_n be the exact solutions of the fractional attractor system (1). The resual functions are represented as the folowing forms

$$|x\left(t_n\right) - x_n| \;=\; \mathcal{O}\left(h^{\min\{\alpha+1,2\}}\right),$$

$$|y\left(t_n\right) - y_n| \;=\; \mathcal{O}\left(h^{\min\{\alpha+1,2\}}\right),$$

$$|z\left(t_n\right) - z_n| \;=\; \mathcal{O}\left(h^{\min\{\alpha+1,2\}}\right).$$

The convergence of the method follows from the convergence to 0 of the step size h. This point closes the numerical procedure. In the following sections, the numerical process described in this section will be used in the bifurcation diagrams, the Lyapunov exponents, the initial conditions, and the coexistence attractors.

APPLICATION OF THE NUMERICAL SCHEME

In this first part, we illustrate the predictor-corrector scheme described in the previous section. We consider different values of the fractional-order derivative to see their impacts on the behaviors of the solutions. The fractional-orders of the Caputo derivative are the following $\alpha = 0.85$, $\alpha = 0.89$, $\alpha = 0.95$, $\alpha = 0.99$.

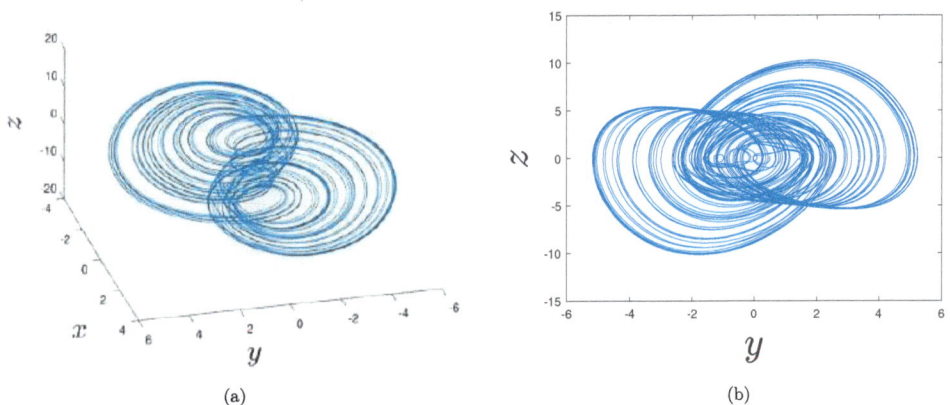

(a)   (b)

Fig. (1). Solutions in planes (x, y, z), (y, z) at order $\alpha = 0.99$.

In the first case, we consider the order $\alpha = 0.99$. We represent, the dynamics of the solutions in different planes (x, y, z), (y, z), (x, z) and (x, y) (Figs. **1a, b** and **2a, b**).

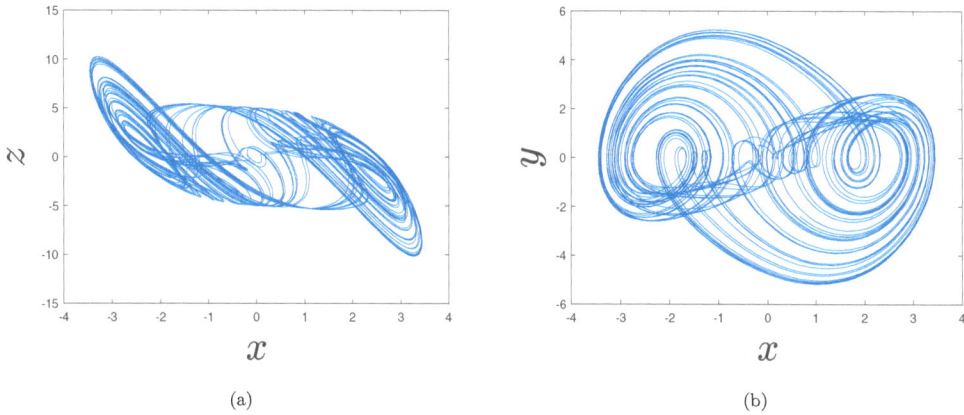

(a)   (b)

Fig. (2). Solutions in planes (x, z), (x, y) at order $\alpha = 0.99$.

The Lyapunov exponents associated with this fractional-order will be calculated in the next section. We now consider the order of the Caputo derivative given by $\alpha = 0.95$, the (Figs. **3a, b** and **4a, b**).

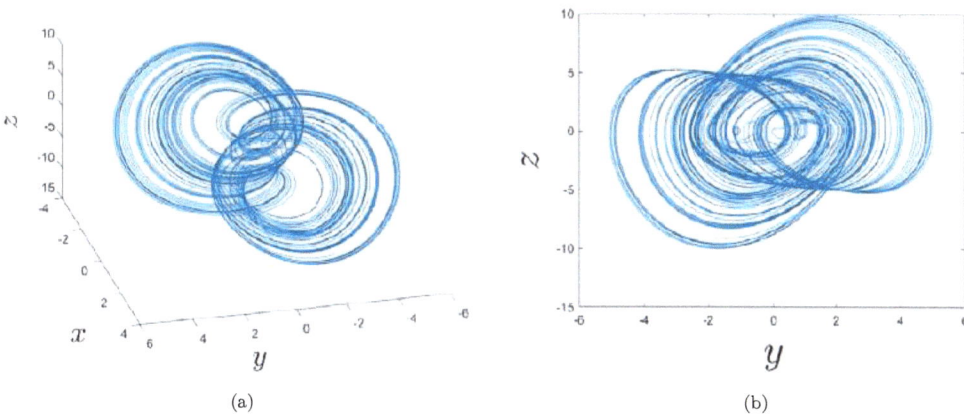

(a)   (b)

Fig. (3). Solutions in planes (x, y, z), and (y, z) at order $\alpha = 0.95$.

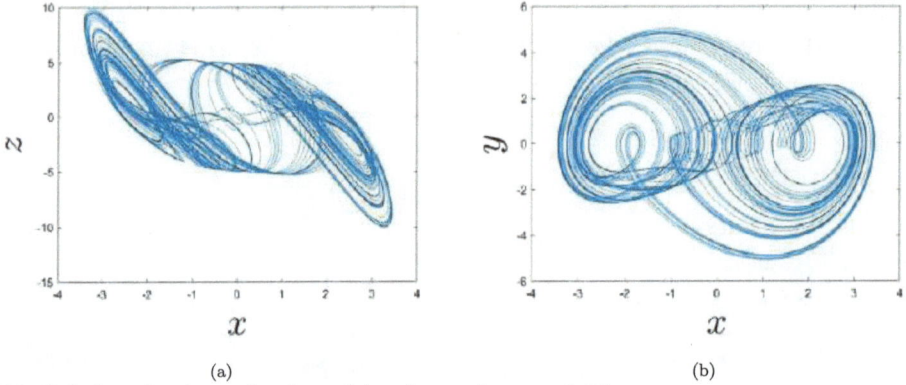

(a) (b)

Fig. (4). Solutions in planes (x, z), and (x, y) at order $\alpha = 0.95$.

In this new case, we consider the order of the Caputo derivative given by $\alpha = 0.89$, and the phase portraits in different planes are depicted in the coming (Figs. **5a, b** and **6a, b**).

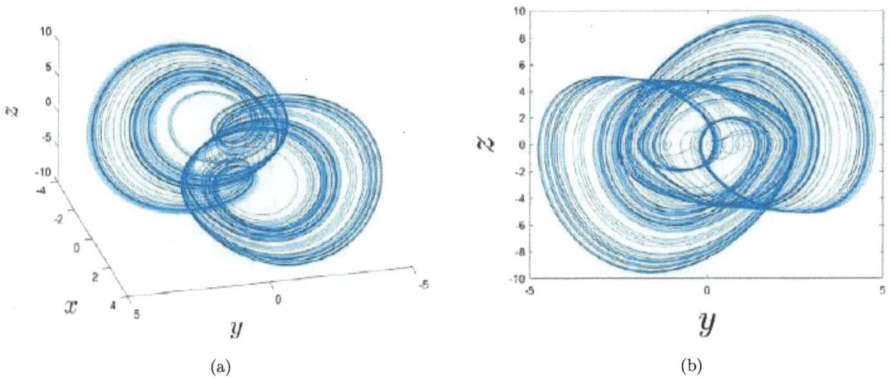

(a) (b)

Fig. (5). Solutions in planes (x, y, z), and (y, z) at order $\alpha = 0.89$.

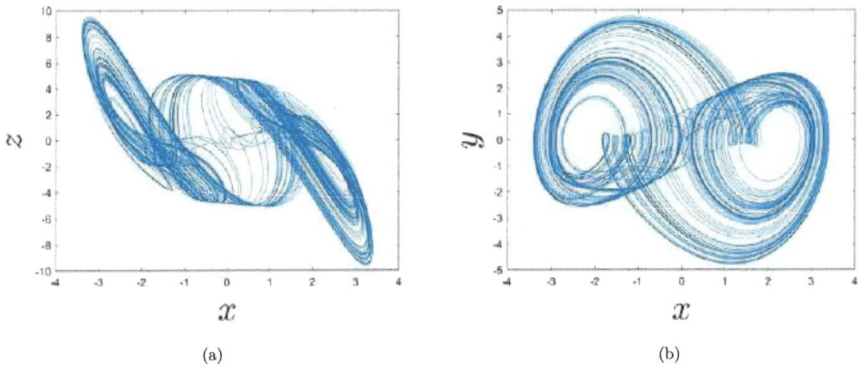

(a) (b)

Fig. (6). Solutions in planes (x, z), and (x, y) at order $\alpha = 0.89$.

In the last case of the portraits of phase (Figs. **7a, b** and **8a, b**). of the considered fractional attractor (1), we consider the order of the Caputo derivative is given by $\alpha = 0.86$.

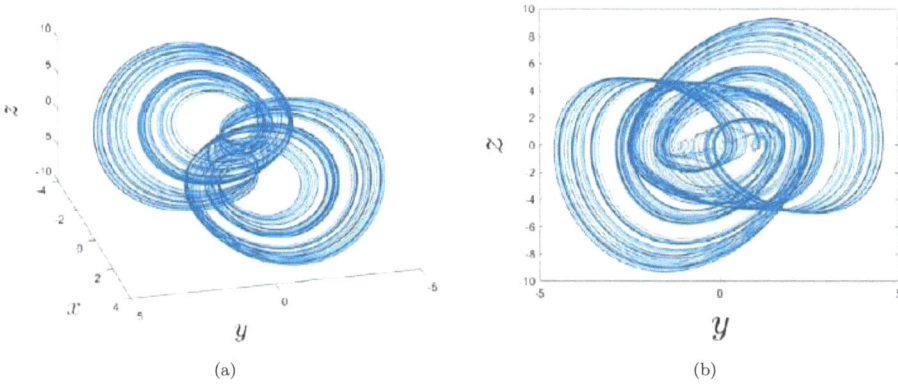

(a)	(b)

Fig. (7). Solutions in planes (x, y, z), and (y, z) at order $\alpha = 0.86$.

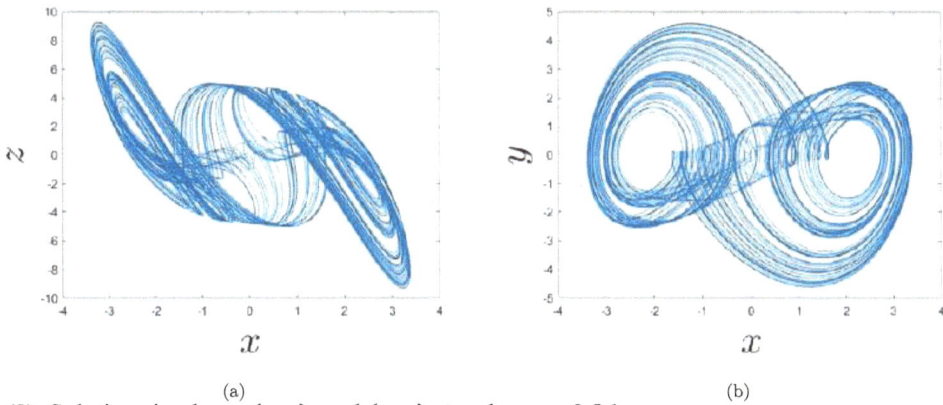

(a)	(b)

Fig. (8). Solutions in planes (x, z), and (x, y) at order $\alpha = 0.86$.

The nature of the chaos in different Caputo derivative orders will be characterized *via* Lyapunov exponents in the next section. It will confirm the impact of the fractional-order derivative. Referring to the figures considered in this section, we can observe the chaos's behaviors are not the same in all the orders considered in this section. The results assigned in this section will be confirmed in real worlds applications, where we will draw the fractional-order system (1) using capacitors and resistors tools. The primary model will be rearranged to avoid frequency problems. In other words, the system will be scaled in an adequate interval for the planes x, y, and z.

LYAPUNOV EXPONENTS AND BIFURCATION MAPS

In this part, we study the impact of the model's parameters on the fractional system dynamics. We give the nature of the chaos at the different order used in the previous section using the fractional Lyapunov exponents and their associated Kaplan Yorke dimension. The method for determining the Lyapunov exponents in the context of the fractional operator is described in [27]s.

We first begin with the Lyapunov exponents and their associated Kaplan Yorke dimensions. The first order of the Caputo derivative is $\alpha = 0.99$. The values of the Lyapunov exponents are described as the following forms

$$Lyap1 = 0.5771, \qquad Lyap2 = -0.3461, \qquad Lyap3 = -1.3229.$$

We confirm that the fractional-order system has chaotic behaviors because there is one positive Lyapunov exponent. We add that the sum of all the Lyapunov exponents is negative. The system is dissipative, and the sum between the first and the second Lyapunov exponent is positive, then we can calculate the Kaplan Yorke dimension. Then for the considered order, its dimension is

$$dim(Lyap) = 2.1746.$$

It is fractional and confirms as well the chaotic behaviors of the system. We now consider the second case with the order $\alpha = 0.95$, and we repeat the same procedure. The values of the Lyapunov exponents are described as the following forms

$$Lyap1 = 0.5811, \qquad Lyap2 = -0.3578, \qquad Lyap3 = -1.5249.$$

We confirm that the fractional-order system has chaotic behaviors due to the existence of one positive Lyapunov exponent. Furthermore, the sum of all the Lyapunov exponents is negative; thus, the system is dissipative. The sum between the first and the second Lyapunov exponent is positive; then, we can calculate the Kaplan Yorke dimension. Then for the considered order, its dimension is

$$dim(Lyap) = 2.1464.$$

The third order of the Caputo derivative is $\alpha = 0.89$. The values of the Lyapunov exponents are described as the following forms

$$Lyap1 = 0.4825, \qquad Lyap2 = -0.2411, \qquad Lyap3 = -1.9366.$$

We confirm that the fractional-order system has chaotic behaviors because there exists one positive Lyapunov exponent. The sum of all the Lyapunov exponents is negative, and the sum between the first and the second Lyapunov exponent is positive, then we can calculate the Kaplan Yorke dimension. For the considered order, the dimension proposed by Kaplan Yorke is

$$dim(Lyap) = 2.1246.$$

The last order of the Caputo derivative is $\alpha = 0.86$. The values of the Lyapunov exponents are described as the following forms

$$Lyap1 = 0.6283, \qquad Lyap2 = -0.3018, \qquad Lyap3 = -2.2620.$$

We confirm that the fractional-order system has chaotic behaviors because there exists one positive Lyapunov exponent. The sum of all the Lyapunov exponents is negative; thus, the system is dissipative. The sum between the first and the second Lyapunov exponent is positive; then, we can calculate the Kaplan Yorke dimension. For the considered order, the dimension proposed by Kaplan Yorke is

$$dim(Lyap) = 2.1443.$$

The final confirmation is for all orders considered in (0.85,1); the behaviors are chaotic. Then we have, in our context, the fractional-order chaotic system. And the chaos is more complex at the order $\alpha = 0.86$; it can be explained by the fact the maximal Lyapunov exponent is obtained to this order.

The second part of this section analyzes the behaviors of the system when the parameter varies. Then we will use the bifurcation map related to the parameter a. The bifurcation diagram is represented in the following Fig. (**9**):

When the parameter a decreases into $(-3, -1)$, the trajectories of the system converge to a stable equilibrium point. When the parameter a decreases into $(-5,5, -3)$ the system enters chaotic behaviors by period-doubling bifurcation and become chaotic for the rest of the variation of the parameter a. Note the system has chaotic behaviors when a decreases in the interval $(-5,5, -3.5)$. For an illustration of the chaotic region, we consider $a = -5$, and we depict the phase portraits of the fractional-order system. We have the following (Figs. **10a, b** and **11a, b**).

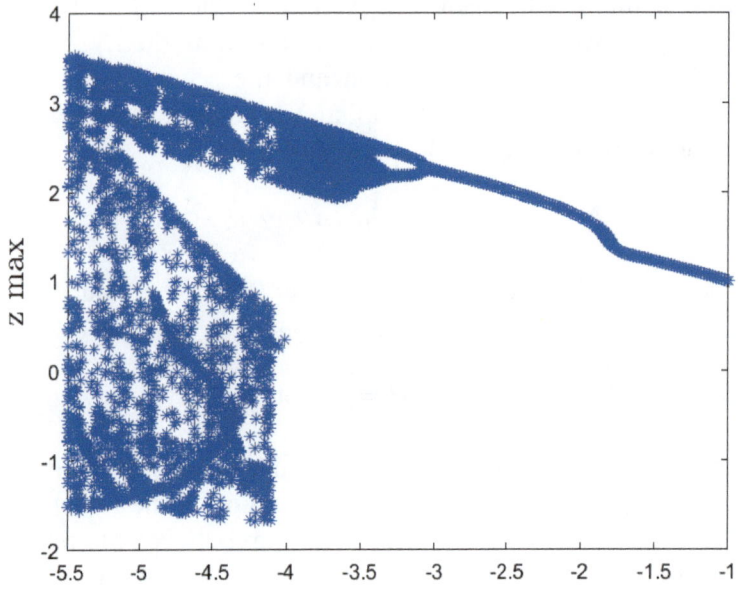

Fig. (9). Bifurcation map according to $a \in (-5.5, -1)$.

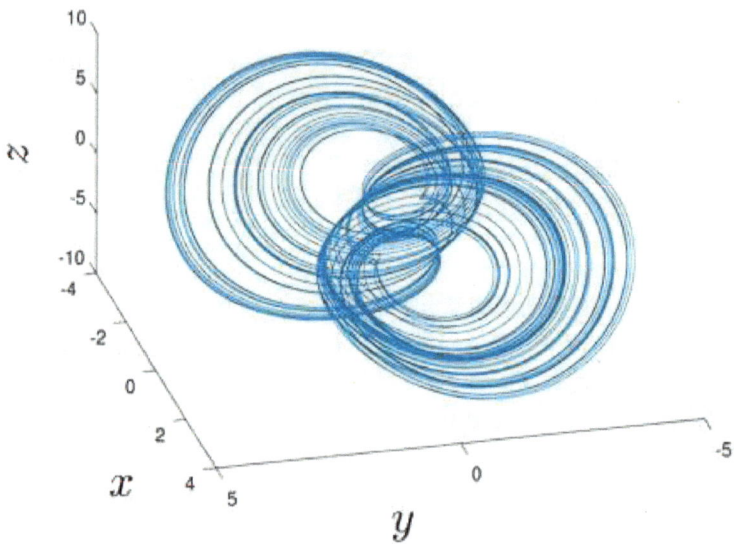

(a)

(b)

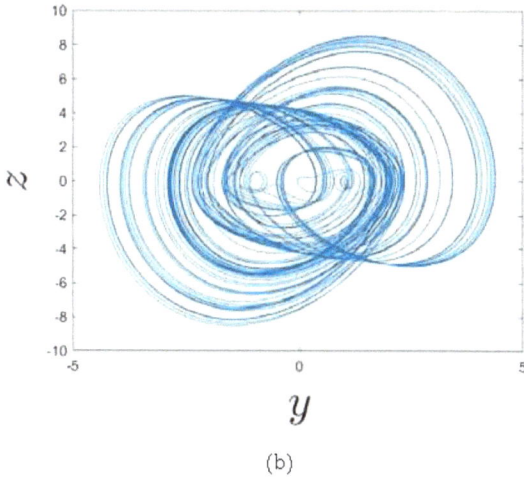

Fig. (10). Solutions in planes (x, y, z), and (y, z) at order $\alpha = 0.99$ with $a = -5$.

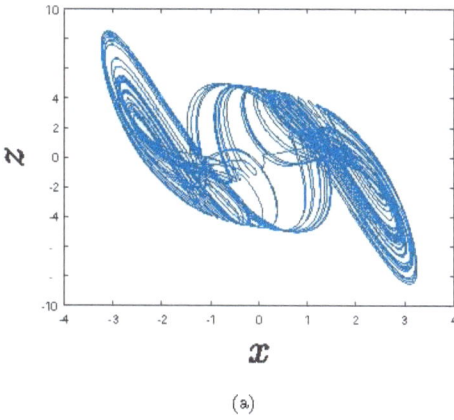

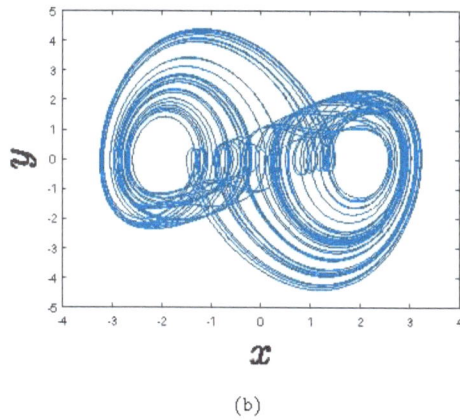

(a)   (b)

Fig. (11). Solutions in planes (x, z), and (x, y) at order $\alpha = 0.99$ with $a = -5$.

STABILITY ANALYSIS

In this section, we focus on the local asymptotic stability of the equilibrium points of the fractional-order chaotic system (1). Stability is crucial in chaotic systems because it is chaotic or hyperchaotic if all the considered system's equilibrium points failed to be stable. Note that when all the fractional-order system's equilibrium points are stable, the system failed to be chaotic or hyperchaotic. This property is important for readers to know it. For the rest of the paper, we use the condition proposed by Matignon, which is given by

$$|\arg(\lambda(J))| > \alpha\pi/2, \tag{5}$$

which is used in the stability analysis in the fractional context [39, 40]. The matrix J denotes the classical Jacobian matrix of (1). In our context, it is in the form

$$J = \begin{pmatrix} 0 & 1 & 0 \\ 0 & 0 & 1 \\ -a + 3x^2 & -3.5 & -1 \end{pmatrix}.$$

The equilibrium points of our considered model (1) are given by $A(0,0,0)$, $B(2.34,0,0)$ and $C(-2.34,0,0)$. For the local stability of the first point, the Jacobian matrix at point A is given by the following form

$$J = \begin{pmatrix} 0 & 1 & 0 \\ 0 & 0 & 1 \\ 5.5 & -3.5 & -1 \end{pmatrix}.$$

The eigenvalues of the above matrix are given by $\lambda_1 = 1$, $\lambda_2 = -1 + 2.12i$ and $\lambda_3 = -1 - 2.12i$. The second and the last eigenvalue satisfy clearly the condition $|\arg(\lambda(J))| = \pi > \alpha\pi/2$, but the first eigenvalue does not confirm the condition (5) $|\arg(\lambda(J))| = 0 < \alpha\pi/2$, for all orders α of the Caputo derivative. Thus the point A is not stable.

For the local stability of the second point. The Jacobian matrix at the equilibrium point $B(2.34,0,0)$ gives the following matrix

$$J = \begin{pmatrix} 0 & 1 & 0 \\ 0 & 0 & 1 \\ -10.92 & -3.5 & -1 \end{pmatrix}.$$

The eigenvalues of the above matrix are given by $\lambda_1 = -2$, $\lambda_2 = 0.5 + 2.29i$ and $\lambda_3 = -1 - 2.12i$. The first eigenvalue satisfies the condition $|\arg(\lambda(J))| = \pi >$

$\alpha\pi/2$,. The second and the last have the same argument satisfying $|\arg(\lambda(J))| = 13\pi/20 > \alpha\pi/2$, means when the order $\alpha < 0.85$. ..., point B is locally stable; else, it is not a stable point.

The same conditions are obtained with the equilibrium point $C(-2.34,0,0)$. For the local stability of the second point. The Jacobian matrix at the equilibrium point $C(-2.34,0,0)$ is given by the following matrix

$$J = \begin{pmatrix} 0 & 1 & 0 \\ 0 & 0 & 1 \\ -10.92 & -3.5 & -1 \end{pmatrix}.$$

The eigenvalues of the above matrix are given by $\lambda_1 = -2$, $\lambda_2 = 0.5 + 2.29i$ and $\lambda_3 = -1 - 2.12i$. The first eigenvalue satisfies the condition $|\arg(\lambda(J))| = \pi > \alpha\pi/2$,. The second and the last have the same argument satisfying $|\arg(\lambda(J))| = 13\pi/20 > \alpha\pi/2$, means when the order $\alpha < 0.85$. ..., the point C is locally stable, else, it is not stable.

The conclusion is the fractional-order system considered in this section is chaotic when the order of the Caputo derivative satisfies the condition $\alpha > 0.85$. This order explains the choice of the fractional-order derivative into the interval $(0.85,1)$ for the phase portraits in the previous sections.

CHAOS AND INITIAL CONDITION IMPACTS

In this section, we analyze all the sensitivity of the initial conditions. We first perturb the initial conditions, and the second part will be to explore the coexisting attractors. For the rest of the paper, we consider the order $\alpha = 0.86$. Some innovative properties in chaos theory are addressed in this part.

To see the impacts of the initial conditions obtained after the perturbation of x_0, we depict the fractional chaotic model following the variable x at $(-0.2,0.5,0.2)$ and $(-0.2001,0.5,0.2)$. We can observe when the condition x_0 has small changes, then the behaviors of the model are not significantly affected as it can be seen in the bifurcation map associated with $(-0.2,0.5,0.2)$ and $(-0.2001,0.5,0.2)$. We also notice small changes in dynamic behaviors (See Figs. **12a, b**):

(a)   (b)

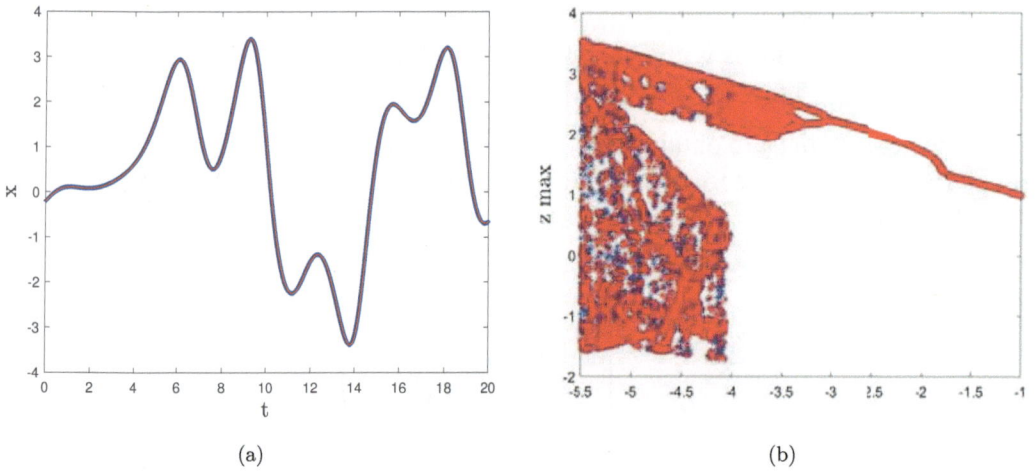

Fig. (12). Sensitivity related to the perturbation of x_0 and its associated bifurcation maps.

The same behaviors are noticed when the other variables are perturbed. Finally, the small perturbations in the initial conditions do not damage our model's chaotic behaviors for our considered model.

In the second part, we study the coexistence of the fractional-order chaotic model considered in this paper. We fix the parameter to $a = -4.5$ and two different initial conditions given by $(-0.2, 0.5, 0.2)$ (blue) and $(0.2, -0.5, -0.2)$. (red). In the following (Figs. **13a, b** and **14a, b**), we represent new attractors, the coexistence on the different planes (x, y), (x, z) and (y, z).

The coexistence is well confirmed by the bifurcation map Fig. (**14b**). Comparing the bifurcation in Figs. (**12b** and **14b**), we notice in both of them the initial conditions are influenced. The difference is in the direction of the perturbations. For the first Fig. (**12b**), the perturbation is unidirectional because one of the initial conditions has a small perturbation; thus, we do not notice a significant change in the bifurcation. But in Fig. (**14b**), it is bifurcation attesting to the existence of pair attractors; the initial conditions were perturbed significantly. We can observe that we have new types of attractors. For innovation, we now consider the pair of two fractional-order attractors in different planes. We fix the parameter to $a = -4.5$ and two different initial conditions given by $(-0.2, 0.5, 0.2)$ and $(0.2, -0.5, -0.2)$.

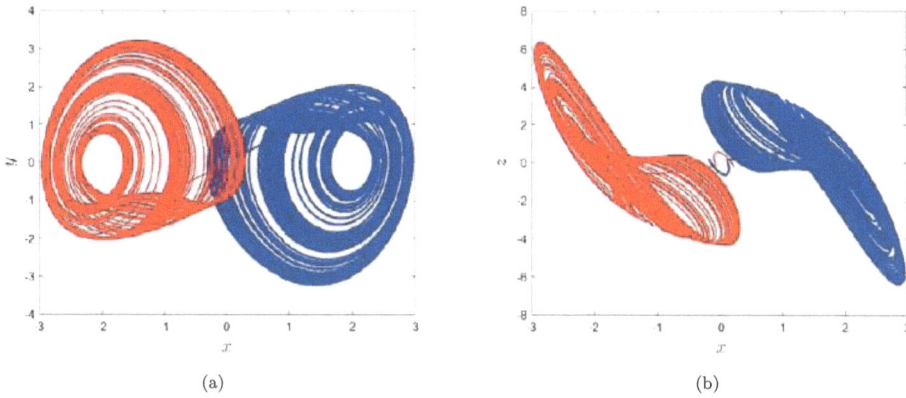

(a)   (b)

Fig. (13). Coexistence of fractional attractors in planes (x, y), and (x, z) at order $\alpha = 0.86$.

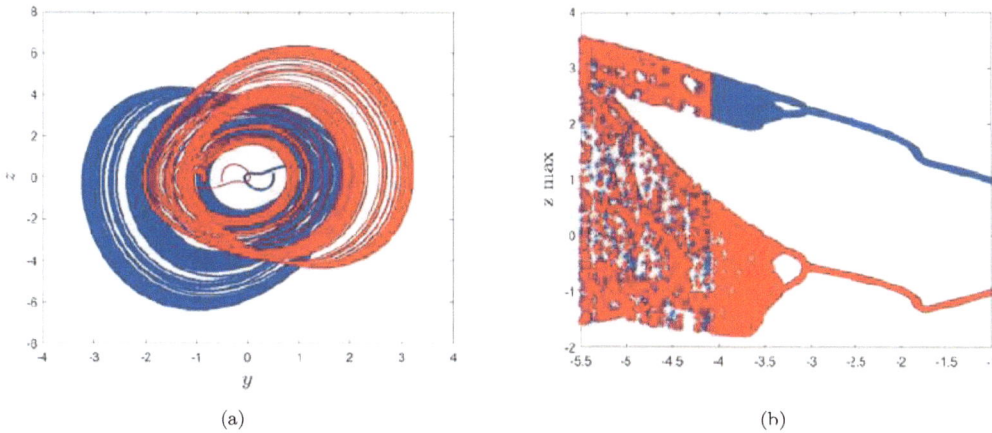

(a)   (b)

Fig. (14). Coexistence of attractors in planes (y, z), and associated bifurcation maps.

This new method to create some attractors will be developed further in the literature. It is also important to mention that the coexistence of two attractors is sensitive to the initial conditions and the variation of the considered fractional model's parameter (Figs. **15a, b** and **16a, b**). We can observe in this chapter the modification of the second initial condition $(0.2, -0.5, -0.2)$ does not modify the behaviors; the system stays to be chaotic, see the values of the Lyapunov exponents

$$Lyap1 = 0.6283, \qquad Lyap2 = -0.3018, \qquad Lyap3 = -2.2620.$$

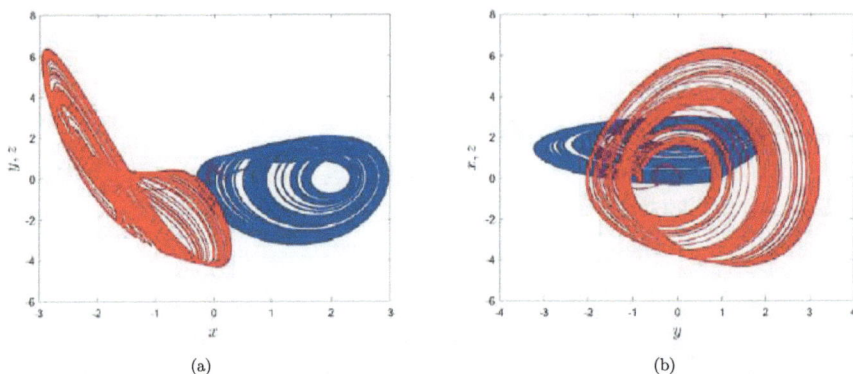

(a) (b)

Fig. (15). Pair of two attractors in planes (x, y)(blue) and (x, z)(red), and pair of two attractors in planes (y, x)(blue) and (y, z)(red).

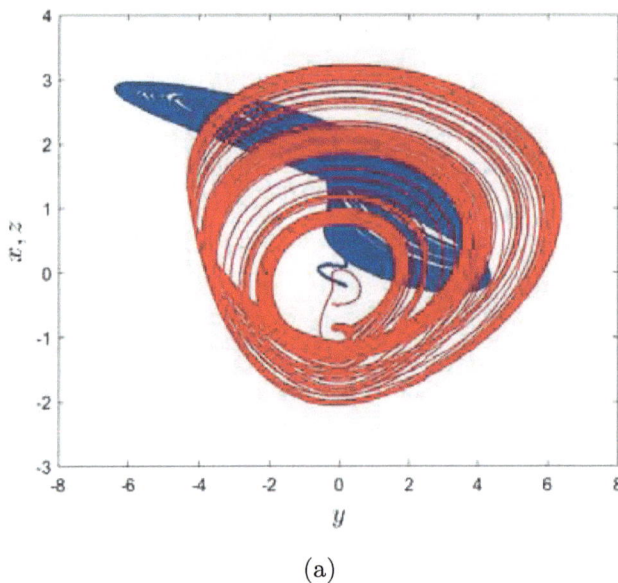

(a)

Fig. (16). Pair of two attractors in planes (z, x)(blue) and (z, y)(red).

Circuit Implementations of the Chaotic System

In this section, we present the schematic circuit of the considered chaotic system. We begin in this section, with the order $\alpha = 1$. This section will confirm the previous results addressed in this chapter. The initial equation was scaled and is now represented as the following differential equation

$$x' = y,$$

$$y' = z,$$

$$z' = -ax - 3.5y - z - 16x^3,$$

where x in initial equation is replaced by $4x$, y in initial equation is $4y$, and z in initial equation is $4z$. Introducing the resistors and capacitors in the modeling, we have to consider the following differential equations

$$x' = \frac{1}{R_1 C_1} y,$$

$$y' = \frac{1}{R_2 C_2} z, \tag{6}$$

$$z' = -\frac{1}{R_3 C_3} x - \frac{1}{R_4 C_3} y - \frac{1}{R_5 C_3} z - \frac{1}{10 R_6 C_3} x^3.$$

For the electrical circuit's schematic associated with the differential equations in (6), we will use ten resistors, three capacitors, two analog multipliers, three ground points, and six VCC points. We fix the resistors and the capacitors as follows $R_1 = R_2 = R_5 = 250k\Omega$, $R_3 = 45.45k\Omega$, $R_4 = 71.42k\Omega$, and $R_6 = 1.5625k\Omega$, and $C_1 = C_2 = C_3 = 1nF$. Now we simulate the schematic circuit by considering the real values of the resistors and capacitors. Note that in the modeling, two multipliers have been utilized to model x^3. The output gain for the multiplier is $1V$, and the output offset $0V$. In the simulation, we influenced the multipliers and the Capacitors' values to obtain phase portraits of good quality. After the simulations, the results are depicted in the following (Figs. **17a, b** and **18a**). We obtain the following phase portraits (**17a, b** and **18a**) for our considered chaotic system.

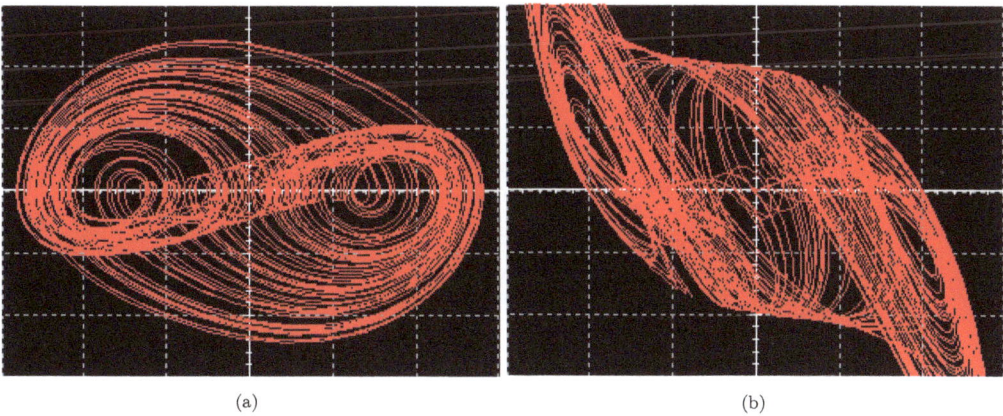

(a)   (b)

Fig. (17). Dynamics evolution in the oscilloscope trough (x, y) and (x, z) planes.

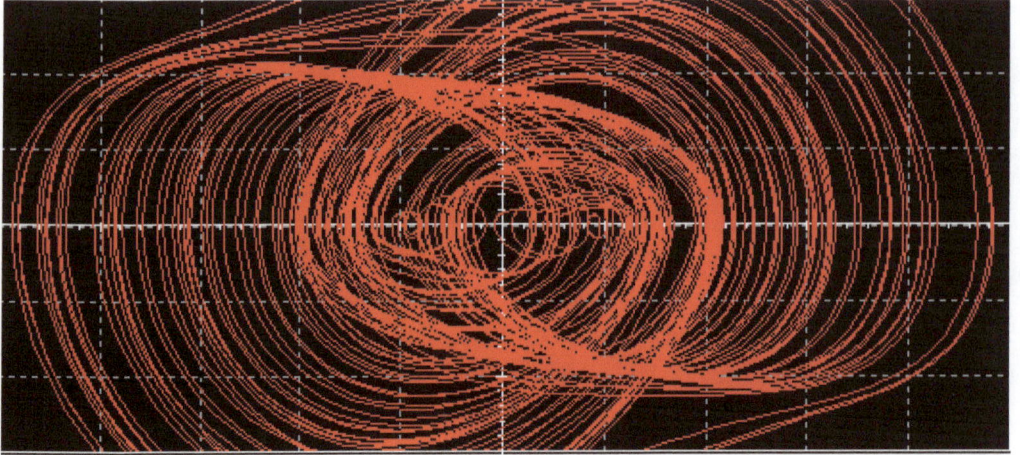

(a)

Fig. (18). Dynamics evolution in the oscilloscope trough (y, z) plane.

We can observe the MATLAB results, and the results in the Figs. (**17a, b** and **18a**) obtained after Multisim simulation are in good agreement. In the second part, we consider incommensurate order; namely, we try to draw the circuit schematic by considering the fractional differential equation

$$D_c^{\alpha^1} x = y,$$

$$D_c^{\alpha^2} y = z, \tag{7}$$

$$D_c^{\alpha^3} z = -ax - 3.5y - z - 16x^3,$$

where the order are $\alpha^1 = 1$, $\alpha^1 = 1$ and $\alpha^3 = 0.99$. To draw this circuit, note that in the circuit associated with the integer-order version, the variable x and y will not change. We introduce in the third equation the fractional-order integrator, which approximation is given by

$$\frac{1}{s^{0.99}} \approx \frac{1.073\,(s+1.0235)\,(s+107.2)}{(s+0.0102)\,(s+1.072)\,(s+112.3)}. \tag{8}$$

The transfer function can be written as the form

$$Z(s) = \frac{\frac{1}{C_1}}{s + \frac{1}{R_1 C_1}} + \frac{\frac{1}{C_2}}{s + \frac{1}{R_2 C_2}} + \frac{\frac{1}{C_3}}{s + \frac{1}{R_3 C_3}}. \tag{9}$$

Comparing Eq. (8) and Eq. (9), we arrive at the values of the resistors, and the capacitors in the fractional integrator, and they should be $C_4 = 20.367nF$, $C_5 = 21.23nF$, $C_6 = 1.0238nF$ and $R_9 = 0.1748k$, $R_{14} = 17.56k$, $R_{15} = 38304k$. For comparison, we first represent the phase portraits with MATLAB with the orders $\alpha^1 = 1$, $\alpha^2 = 1$ and $\alpha^3 = 0.99$ for model (7), we have the following (Figs. **19a, b** and **20a, b**).

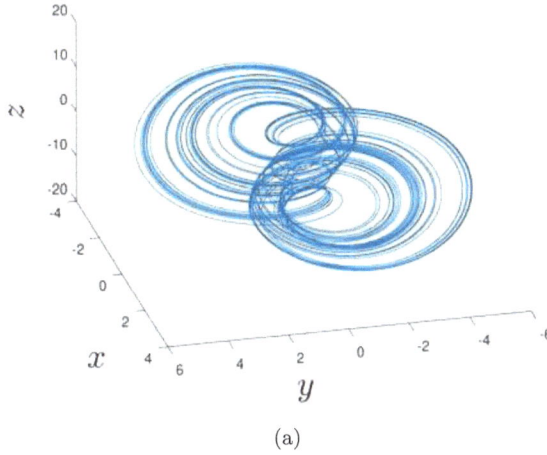

(a)

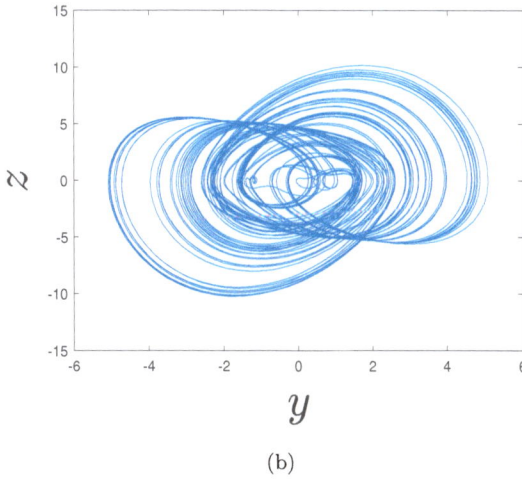

(b)

Fig. (19). Solutions in planes (x, y, z), and (y, z) at orders $\alpha^1 = 1$, $\alpha^2 = 1$ and $\alpha = 0.99$.

To see more, the influence of the orders of the Caputo derivative in the dynamic of the incommensurate fractional-order chaotic system, we consider the phase portraits at the orders $\alpha^1 = 1$, $\alpha^2 = 0.94$ and $\alpha = 0.94$ in Figs. (**24a, b** and **25a, b**).

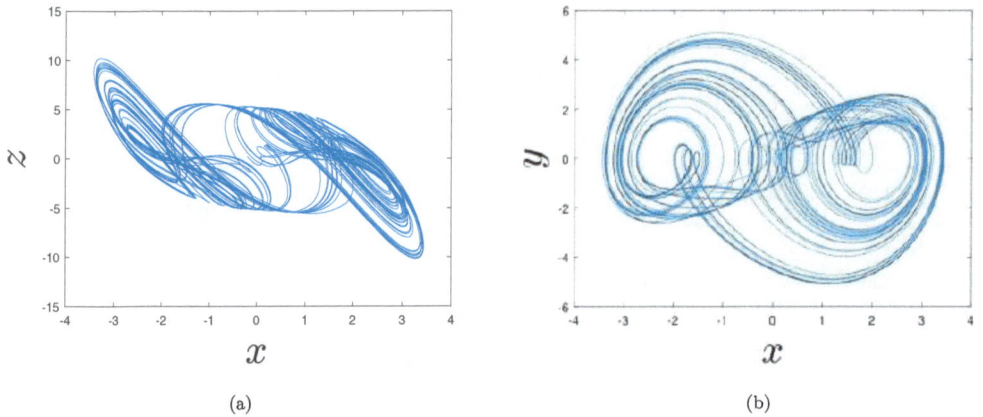

(a)   (b)

Fig. (20). Solutions in planes (x, z), and (x, y) at orders $\alpha^1 = 1$, $\alpha^2 = 1$ and $\alpha = 0.99$.

With the phase portraits obtained in the oscilloscopes (Figs. **21-23**), it has been observed that the theoretical results, and the experimental results are in good agreements.

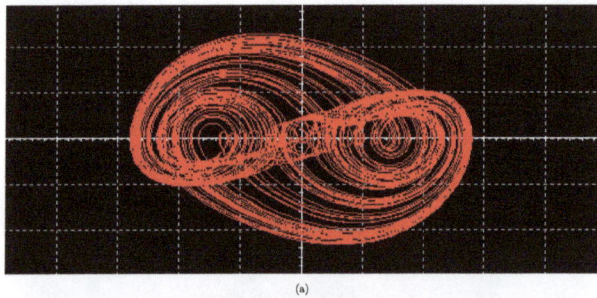

(a)

Fig. (21). Dynamics evolution in the oscilloscope thought (x, y) plane.

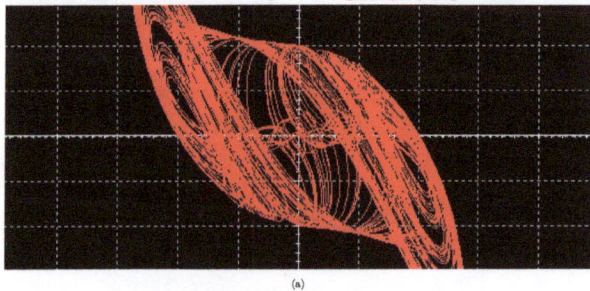

(a)

Fig. (22). Dynamics evolution in the oscilloscope thought (x, z) plane.

To see more, the influence of the orders of the Caputo derivative in the dynamic of the incommensurate fractional-order chaotic system, we consider the phase portraits at the orders $\alpha^1 = 1$, $\alpha^2 = 0.94$ and $\alpha = 0.94$ in Figs. (**24a, b** and **25a, b**).

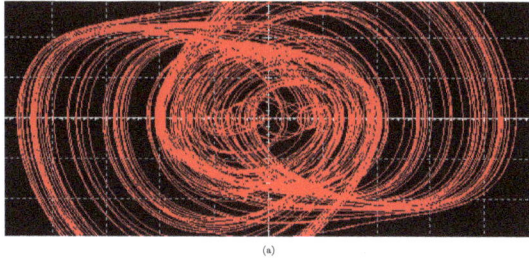

Fig. (23). Dynamics evolution in the oscilloscope thought (y, z) plane.

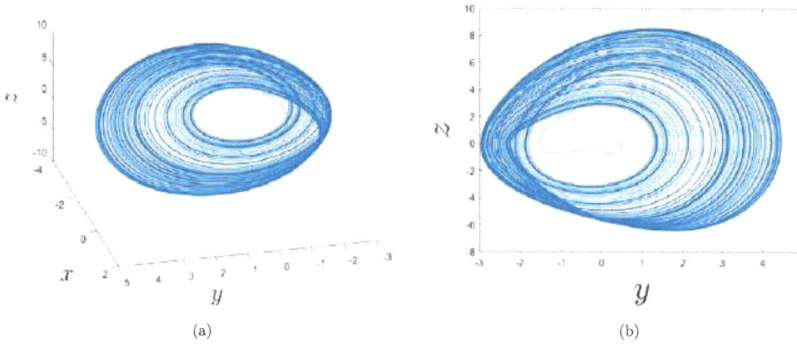

Fig. (24). Solutions in planes (x, y, z), and (y, z) at orders $\alpha^1 = 1$, $\alpha^2 = 0.94$ and $\alpha = 0.94$.

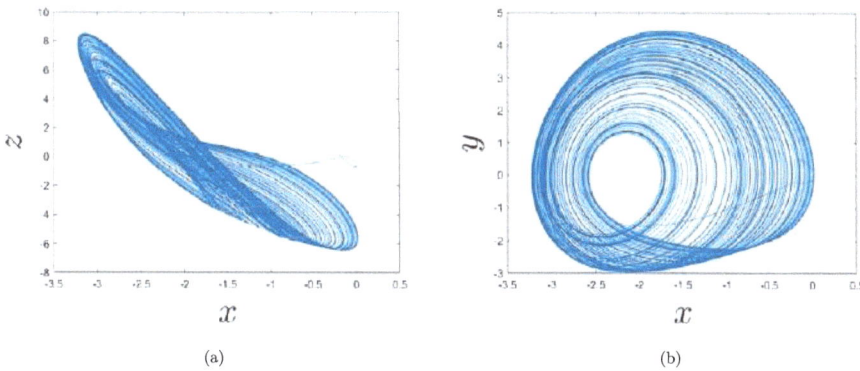

Fig. (25). Solutions in planes (x, z), and (x, y) at orders $\alpha^1 = 1$, $\alpha^2 = 0.94$ and $\alpha = 0.94$.

CONCLUSION

Many points such as the bifurcation diagram with parameters evolutions, bifurcation for coexistence, the coexistence, the phase portraits of the fractional-order chaotic system, the numerical procedure to get the graphical representations, the application of the model in real data and the Lyapunov exponents in the context of the Caputo derivative have been discussed in this chapter. The central importance of this chapter is the schematic of the circuit proposed for our fractional-order chaotic system. The discretization procedure, including the discretization of the Riemann-Liouville derivative, is highly beneficial in our investigations in this present chapter. Future research directions advise investigators to check the same model and study the same properties as described in this chapter by considering the incommensurate orders notably, how to get the Lyapunov exponents when the fractional order incommensurate system are considered. Via bifurcation, Lyapunov exponents, and the stability analysis, we have detected in this chapter the so-called chaotic regions. The advantage of the used numerical scheme is the numerical discretization of the Riemann-Liouville fractional integral. The second advantage is the stability and the convergence of the utilized numerical, obtained when the functions used to model the chaotic system are Lipschitz continuous. The last advantage of the numerical scheme is the implementation in MATLAB to depict the phase portrait that has no inconveniences.

CONSENT FOR PUBLICATION

Not applicable.

CONFLICT OF INTEREST

The author declares no conflict of interest, financial or otherwise.

ACKNOWLEDGEMENT

Declared none.

REFERENCES

[1]   A. Akgul, C. Li, and I. Pehlivan, "Amplitude Control Analysis of a Four-Wing Chaotic Attractor, its Electronic Circuit Designs and Microcontroller-Based Random Number Generator", *J. Circuits Syst. Comput.,* vol. 26, no. 12, p. 1750190, 2017.
      http://dx.doi.org/10.1142/S0218126617501900
[2]   S. Vaidyanathan, C. Volos, and V.T. Pham, "Hyperchaos, adaptive control and synchronization of a novel 5-D hyperchaotic system with three positive Lyapunov exponents and its SPICE implementation", *Archives of Control Sciences,* vol. 24, no. 4, pp. 409-446, 2014.
      http://dx.doi.org/10.2478/acsc-2014-0023
[3]   K. Rajagopal, A. Akgul, S. Jafari, A. Karthikeyan, U. Cavusoglu, and S. Kacar, "An exponential jerk system, its fractional-order form with dynamical analysis and engineering application", *Soft Comput.,* vol. 24, no. 10, pp. 7469-7479, 2020.
      http://dx.doi.org/10.1007/s00500-019-04373-w
[4]   S. Jafari, and J.C. Sprott, "Simple chaotic flows with a line equilibrium", *Chaos Solitons Fractals,* vol. 57, pp. 79-84, 2013.
      http://dx.doi.org/10.1016/j.chaos.2013.08.018

[5]   J.C. Sprott, S. Jafari, A.J.M. Khalaf, and T. Kapitaniak, "Megastability: Coexistence of a countable infinity of nested attractors in a periodically-forced oscillator with spatially-periodic damping", *Eur. Phys. J. Spec. Top.*, vol. 226, no. 9, pp. 1979-1985, 2017.
        http://dx.doi.org/10.1140/epjst/e2017-70037-1

[6]   A.A. Kilbas, H.M. Srivastava, and J.J. Trujillo, "Theory and applications of fractional differential equations", *North-Holland Mathematics Studies,* vol. 204, Elsevier: Amsterdam, The Netherlands, 2006.

[7]   I. Podlubny, "Fractional Differential Equations", *Mathematics in Science and Engineering,* vol. 198, Academic Press: New York, NY, USA, 1999.

[8]   A. Atangana, and D. Baleanu, "New fractional derivatives with nonlocal and non-singular kernel: Theory and application to heat transfer model", *Therm. Sci.,* vol. 20, no. 2, pp. 763-769, 2016.
        http://dx.doi.org/10.2298/TSCI160111018A

[9]   M. Caputo, and M. Fabrizio, "A new definition of fractional derivative without singular kernel", *Progr. Fract. Differ. Appl,* vol. 1, no. 2, pp. 1-15, 2015.

[10]   N. Sene, "Solutions for some conformable differential equations", *Progr. Fract. Differ. Appl,* vol. 4, no. 4, pp. 493-501, 2020.

[11]   F. Mansal, and N. Sene, "Analysis of fractional fishery model with reserve area in the context of time-fractional order derivative", *Chaos Solitons Fractals,* vol. 140, p. 110200, 2020.
        http://dx.doi.org/10.1016/j.chaos.2020.110200

[12]   F. Özköse, M.T. Şenel, and R. Habbireeh, "Fractional-order mathematical modelling of cancer cells-cancer stem cells-immune system interaction with chemotherapy", *Mathematical Modelling and Numerical Simulation with Applications,* vol. 1, no. 2, pp. 67-83, 2021.
        http://dx.doi.org/10.53391/mmnsa.2021.01.007

[13]   H. Joshi, and B.K. Jha, "Chaos of calcium diffusion in Parkinson s infectious disease model and treatment mechanism via Hilfer fractional derivative", *Mathematical Modelling and Numerical Simulation with Applications,* vol. 1, no. 2, pp. 84-94, 2021.
        http://dx.doi.org/10.53391/mmnsa.2021.01.008

[14]   P.A. Naik, Z. Eskandari, and H.E. Shahraki, "Flip and generalized flip bifurcations of a two-dimensional discrete-time chemical model", *Mathematical Modelling and Numerical Simulation with Applications,* vol. 1, no. 2, pp. 95-101, 2021.
        http://dx.doi.org/10.53391/mmnsa.2021.01.009

[15]   T. Mekkaoui, Z. Hammouch, D. Kumar, and J. Singh, "A new approximation scheme for solving ordinary differential equation with gomez-atangana-caputo fractional derivative", *Methods of Mathematical Modelling,* vol. 15, 2019.
        http://dx.doi.org/10.1201/9780429274114-4

[16]   P.A. Naik, K.M. Owolabi, M. Yavuz, and J. Zu, "Chaotic dynamics of a fractional order HIV-1 model involving AIDS-related cancer cells", *Chaos Solitons Fractals,* vol. 140, p. 110272, 2020.
        http://dx.doi.org/10.1016/j.chaos.2020.110272

[17]   M. Yavuz, "Characterizations of two different fractional operators without singular kernel", *Math. Model. Nat. Phenom.,* vol. 14, no. 3, p. 302, 2019.
        http://dx.doi.org/10.1051/mmnp/2018070

[18]   N. Sene, "Second-grade fluid with Newtonian heating under Caputo fractional derivative: analytical investigations via Laplace transforms", *Mathematical Modelling and Numerical Simulation with Applications,* vol. 2, no. 1, pp. 13-25, 2022.
        http://dx.doi.org/10.53391/mmnsa.2022.01.002

[19]   Z. Hammouch, M. Yavuz, and N. Özdemir, "Numerical solutions and synchronization of a variable-order fractional chaotic system", *Mathematical Modelling and Numerical Simulation with Applications,* vol. 1, no. 1, pp. 11-23, 2021.
        http://dx.doi.org/10.53391/mmnsa.2021.01.002

[20]   B. Daşbaşı, "Stability analysis of an incommensurate fractional-order SIR model", *Mathematical Modelling and Numerical Simulation with Applications,* vol. 1, no. 1, pp. 44-55, 2021.
        http://dx.doi.org/10.53391/mmnsa.2021.01.005

[21]   P.A. Naik, M. Yavuz, S. Qureshi, J. Zu, and S. Townley, "Modeling and analysis of COVID-19 epidemics with treatment in fractional derivatives using real data from Pakistan", *Eur. Phys. J. Plus,* vol. 135, no. 10, p. 795, 2020.
        http://dx.doi.org/10.1140/epjp/s13360-020-00819-5 PMID: 33145145

[22]   Z.A. Zafar, N. Sene, H. Rezazadeh, and N. Esfandian, "Tangent nonlinear equation in context of fractal fractional operators with nonsingular kernel", *Math. Sci.,* pp. 1-11, 2021.
       http://dx.doi.org/10.1007/s40096-021-00403-7 PMID: 35673627

[23]   A. Yokuş, "Construction of different types of traveling wave solutions of the relativistic wave equation associated with the Schrödinger equation", *Mathematical Modelling and Numerical Simulation with Applications,* vol. 1, no. 1, pp. 24-31, 2021.
       http://dx.doi.org/10.53391/mmnsa.2021.01.003

[24]   P. Veeresha, "A numerical approach to the coupled atmospheric ocean model using a fractional operator", *Mathematical Modelling and Numerical Simulation with Applications,* vol. 1, no. 1, pp. 1-10, 2021.
       http://dx.doi.org/10.53391/mmnsa.2021.01.001

[25]   I. Petras, *Fractional-Order Chaotic Systems. Nonlinear Physical Science,* Springer book, 2011, pp. 103-184.

[26]   I. Petráš, "A note on the fractional-order Chua's system", *Chaos Solitons Fractals,* vol. 38, no. 1, pp. 140-147, 2008.
       http://dx.doi.org/10.1016/j.chaos.2006.10.054

[27]   M.F. Danca, and N. Kuznetsov, "Matlab Code for Lyapunov Exponents of Fractional-Order Systems", *Int. J. Bifurcat. Chaos,* vol. 28, no. 5, p. 1850067, 2018.
       http://dx.doi.org/10.1142/S0218127418500670

[28]   S. Ren, S. Panahi, K. Rajagopal, A. Akgul, VT. Pham, and S. Jafari, *A New Chaotic Flow with Hidden Attractor: The First Hyperjerk System with No Equilibrium,* aop, 2018.
       http://dx.doi.org/10.1515/zna-2017-0409

[29]   K. Rajagopal, A. Karthikeyan, and P. Duraisamy, "Hyperchaotic Chameleon: Fractional Order FPGA Implementation", *Complexity,* vol. 2017, pp. 1-16, 2017.
       http://dx.doi.org/10.1155/2017/8979408

[30]   V.T. Pham, C. Volos, S. Jafari, and T. Kapitaniak, "Coexistence of hidden chaotic attractors in a novel no-equilibrium system", *Nonlinear Dyn.,* vol. 87, no. 3, pp. 2001-2010, 2017.
       http://dx.doi.org/10.1007/s11071-016-3170-x

[31]   J. Lü, G. Chen, and D. Cheng, "G. chen, and D. cheng, A new chaotic system and beyond: the generalized Lorenz-like system", *Int. J. Bifurcat. Chaos,* vol. 14, no. 5, pp. 1507-1537, 2004.
       http://dx.doi.org/10.1142/S021812740401014X

[32]   T. M. Shahiri, N. A. Ranjbar, R. Ghaderi, S.H. Hosseinnia, and S. Momani, "Control and Synchronization of Chaotic Fractional-Order Coullet System via Active Controller. (2)",

[33]   M. Diouf, and N. Sene, "Analysis of the Financial Chaotic Model with the Fractional Derivative Operator", *Complexity,* vol. 2020, pp. 1-14, 2020.
       http://dx.doi.org/10.1155/2020/9845031

[34]   N. Sene, "Analysis of a Four-Dimensional Hyperchaotic System Described by the Caputo–Liouville Fractional Derivative", *Complexity,* vol. 2020, pp. 1-20, 2020.
       http://dx.doi.org/10.1155/2020/8889831

[35]   N. Sene, and A. Ndiaye, "On Class of Fractional-Order Chaotic or Hyperchaotic Systems in the Context of the Caputo Fractional-Order Derivative", *Journal of Mathematics,* 2020.
       http://dx.doi.org/10.1155/2020/8815377

[36]   N. Sene, "Analysis of a fractional-order chaotic system in the context of the Caputo fractional derivative via bifurcation and Lyapunov exponents", *Journal of King Saud University-Science,* vol. 101275.
       http://dx.doi.org/10.1016/j.jksus.2020.101275

[37]   N. Sene, "Introduction to the fractional-order chaotic system under fractional operator in Caputo sense", *Alex. Eng. J.,* vol. 60, no. 4, pp. 3997-4014, 2021.
       http://dx.doi.org/10.1016/j.aej.2021.02.056

[38]   R. Garrappa, "Numerical solution of fractional differential equations: A survey and a software tutorial 2018", *Mathematics,* vol. 6, no. 2, p. 16, 2018.
       http://dx.doi.org/10.3390/math6020016

[39]   D. Matignon, "Stability results on fractional differential equations to control processing", *IMACS,* IEEE-SMC: Lille, France, 1996, 2, 963-968.

[40]   N. Sene, "Global asymptotic stability of the fractional differential equations", *J. Nonlinear Sci. Appl.,* vol. 13, pp. 171-175, 2020.

CHAPTER 2

A New Method of Multistage Optimal Homotopy Asymptotic Method for Solution of Fractional Optimal Control Problem

Oluwaseun O. Okundalaye[1]*, Necati Özdemir[2] and Wan A. M. Othman[3]

[1] *Department of Mathematical Sciences, Faculty of Science, Adekunle Ajasin University, Akungba-Akoko, Nigeria*

[2] *Department of Mathematics, Faculty of Science and Arts, Balikesir University, Balikesir, Turkey*

[3]*Institute of Mathematical Sciences, University of Malaya, 50603 Kuala Lumpur, Malaysia*

Abstract: This paper deals with a recent approximate analytical approach of the multistage optimal homotopy asymptotic method (MOHAM) for fractional optimal control problems (FOCPs). In this paper, FOCPs are developed in terms of a conformable derivative operator (CDO) sense. It is validated that the right CDO appears naturally in the formulation even when the system dynamics are described with the left CDO only. The CDO is employed to enlarge the stability region of the dynamical systems of the optimal control problems (OCPs). The necessary and transversal conditions are achieved using a Hamiltonian technique. The results demonstrated that as the fractional-order solution derivative tends to integer-order 1, the formulations lead to integer-order system solutions. Numerical results and a comparison with the exact solution and other approximate analytical solutions in fractional order are given to validate the efficiency of the MOHAM. Some numerical examples are included to demonstrate the effectiveness and applicability of the new technique.

Keywords: Approximate analytical solution, Convergence analysis, Conformable derivative operator, Fractional calculus, Fractional Hamiltonian approach, Fractional optimal control problems.

INTRODUCTION

The global definition of an optimal control problems depends on the minimization

---
*Corresponding author Oluwaseun O. Okundalaye: Department of Mathematical Sciences, Faculty of Science, Adekunle Ajasin University, Akungba-Akoko, Nigeria;
E-mail: okundalaye.oluwaseun@aaua.edu.mg

Mehmet Yavuz & Necati Özdemir (Eds.)

of an objective function of the state and control inputs of the system over a set of relevant control functions. The OCPs usually emerge in diverse areas of applied science, and well-founded studies have been done in the classical derivatives dynamic systems.

A non-linear constrained OCPs can be of various types, relying on the conditions constrained on the final time and state. It can be grouped as fixed final state-fixed final time, free final state-fixed final time, fixed final state-free final time, and free final state-free final time. Most computing techniques for the solution of OCPs conveniently solved the unconstrained problem, but inequality constraints often resulted in both exact and numerical computational difficulties. In control, the state-space illustrations are an effective approach for stability analysis and OCPs formulation. Fractional calculus (FC) is the generalization of traditional calculus and has drawn the attention of several authors in the areas of applied science to described more precisely the dynamics of many systems using fractional calculus. It has been divulged in the literature that systems represented using FC give more interesting behaviour [1-5]. Also, it has been demonstrated that the substances with memory, genetic properties, heat conduction, and gas diffusion can be modeled more accurately with FC [6-9]. Many definitions of fractional-order derivative operator (FODO) can be seen [10-16]. The Riemann-Liouville (RL) derivative operator is not consistently usable for modeling physical systems because the RL solution requires unnatural initial conditions [17-19]. In contrast, the Caputo fractional derivative operator (CFDO) accepts initial conditions like the integer-order systems. Thus, CFDO is good for modeling physical systems [20-23], epidemiological analysis of COVID-19 [24], discretization method for an expanded family of distributions [25], fractional-order COVID-19 epidemic model [26], the transmission of COVID-19 dynamic system [27], fractional Burger–Fisher equations [28], time-fractional Burgers-Coupled equations [29-31], and time-fractional Fisher's equations [32, 33]. The fractional-order derivative operators have been applied to many problems in the area of OCPs. We refer the researchers who are interested in the theory and applications of FC to these books [34-39] with some papers on FOCPs [40-43]. The FOCPs of several cases have been constructed and considered using various variations of FC: Riemann-Liouvile for FOCPs [44, 45], Mittag-Leffler for FOCPs [46] Caputo for FOCPs [47], and Atangana-Baleanu for FOCPs [48]. Agrawal gave a general formulation and solution scheme for a class of FOCPs in terms of the RL [49]. Recently, fractional conservation laws for FOCPs with RL fractional derivatives (RLFDs) were studied [50], FOCPs in Caputo sense were addressed [51], one state and one control variable, and one fractional state equations [52], and Hamiltonian equations for fractional variational problems [53]. The FOCPs are OCPs in which the differential equations (DEs) governing the dynamics system exhibit at least one FODO [54]. Authors in [55], gave a pseudo-state-space-based FOCPs formulation and a solution scheme. Fixed and free final-time FOCPs are considered in the study [56], second-order necessary optimality condition for FOCPs in the Caputo sense [57], and FOCPs of an HIV-immune system in terms of the Caputo sense [58]. Recent approximate analytical methods (AAM) are: modified Adomian decomposition method for (FOCPs) [59], variational iteration method for optimal solutions FOCPs [60], conformable fractional optimal control problem of heat conduction equations using Laplace and finite Fourier sine transforms [61], spectral Galerkin approximation [62], new approximate-analytical solutions for PDEs [63], transcription methods for FOCPs [64], but the methods mentioned above lack convergence criteria and interval of convergent.

In 1992, Liao proposed the homotopy analysis method (HAM), independent of any small or large physical parameters [65]. Homotopy methods are robust mathematical tools for obtaining a solution to many non-linear problems. Contrary to all other approximate analytic methods, it gives us a convenient means to guarantee the convergence of the series solution of non-linear problems by putting in an auxiliary parameter, called the convergence-control parameter (CCP), and offers a solution to this problem where the exact solution is not available. Using this parameter, we can easily control and possibly extend the convergence region of the solution obtained; this merit makes HAM an excellent method for applied mathematicians [66]. It has been verified that the HAM solution with this CCP is a Taylor series expansion of the analytical solution at some point [67]. In 2003, the elementary concepts of HAM and some applications largely related to non-linear ODEs were described systematically by the author in the book "Beyond Perturbation" [68]. The HAM has attracted many researchers in nearly a score of nations. It has been intensely employed to resolve many non-linear problems in a scientific discipline, finance, and technology [69] for solving the non-linear problem, which was later advanced to OHAM for non-integer order [70], new fractional homotopy method for OCPs [71], optimal control of a constrained fractionally damped elastic beam [72], and comparisons of OHAM [73]. But MOHAM has never been used to solve FOCPs, which drives this research work. The present work aims to find the approximate analytical solutions for FOCPs using a new novel technique called MOHAM. Our focus in this paper is to widen the application of MOHAM to obtain accurate solution of FOCPs. We provide answers to convergence criteria for accurate optimal solutions, the convergence of series solutions, and the solution using MOHAM with an optimization technique of the Galerkin method. The merit of this paper is that CDO is employed to enlarge the stability region of the dynamical. The scope covers the limitations in approximate analytical methods regarding control-convergence parameters and utilized the simplicity and effectiveness of the latest conformable fractional-order derivative. The objective is to develop a conformable fractional mathematical model for non-linear FOCPs. It is hoped that the simplicity of this formulation will open a new development in the area of FOCPs.

We arrange the paper as follows: In Section 2, a brief introduction to the fractional calculus and the necessary optimality conditions are discussed. In Section 3, we formulate MOHAM with FOCPs and show the convergence theorem of the technique. We present numerical examples and results in Section 4. Finally, in Section 5, we present the conclusion.

FRACTIONAL CALCULUS

Fractional calculus, as the generalization of traditional calculus, has contributed a substantial part to mathematics, engineering, and different sciences. Literature review that fractional calculus provides more exact models of many applications and shows the behaviour of the dynamic system in sciences than traditional calculus [74, 75]. There exist several definitions regarding the FC, and some basic and common definitions are Riesz, Riemann–Liouville (RL), Hadamard, Grünwald–Letnikov (GL), Caputo–Fabrizio (CF), generalized Caputo FDE [76], Mittag-Leffler kernels [77] and Atangana–Baleanu (AB) in the literature. Furthermore, many researchers studied new fractional operators with local, nonlocal, singular, and non-singular kernels [78-81]. The conformable derivative was introduced in [82] based on the concept of the local derivative with fractional components. This derivative allows for many extensions of some fundamental theorems in calculus (i.e., the product rule, Rolle's theorem, chain rule, and mean value theorem). It can be found that many authors focus on conformable derivative operators of a real-life problem [83-88] with the definition of a conformable derivative operator (CDO) that conserves many properties of non-integer order derivative [89-91], and generalized conformable for mean value theorem [92].

The definition of conformable derivative (CFD) preserves many properties of classical order derivatives [93]. Some features that we will adopt are:

Definition 2.1. A (left) fractional derivative starting from s of a function ξ: $[s, \infty) \to \Re$ of order $\alpha \in (m-1, m), m \in N$ is defined by

$$T_s^\alpha \xi(t) = \xi^{(\alpha)}(t) = \lim_{\epsilon \to 0} \frac{\xi^{(m-1)}\left(t+\epsilon(t-s)^{(m-\alpha)}-\xi^{(m-1)}t\right)}{\epsilon}, t > s, \qquad (1)$$

$$T_s^\alpha \xi(s) = \lim_{x \to s^+} T_s^\alpha \xi(t), \tag{2}$$

Provided the limits exist, and $\xi(t)$ is $(m-1)-$differentiable at $t > s$.

We can define the (right) fractional derivative terminating at s of a function $\xi:(-\infty, s] \to \Re$ of order $\alpha \in (m-1, m), m \in N$ is defined by

$$_s^\alpha T\xi(t) = \xi^{(\alpha)}(t) = (-1)^m \lim_{\epsilon \to 0} \frac{\xi^{(m-1)}\left(t + \epsilon(s-t)^{(m-\alpha)} - \xi^{(m-1)}t\right)}{\epsilon}, t < s, \tag{3}$$

$$_s^\alpha T\xi(s) = \lim_{x \to s^-} {}_s^\alpha T\xi(t), \tag{4}$$

Provided the limits exist, and $\xi(t)$ is $(m-1)$-differentiable at $t < s$.

If $T_s^\alpha \xi(t)$ exists on $t > s$, then we say that ξ is left $\alpha-$differentiable on $t > s$ whereas ξ is right $\alpha-$differentiable on $t < s$.

Definition 2.2. A (left) fractional integral starting from s of a function $\xi:[s, \infty) \to \Re$ of order $\alpha \in (m-1, m), m \in N$ is defined by

$$I_s^\alpha \xi(t) = \frac{1}{(m-1)!} \int_s^t \frac{(t-x)^{m-1}\xi(x)}{(x-s)^{m-\alpha}} dx, \alpha > 0, t > s, \tag{5}$$

$$I_s^0 \xi(x) = \xi(x). \tag{6}$$

We can define the (right) fractional integral terminating at s of a function $\xi:(-\infty, s] \to \Re$ of order $\alpha \in (m-1, m), m \in N$ as follows:

$$_s^\alpha I\xi(t) = \frac{1}{(m-1)!} \int_t^s \frac{(x-t)^{m-1}\xi(x)}{(s-x)^{m-\alpha}} dx, \alpha > 0, t < s, \tag{7}$$

$$_s^0 I\xi(x) = \xi(x). \tag{8}$$

It is worth mentioning here that $I_s^\alpha I_s^\beta \xi(t) \neq I_s^\beta I_s^\alpha \xi(t)$ and $_s^\alpha I_s^\beta I\xi(t) \neq_s^\beta I_s^\alpha I\xi(t)$.

Lemma 2.1. If $\alpha \in (m-1, m), m \in N$ and $\xi:[s, \infty) \to \Re$ is $(m-1)$-differentiable, then

(1) $T_s^\alpha I_s^\alpha \xi(t) = \xi(t),$

(2) $I_s^\alpha T_s^\alpha \xi(t) = \xi(t) - \sum_{k=0}^{m-1} \xi^{(k)}(s) \frac{(t-s)^k}{k!}, t > s.$

Lemma 2.2. If $\alpha \in (m-1, m)$, $m \in N$ and $\xi:[-\infty, s) \to \Re$ is $(m-1)$-differentiable, then

(1) $_s^\alpha T_s^\alpha I \xi(t) = \xi(t)$,

(2) $_s^\alpha I_s^\alpha T \xi(t) = \xi(t) - \sum_{k=0}^{m-1} (-1)\xi^{(k)}(s) \frac{(s-t)^k}{k!}, t < s.$

THE HAMILTONIAN OPTIMALITY CONDITIONS FORMULATION

We start by looking into the following non-linear constrained dynamic system

$$T_s^\alpha \xi = A(t)\xi + B(t)u, \tag{9}$$

$$\xi(a) = \xi_a, \tag{10}$$

with control signal $u(t)$ which minimizes the cost function

$$J = \frac{1}{2} \int_a^T [\xi^T Q(t)\xi + u^T R(t)u]dt, \tag{11}$$

where $u(t)$ is the state variable and $0 < \alpha < 1$. $Q(t)$ and $R(t)$ are selected to be positive-semi definite and positive-definite matrices, respectively. We obtain the necessary equations by defining the Hamiltonian

$$H(p,t) = \frac{1}{2}\xi^T Q(t)\xi + \frac{1}{2}u^T R(t)u + p^T[A(t)\xi + B(t)u], \tag{12}$$

where p is the vector of the Lagrange multiplier, and so we define a performance index as

$$J = \int_a^T [H(\xi, u, p, t) - p^T T_s^\alpha \xi]dt, \tag{13}$$

Using the method of integration by parts, we have

$$J = [\xi(t)I_\alpha p(t)]_a^T + \int_a^T [H(\xi, u, p, t) - \xi_t T_s^\alpha p]dt, \tag{14}$$

taking the condition $\xi(t_0)$ is fixed, and $\xi(T)$ is unspecified. Therefore, the vectors ξ and u should satisfy equation $J = 0$ with the relations below

$$\frac{\partial H}{\partial \xi} = Q(t)\xi + A^T(t)p = T_s^\alpha p, \tag{15}$$

$$\frac{\partial H}{\partial p} = A(t)\xi + B^T(t)u =_s^\alpha T\xi, \tag{16}$$

$$\frac{\partial H}{\partial u} = 0 = R(t)u + B^T(t)p, \tag{17}$$

$$p(T) = 0, \tag{18}$$

From relation Eq. (17) we have

$$u(t) = -R^{-1}(t)B^T(t)p. \tag{19}$$

The Hamiltonian equations for FOCP Eqs. (15)-(17) represent the necessary conditions for the optimality of FOCP. The difference between arbitrary and integer OCP is that the arbitrary-order FOCP consists of the left and the right FD. Examine that Eq. (15) contains the left conformable derivative, whereas Eq. (16) contains the right conformable derivative. This demonstrates that the solution of OCP requires the insight of not only forward derivatives (FD) but also backward derivatives (BD) to account for end conditions. In integer-order OCP theories, this issue is either not addressed or not highlighted. This is broad because the BD of order 1 turns out to be negative of the FD of order 1. It can be shown that the $\lim_{\alpha \to 1}$ Eqs. (15)-(16) goes back to those obtained using standard methods.

THE OHAM FORMULATION WITH FRACTIONAL OPTIMAL CONTROL PROBLEMS

We exemplify the basic concept of the OHAM by considering the following general equations

$$T_s^\alpha \xi_k(t) + L_k(\xi_k(t)) + N_k(\xi_k(t)) - g_k(t) = 0 \ \ t \in [0,1] \ \ k = 1,2\ldots m, \tag{20}$$

with initial conditions

$$\xi_k(b) = a_i, \tag{21}$$

where T_s^α is the CFD, L_k is a linear operator, N_k is a non-linear operator, t is an independent variable, $x_k(t)$ is an unknown function, φ is the problem domain, and $g_k(t)$ is a known function. According to OHAM, we formulate a homotopy map $H_k(\phi_i(t,p)): \varphi \times [0,1] \to \varphi$ which satisfies Eq. (20) can be constructed using OHAM as.

$$(1 - \ell)[T_s^\alpha(\aleph_k(t, \ell))] = H_k(\ell)[T_s^\alpha \aleph_k(t, \ell) + N_k \aleph_k(t, \ell) + L_k \aleph_k(t, \ell) + g_k(t)], \qquad (22)$$

where embedding parameter (ℓ) is $0 \leq \ell \leq 1$, auxiliary function $H_k(\ell)$ $\forall$ $\ell \neq 0$, unknown function $(\aleph_k(t, \ell))$ and $H(0) = 0$. When $\ell = 0$ and $\ell = 1$, it holds that $\aleph_k(t, 0) = \psi_{k,0}(t)$ and $\aleph_k(t, 1) = \psi_k(t)$ respectively. Thus as ℓ moves from 0 to 1, the solution $\aleph_k(t, \ell)$ approach from $\psi_{k,0}(t)$ to $\psi_k(t)$, where initial guess $\psi_{k,0}(t)$ satisfies the linear operator generated from Eq. (20) for $\ell = 0$ as

$$T_s^\alpha(\psi_{k,0}(t)) = 0.\, \psi_{k,0}(b) = 0. \qquad (23)$$

The $H_k(\ell)$ is given as

$$H_k(\ell) = \sum_{j=1}^n \ell^j C_j, \qquad (24)$$

where C_j^s can be known later. We get an approximate solution by expanding $\aleph_k(t, \ell, C_j)$ in Taylor's series in terms of ℓ,

$$\aleph_k(t, \ell, C_j) = \psi_{k,0}(t) + \sum_{k \geq 1} \psi_{i,k}(t, C_j)\ell^i \quad j = 1, 2, \ldots, n, \qquad (25)$$

using above in Eq. (20) with collections of the coefficient like the power of ℓ gives the governing equations $\psi_{i,0}(t)$ in a linear form in Eq. (23). Then 1st problems are given as

$$T_s^\alpha(\psi_{k,1}(t)) + g_k(t) = C_1 N_0(\psi_{k,0}(t)), \psi_{k,1}(b) = 0, \qquad (26)$$

the general governing equations for $\psi_{k,i}(t)$ is

$$T_s^\alpha\left(\psi_{k,i}(t)\right) - T_s^\alpha\left(\psi_{k,i-1}(t)\right) = C_i N_{k,0}\left(\psi_{k,0}(t)\right) +$$
$$\sum_{m=1}^{i-1} C_{j,m}\left[T_s^\alpha\left(\psi_{k,i-m}(t)\right) + \right.$$
$$\left. N_{k,i-m}\left(\psi_{k,i-1}(t)\right)\right] \qquad (27)$$

$$\psi_{k,i}(b) = 0, \quad i = 2, 3, \ldots m,$$

where $N_{k,m}(\psi_0(t), \psi_{k,1}(t), \ldots, \psi_{k,m}(t))$ is the coefficient of ℓ^m, produce by expanding $N_k(\aleph_k(t, \ell, C_j))$ in series relating to ℓ

$$N_k(\aleph_k(t, \ell, C_j)) = N_{k,0}(\psi_{k,0}(t)) + \sum_{m \geq 1} N_{k,m}(\psi_0, \psi_1, \ldots \psi_m)\ell^m, \qquad (28)$$

The convergence of series solution Eq. (29) relies on C_j^S. If it's convergent at $\ell = 1$ gives solution to Eq. (20) as

$$\psi_k(t, C_j) = \psi_{k,0}(t) + \sum_{k\geq 1}^{m} \psi_{i,k}(t, C_j), j = 1, 2, \dots, n, \tag{29}$$

using Eq. (30) in Eq. (20), we have an expression for the residual error as

$$R_k(t, C_j) = T_s^\alpha(\psi_k(t, C_j)) + L_k(\psi_k(t, C_j)) + N_k(\psi_k(t, C_j) - g_k(t)). \tag{31}$$

If

$$R_k(t, C_j) = 0, \tag{32}$$

then $\tilde{\psi}_k(t, C_j)$ is the exact solution. Usually, such a case does not occur. We adopt optimization techinque of Galerkin method to find the optimal values C_j^S as given below

$$\ell_k = \frac{\partial \tilde{\psi}_k(t, C_j)}{\partial C_j} = 0 \quad k = 1, 2, \dots m, \tag{33}$$

minimize the functional

$$\Delta_k(C_j) = \int_a^b \ell_k \times R_k(t, C_j) dt. \tag{34}$$

Where the values of a and b depend on the given problem. With these known C_j^S, the approximate analytical solution Eq. (20) is well known.

If the interval of the time variable is long, then OHAM fails to reach accurate solutions. MOHAM overcome this shortcoming by partitioning the time interval $[t_0, T]$, into N subintervals $[t_0, t_1) \dots, [t_{\gamma-1}, t_\gamma]$ where $t_\gamma = T$ and OHAM will be utilized over each subinterval. The endpoint in each sub-interval denotes an initial approximation to the solution over the following interval. The procedure will continue until we obtain a pre-assigned time (T). Utilization of MOHAM is relative to OHAM, with some minor changes from C_i to $C_{i,j}$ respectively. Also, initial approximation in $[t_{y-1}, t_y]$, $\gamma = 0, 1, \dots, N - 1$, will be considered as

$$u_{0,j}(t_j) = \alpha_j, \quad j = 1..N. \tag{35}$$

In addition, the deformation equation in each subinterval will change to the following

$$(1-p)[L_j(u_j(t,p)) - u_{0,j}(t)] = H_j(P,t)[L_j(u_j(t,P)] + f(t) + N_j(u_j(t,P))], \tag{36}$$

$H(p,t)$ will be generalized as follows,

$$H_j(P,t) = (C_{1,j} + C_{2,j}t + C_{3,j}t^2 + \ldots)P, \quad j = 1,\ldots,N. \tag{37}$$

For $i = 1,2,\ldots m$, and $j = 1,2,\ldots N$, we have

$$u_j(t,C_{i,j}) = u_{0,j}(t) + \sum_{k=1}^{m} u_{k,j}(t,C_{i,j}), \tag{38}$$

$$R_j(t,C_{i,j}) = T_s^\alpha(u_j(t,C_{i,j})) + L_j(u_j(t,C_{i,j})) + N_j(u_j(t,C_{i,j})) - g_j(t), \tag{39}$$

$$J_j(C_{i,j}) = \int_{t\gamma}^{t\gamma+h} R_j^2(s,C_{i,j})ds \quad \gamma = 0,1,\ldots N-1. \tag{40}$$

The length of the subinterval $[t_\gamma, t_{\gamma+1}]$ is h, and the number of subintervals is $N = \lfloor\frac{T}{N}\rfloor$. Now we consider the derivatives Eq. (40) off for $C_{i,j}$ to zero. We define $\alpha_j = u_j(t_j)$ in each subinterval $[t_\gamma, t_{\gamma+1})$. Therefore, the convergence control parameters can be determined from the solution of the following system of equations.

$$\frac{\partial J_j}{\partial C_{1,j}} = \frac{\partial J_j}{\partial C_{2,j}} = \ldots = \frac{\partial J_j}{\partial C_{m,j}} = 0. \tag{41}$$

We calculate the approximate analytical solutions on each subinterval as follows.

$$u(t) = \begin{pmatrix} u_1(t), & t_0 \le t < t_1, \\ u_2(t), & t_1 \le t < t_2, \\ \cdot \\ \cdot \\ \cdot \\ u_N(t), & t_{N-1} \le t \le T. \end{pmatrix} \tag{42}$$

We calculate the correctness of OHAM by

(1) Error norm L_2

$$L_2 = ||\Psi^{exact} - \Psi_N|| \approx \sqrt{\frac{b-a}{N} \sum_{k=0}^{N} |\psi_k^{exact} - (\psi_N)_k|^2},\qquad (43)$$

(2) Error norm L_∞

$$L_\infty = ||\Psi^{exact} - \Psi_N||_\infty \approx max_k |\psi_k^{exact} - (\psi_N)_k|.\qquad (44)$$

CONVERGENT THEOREM

We prove the convergent theorem of MOHAM by following the approach used in [85].

Theorem. For the stage j, if the series Eq. (38) converges to $u_j(t)$, where $u_{j,k}(t) \in L(R^+)$ is produced by Eq. (26) and the k-order deformation Eq. (27), then $u_i(t)$ is the exact solution of Eq. (20).

Proof. Since the series $\sum_{k=1}^{\infty} u_{j,k}(t, C_{i,j})$ is convergent, we have $u_j(t, C_{i,j})$ then

$$\lim_{k \to \infty} u_{j,k}(t, C_{i,j}) = 0 \quad \forall \quad k = 1,2\ldots n.\qquad (45)$$

from Eq. (27), we can write

$$\sum_{i=1}^{\infty} [C_i N_{k,0}(u_{j,0}(t)) + \sum_{m=1}^{i-1} C_{j,m}[T_s^\alpha(u_{j,i-m}(t)) + N_{k,i-m}(u_{j,i-1}(t))]$$

$$= \sum_{k=1}^{\infty} [T_s^\alpha(u_{j,k}(t)) - T_s^\alpha \xi(u_{j,k-1}(t))],\qquad (46)$$

$$= \lim_{n \to \infty} \sum_{k=1}^{n} T_s^\alpha(u_{j,k}(t)) - T_s^\alpha(u_{j,k-1}(t)),\qquad (47)$$

$$= T_s^\alpha u_{11}(t) + (T_s^\alpha u_{22}(t) - T_s^\alpha u_{21}(t)) + .. + (T_s^\alpha u_{nn}(t) - T_s^\alpha u_{n(n-1)}(t)),\ (48)$$

$$= T_s^\alpha [\lim_{n \to \infty} \sum_{m=1}^{n} u_{nn}(t)] = T_s^\alpha [\lim_{n \to \infty} u_{nn}(t)] = 0,\qquad (49)$$

equating the RHS of Eq. (49) with the equation below

$$0 = \sum_{m=1}^{\infty} [T_s^\alpha u_{j(m-1)} + L_j u_{j(m-1)} + N_j u_{j(m-1)} - g_j(t)],\qquad (50)$$

$$\sum_{m=1}^{\infty} [T_s^{\alpha} u_{j(m-1)} + L_j u_{j(m-1)} + N_j u_{j(m-1)})] = g_j(t), \tag{51}$$

$$T_s^{\alpha} u_j(t, C_{i,j}) + L_j u_j(t, C_{i,j}) + N_j u_j(t, C_{i,j}) - g_j(t) = 0 \quad \forall \ j = 1,2..n. \tag{52}$$

If the $C_{i,j}$ is chosen properly, then Eq. (52) leads to the solution of Eqs. (20)-(21).

NUMERICAL EXAMPLE AND RESULTS

Example 1. Find the control $u(t)$ that minimizes the quadratic performance index

$$J(u) = \frac{1}{2} \int_0^1 [z^2(t) + u^2] dt, \tag{53}$$

subject to system dynamics

$$T_s^{\alpha} = -z + u,$$

and the initial condition

$$z(0) = 1. \tag{54}$$

The exact solution for the system for $\alpha = 1$ is given as

$$z(t) = cosh(\sqrt{2}t) + \beta sinh(\sqrt{2}t),$$

$$u(t) = (1 + \sqrt{2\beta})cosh(\sqrt{2}t) + (\sqrt{2} + \beta)sinh(\sqrt{2}t),$$

where

$$\beta = \frac{cosh(\sqrt{2} + \sqrt{2}sinh(\sqrt{2})}{\sqrt{2cosh(\sqrt{2} + sinh(\sqrt{2})}} \approx -0.98. \tag{55}$$

Following Eqs. (8)-(11) procedure, we have

$$T_s^{\alpha} z = -z + u,$$

$${}_s^{\alpha} T u = -z - u, \tag{56}$$

and

$$z(0) = 1, u(1) = 0. \tag{57}$$

Using OHAM, the solutions for fractional-order are acquired. We determine the linear and non-linear operators as

$$L_1[\aleph_1(t,\ell)] = T_s^\alpha \aleph_1(t,\ell), \tag{58}$$

$$L_2[\aleph_2(t,\ell)] = {}_s^\alpha T\aleph_2(t,\ell), \tag{59}$$

$$N_1[\aleph_1(t,\ell)] = T_s^\alpha \aleph_1(t,\ell) + \aleph_1(t,\ell) + \aleph_2(t,\ell), \tag{60}$$

$$N_2[\aleph_2(t,\ell)] = {}_s^\alpha T\aleph_2(t\ell) - (\aleph_1(t,\ell) - \aleph_2(t,\ell), \tag{61}$$

using homotopy in Eq. (22)

$$(1-\ell)T_s^\alpha \aleph_1(t,\ell) = H_k(\ell)[T_s^\alpha \aleph_1(t,\ell) + \aleph_1(t,\ell) + \aleph_2(t,\ell)], \tag{62}$$

$$(1-\ell){}_s^\alpha T\aleph_2(t,\ell) = H_k(\ell)[{}_s^\alpha T\aleph_2(t,\ell) - (\aleph_1(t,\ell) - \aleph_2(t,\ell], \tag{63}$$

where

$$\aleph_1(t,\ell) = z_0(t) + \sum_{j\leq 1} z_{1,j}(t)\ell^j, \tag{64}$$

$$\aleph_2(t,\ell) = u_0(t) + \sum_{j\leq 1} u_{1,j}(t)\ell^j, \tag{65}$$

$$H_k(\ell) = \ell C_1 + \ell^2 C_2 + \ell^3 C_3 + \dots \quad k = 1,2\dots m. \tag{66}$$

substitute $\aleph_1(t,\ell), \aleph_2(t,\ell)$, and $H_k(\ell)$ into Eq. (64)-Eq. (66), and equating the coefficient of likes power of ℓ, gives linear FDEs as,

$$\ell^0: T_s^\alpha z_0(t) = 0, \tag{67}$$

$$\ell^0: {}_s^\alpha T u_0(t) = 0, \tag{68}$$

$$\ell^1: T_s^\alpha z_1(t) = T_s^\alpha z_0(t)C_1 - T_s^\alpha z_0(t) + z_0(t)C_1 + u_0(t)C_1 = 0, \tag{69}$$

$$\ell^1: {}_s^\alpha T u_1(t) = {}_s^\alpha T u_0(t)C_1 - T_s^\alpha u_0(t) + z_0(t)C_1 + u_0(t)C_1 = 0, \tag{70}$$

$$\ell^2: T_s^\alpha z_2(t) = T_s^\alpha z_0(t)C_2 + T_s^\alpha z_1(t)C_1 - T_s^\alpha z_1(t) + z_0(t)C_2 + z_1(t)C_1$$
$$+ u_0(t)C_2 + u_1(t)C_1 = 0, \tag{71}$$

$$\ell^2: {}_s^\alpha T u_2(t) = {}_s^\alpha T u_0(t)C_2 + {}_s^\alpha T u_1(t)C_1 - {}_s^\alpha T u_1(t) - u_0(t)C_2 - u_1(t)C_1$$

$$+z_0(t)C_2 + z_1(t)C_1 = 0. \tag{72}$$

Using the operator Lemmas (2.1) and (2.2) on the above equations with the initial condition gives

$$z_0(t) = 0, \tag{73}$$

$$u_0(t) = 0, \tag{74}$$

$$z_1(t, C_1) = -5t^{\frac{1}{5}}C_1 + 1, \tag{75}$$

$$u_1(t, C_1) = 5t^{\frac{1}{5}}C_1, \tag{76}$$

$$z_2(t, C_1, C_2) = 5t^{\frac{1}{5}}C_1^2 - 2C_1 + C_2 + 1, \tag{77}$$

$$u_2(t, C_1, C_2) = -25t^{\frac{2}{5}}C_1^2 - 5t^{\frac{1}{5}}C_1^2 + 10t^{\frac{1}{5}}C_1 + 5C_2t^{\frac{1}{5}}. \tag{78}$$

Summing up the solution from Eq. (73)-Eq. (78), the 3rd -order approximate analytical method generated by OHAM, for $\ell = 1$ are

$$z(t, C_1, C_2) = (5C_1^2 - 15C_1 - 5C_2)t^{\frac{1}{5}} + 3, \tag{79}$$

$$u(t, C_1, C_2) = (15C_1 - 5C_1^2 + 5C_2)t^{\frac{1}{5}} - 25^{\frac{2}{5}}C_1^2. \tag{80}$$

We determine C_1 and C_2 by using the procedure mentioned in Eq. (40)-Eq. (42).

for $z(t)$,

$$C_1 = -0.2553704121, C_2 = 1.408561895,$$

and for $u(t)$,

$$C_1 = -0.1925118854, C_2 = 1.0134328546.$$

The approximate analytical solution given in Eq. (79) and Eq. (80), we have

$$z(t) = -4.591553468t^{\frac{1}{5}} + 3, \tag{81}$$

$$u(t) = -0.9265206505t^{\frac{2}{5}} + 1.994181864t^{\frac{1}{5}}. \tag{82}$$

To derive a solution to Eqs. (56)-(57), for $0 \le t \le 1$ by MOHAM, we consider the following initial approximation

$$z_{0,j}(t_j) = \gamma_j \tag{83}$$

$$u_{0,j}(t_j) = \beta_j \tag{84}$$

We choose the auxiliary function $H_i(p,t)$ as $H(p,t) = (C_{1,j} + C_{2,j}t + C_{3,j}t^2)P$ which gave us the most suitable way to regulate the convergence domain. According to Eq. (38), the first-order MOHAM approximate solution for $m=1$, as

$$\tilde{z}_j(t) = z_{0,j}(t) + z_{1,j}(t). \tag{85}$$

$$\tilde{u}_j(t) = u_{0,j}(t) + u_{1,j}(t), \tag{86}$$

where

$$z_{1,j}(t) = (C_{1,j} + C_{2,j}t)(-z_{0,j}(t) + u_{0,j}(t), \tag{87}$$

$$u_{1,j}(t) = (C_{1,j} + C_{2,j}t)(-z_{0,j}(t) - u_{0,j}(t), \tag{88}$$

$$z_{1,j}(t) = (-\gamma_j + \beta_j)(C_{1,j} + C_{2,j}t), \tag{89}$$

$$u_{1,j}(t) = (-\gamma_j - \beta_j)(C_{1,j} + C_{2,j}t). \tag{90}$$

By substituting Eqs. (89-90) into Eqs. (85-86), regarding Eqs. (83-84), we obtain

$$\tilde{z}_j(t) = \gamma_j - \gamma_j(C_{1,j} + \gamma_j C_{2,j}t) + \beta_j(C_{1,j} + \beta_j C_{2,j}t) \tag{91}$$

$$\tilde{u}_j(t) = \beta_j - \gamma_j(C_{1,j} + \gamma_j C_{2,j}t) - \beta_j(C_{1,j} - \beta_j C_{2,j}t) \tag{92}$$

Table 1. Control-convergence parameters $C_{i,j}^s$ at different values of α.

Variable	z(t)	z(t)	u(t)	u(t)
α	$C_{1,j}$	$C_{2,j}$	$C_{1,j}$	$C_{2,j}$
1.0	-0.154562111	1.296142742	-0.1025358652	0.7138337487

(Table 1) cont.....

| 0.9 | -0.193566642 | 1.246461654 | -0.1536324112 | 0.5135317435 |
| 0.8 | -0.223456220 | 1.195643743 | -0.2154217924 | 0.113937310 |

By substituting the solution of Eqs. (91-92) into Eqs. (40-42) using the results in Table **1** for values of $C_{i,j}$, the second-order MOHAM approximate solution for $m = 3$ are given as

$$z(t) = \begin{cases} -4.591553468t^{\frac{1}{5}} + 2t, & 0 \leq t < 0.2, \\ -3.211631615t^{\frac{1}{5}} + 3t, & 0.2 \leq t < 0.4, \\ -3.031573726t^{\frac{1}{5}} + 4t, & 0.4 \leq t < 0.6, \\ -2.131353454t^{\frac{1}{5}} + 5t, & 0.6 \leq t < 0.8, \\ -1.391543518t^{\frac{1}{5}} + 6t, & 0.8 \leq t \leq 1. \end{cases} \quad (93)$$

$$u(t) = \begin{cases} -0.9265206505t^{\frac{2}{5}} + 1.994181864t^{\frac{1}{5}}, & 0 \leq t < 0.2, \\ -0.7225106715t^{\frac{2}{5}} + 2.852161561t^{\frac{1}{5}}, & 0.2 \leq t < 0.4, \\ -0.5065207501t^{\frac{2}{5}} + 3.903171812t^{\frac{1}{5}}, & 0.4 \leq t < 0.6, \\ -0.4155216514t^{\frac{2}{5}} + 5.244281763t^{\frac{1}{5}}, & 0.6 \leq t < 0.8, \\ -0.2245216516t^{\frac{2}{5}} + 7.844171853t^{\frac{1}{5}}, & 0.8 \leq t \leq 1. \end{cases} \quad (94)$$

The second-order of the MOHAM approximation solution was used for the problem Eqs. (56)-(57) with the step $h = 0.2, t_j = 0, \gamma = 1$, and $t_j = 1, \beta = 0$, starting with $t_0 = 0$, to $t_5 = T = 1$.

Figs. (**1**) and (**2**) show the AAS of the state $z(t)$ and control $u(t)$ at the value of $\alpha = 1$ using MOHAM after the 2nd-order of approximate solution. Table **1** shows the optimal control convergent parameter, Tables **2** & **3** comparisons between MOHAM, OHAM, and the exact, Table **4** optimal values of the cost function. It is observed that MOHAM converges to optimal values faster in CPU times and iterations. From the Figs. (**1** and **2**), we observe that the proposed technique solutions are in good agreement with the analytical solution, and this shows that the method is efficient for accurate approximate solutions.

Table 2. Comparisons of Exact, MOHAM, and OHAM, α=1 at at state z(t).

t_j	Exact z(t)	MOHAM z(t)	OHAM z(t)	Absolute Error
0.0	1.000000000	1.000000000	1.000000000	0.00000000
0.1	0.667144195	0.667203453	0.667606464	5.9258E-05
0.2	0.571814575	0.572490044	0.572490044	0.000675469
0.3	0.512155196	0.512119634	0.513009607	3.5562E-05
0.4	0.469955226	0.463977768	0.470973769	0.005977458
0.5	0.443945101	0.441875124	0.451473134	0.002069977
0.6	0.438392205	0.438156416	0.439567403	0.000235789
0.7	0.414075062	0.414457216	0.415403119	0.000382154
0.8	0.395049669	0.396438920	0.396528946	0.001389251
0.9	0.380080330	0.381524761	0.381710581	0.001444431
1	0.368333813	0.368736762	0.368415754	0.000402949

Table 3. Comparisons of Exact, MOHAM, and OHAM, α=1 at control u(t).

t_j	Exact u(t)	MOHAM u(t)	OHAM u(t)	Absolute Error
0.0	0.000000000	0.00000000	0.0000000	0.00000000
0.1	0.681603987	0.681141712	0.681141711	0.000462275
0.2	1.151495891	1.150820423	1.150820422	0.000675468
0.3	1.538030634	1.537176225	1.537176224	0.000854409
0.4	1.887741076	1.886722536	1.886722533	0.00101854
0.5	2.218155658	2.216980467	2.216980462	0.001175191

(Table 3) cont.....

0.6	2.538065126	2.536737063	2.536737061	0.001328063
0.7	2.852602442	2.851123164	2.851123163	0.001479278
0.8	3.165072917	3.163442665	3.163442664	0.001630252
0.9	3.477758981	3.475977039	3.475977036	0.001781942
1	3.792323188	3.791388125	3.791388123	0.000935063

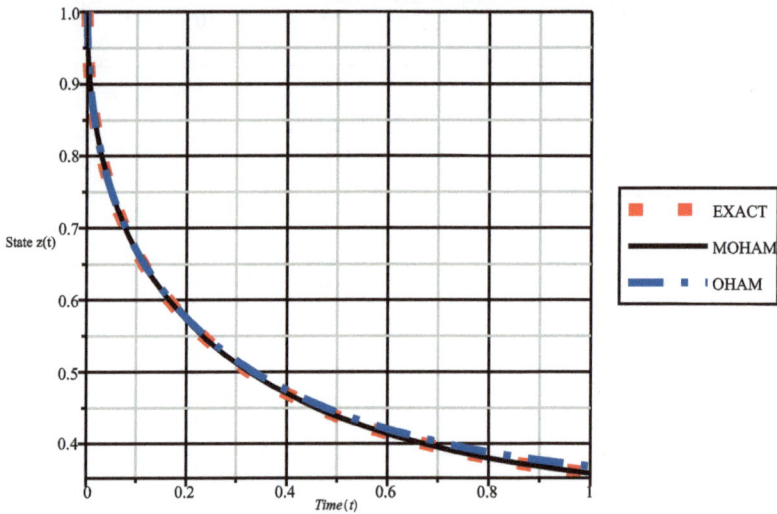

Fig. (1). For value $\alpha = 1$ (Exact=solid, MOHAM=dot, and OHAM=dash dot) at state z(t).

Table 4. Optimal value of J at different choices of α for example 1.

$s\alpha$	MOHAM	OHAM
1.0	0.1929089	0.1929090
0.9	0.17950	0.17152
0.8	0.16706	0.16709

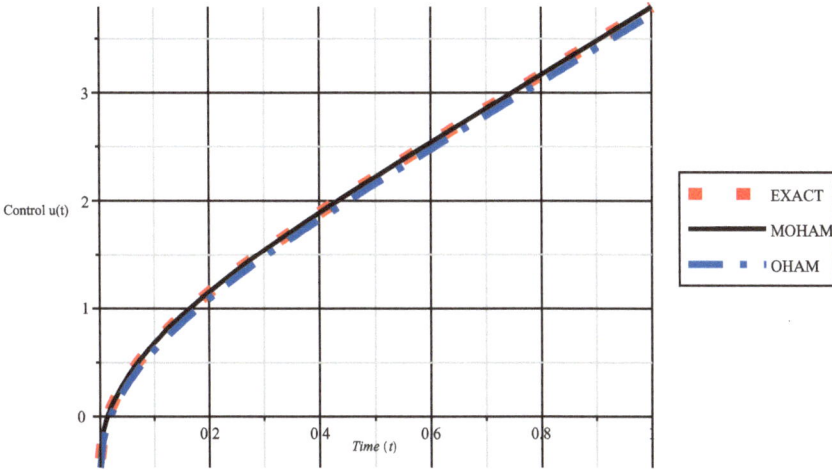

Fig. (2). For the value of $\alpha = 1$ (Exact=solid, MOHAM=dot, and OHAM=dash dot) at control u(t).

Example 2. Considering the following problem of minimizing the functional

$$J(u) = \frac{1}{2}\int_0^1 [z_1(t)^2 + z_2(t)^2 + u(t)^2]dt, \tag{95}$$

Subjected to the dynamic constraints

$$T_s^\alpha z_1(t) = -z_1(t) + z_2(t) + u(t), \tag{96}$$

$$_s^\alpha T z_2(t) = -2z_2(t), \tag{97}$$

and the initial conditions

$$z_1(0) = z_2(0) = 1. \tag{98}$$

The exact solution for this system for $\alpha = 1$ is

$$z_1(t) = \frac{-3}{2}e^{-2t} + 2.48164e^{-\sqrt{2t}} + 0.018352e^{\sqrt{2t}}, \tag{99}$$

$$z_2(t) = e^{-2t}, \tag{100}$$

$$u(t) = \frac{1}{2}e^{-2t} - 1.027934e^{-\sqrt{2t}} + 0.0443056e^{\sqrt{2t}}. \tag{101}$$

from Eqs. (96-98), we have the following system of non-linear equations

$$T_s^\alpha z_1(t) - z_1(t) + z_2(t) + u(t) = 0, \tag{102}$$

$$_s^\alpha T z_2(t) + 2z_2(t) = 0, \tag{103}$$

$$u(t) + p_1(t) = 0, \tag{104}$$

$$T_s^\alpha p_1(t) + p_(t) - z_1(t) = 0, \tag{105}$$

$$_s^\alpha T p_2(t) - 2p_2(t) + p_1(t) + z_2(t) = 0, \tag{106}$$

and

$$z_1(0) = z_2(0) = 1, p_1(1) = p_2(1) = 0. \tag{107}$$

Following the same standard procedure for OHAM with the 3rd-order approximate analytical solution, we have

$$z_1(t) = -2.950819672 \times t^{\frac{1}{10}} + 3.01234 + 6.543676455 \times 10^{-17} \times t^{\frac{1}{5}}, \tag{108}$$

$$z_2(t) = -0.378824654 \times t^{\frac{1}{10}} + 2.036182595 - 1.657357940 \times t^{\frac{1}{5}}, \tag{109}$$

$$u(t) = 3.913043477 \times t^{\frac{1}{10}} - 3.913043477 - 4.327008279 \times 10^{-17} \times t^{\frac{1}{5}}. \tag{110}$$

Following the same standard procedure for MOHAM with the 2nd-order approximate analytical solution, we have the value for the control-convergence parameter as:

$$z_1(t) = \begin{cases} -2.450416561 \times t^{\frac{1}{10}} + 3.01253 + 6.142675490 \times 10^{-17} \times t^{\frac{1}{5}}, & 0 \leq t < 0.2, \\ -1.750319270 \times t^{\frac{1}{10}} + 2.81394 + 5.842476431 \times 10^{-17} \times t^{\frac{1}{5}}, & 0.2 \leq t < 0.4, \\ -1.451314151 \times t^{\frac{1}{10}} + 2.52563 + 5.622356111 \times 10^{-17} \times t^{\frac{1}{5}}, & 0.4 \leq t < 0.6, \\ -1.251829572 \times t^{\frac{1}{10}} + 2.31469 + 5.443886449 \times 10^{-17} \times t^{\frac{1}{5}}, & 0.6 \leq t < 0.8, \\ -0.851718672 \times t^{\frac{1}{10}} + 1.81387 + 4.933501301 \times 10^{-17} \times t^{\frac{1}{5}}, & 0.8 \leq t \leq 1. \end{cases} \tag{111}$$

$$
z_2(t) = \begin{cases}
-0.361424324 \times t^{\frac{1}{10}} + 2.045173486 - 1.546656421 \times t^{\frac{1}{5}}, & 0 \le t < 0.2, \\
-0.271224412 \times t^{\frac{1}{10}} + 1.335072453 - 1.152455551 \times t^{\frac{1}{5}}, & 0.2 \le t < 0.4, \\
-0.221567410 \times t^{\frac{1}{10}} + 1.235164501 - 1.121328560 \times t^{\frac{1}{5}}, & 0.4 \le t < 0.6, \\
-0.176723631 \times t^{\frac{1}{10}} + 0.134163406 - 0.954245521 \times t^{\frac{1}{5}}, & 0.6 \le t < 0.8, \\
-0.077623511 \times t^{\frac{1}{10}} + 0.037254672 - 0.246468320 \times t^{\frac{1}{5}}, & 0.8 \le t \le 1.
\end{cases} \tag{112}
$$

$$
u(t) = \begin{cases}
3.722133423 \times t^{\frac{1}{10}} - 3.713043477 - 4.148118351 \times 10^{-17} \times t^{\frac{1}{5}}, & 0 \le t < 0.2, \\
2.912445697 \times t^{\frac{1}{10}} - 2.912445977 - 3.715125381 \times 10^{-17} \times t^{\frac{1}{5}}, & 0.2 \le t < 0.4, \\
2.534566482 \times t^{\frac{1}{10}} - 2.534566482 - 2.924217262 \times 10^{-17} \times t^{\frac{1}{5}}, & 0.4 \le t < 0.6, \\
1.934348487 \times t^{\frac{1}{10}} - 1.934348487 - 2.126126162 \times 10^{-17} \times t^{\frac{1}{5}}, & 0.6 \le t < 0.8, \\
1.213163281 \times t^{\frac{1}{10}} - 1.213163281 - 1.816125356 \times 10^{-17} \times t^{\frac{1}{5}}, & 0.8 \le t \le 1.
\end{cases} \tag{113}
$$

Figs. (3-5) show the AAS of the states $z_1(t)$, $z_2(t)$, and controls $u(t)$ at the value of $\alpha = 1$ with MOHAM after the 2nd-order of approximate solution. Tables **5** & **6** show the the optimal control convergent parameter, Tables **7-9** comparisons between MOHAM, OHAM, and the exact, Table **10** optimal values of the cost function. It is observed that MOHAM converges to optimal values faster in CPU times and iterations. From the Figures, we observe that the proposed technique solutions are in good agreement with the analytical solution, and this shows that the method is efficient for accurate approximate solutions.

Table 5. Control-convergence parameters $C^s_{i,j}$ at different values of α.

Variable	$z_1(t)$	$z_1(t)$	$z_2(t)$	$z_2(t)$
α	$C_{1,j}$	$C_{2,j}$	$C_{1,j}$	$C_{2,j}$
1.0	0.13936248110^{-7}	0.1761627547	-0.1834567241	0.2099839413
0.9	$0.114327352\ 10^{-7}$	0.1245675474	-0.1553456356	0.1845623225
0.8	$0.126357171\ 10^{-7}$	0.0952576356	-0.1054638215	0.158315202

Table 6. Control-convergence parameters $C_{i,j}^s$ at different values of α.

Variable	u(t)	u(t)
α	$C_{1,j}$	$C_{2,j}$
1.0	-2.571456114 10^{-10}	0.2485236311
0.9	-3.645534245 10^{-10}	0.1745233411
0.8	-4.864682317 10^{-10}	0.1237245932

The second-order of the MOHAM approximation solution was used for the problem Eqs. (102)-(107) with the step $h = 0.2, t_j = 0, \gamma = 1, t_j = 0, \beta = 1, t_j = 1, \tau = 0$, and $t_j = 1, \zeta = 0$ starting with $t_0 = 0$, to $t_5 = T = 1$.

Table 7. Comparisons of (Exact, MOHAM, and OHAM, $\alpha = 1$) at state z1 (t) .

t_j	Exact $z_1(t)$	MOHAM $z_1(t)$	OHAM $z_1(t)$	Absolute Error
0.0	1.0000000	1.00000000	1.0000000	0.00000000
0.1	0.387385984	0.388114021	0.388025392	0.000728037
0.2	0.347524221	0.348021502	0.348055508	0.000497281
0.3	0.360365496	0.360515265	0.360826387	0.000149769
0.4	0.385494492	0.385602123	0.385903334	0.000107631
0.5	0.385494492	0.385601022	0.385903334	0.00010653
0.6	0.411013257	0.411171101	0.411381137	0.000157844
0.7	0.432930139	0.433153330	0.433264531	0.000223191
0.8	0.45013016	0.450224352	0.450436393	9.4192E-05

(Table 7) cont.....

| 0.9 | 0.462652925 | 0.462714085 | 0.462935191 | 6.116E-05 |
| 1 | 0.470997167 | 0.470999812 | 0.470999841 | 2.645E-06 |

Table 8. Comparisons of (Exact, MOHAM and OHAM, $\alpha = 1$) at state z2(t).

t_j	Exact $z_2(t)$	MOHAM $z_2(t)$	OHAM $z_2(t)$	Absolute error
0.0	1.0000000	1.00000000	1.0000000	0.00000000
0.1	0.852143789	0.852745225	0.852996359	0.000601436
0.2	0.726149037	0.727602788	0.727602788	0.001453751
0.3	0.618783392	0.619531418	0.620642529	0.000748026
0.4	0.527292424	0.528314717	0.529405818	0.001022293
0.5	0.449328964	0.450360124	0.451581235	0.00103116
0.6	0.382892886	0.383165125	0.385197149	0.000272239
0.7	0.326279795	0.327361543	0.328571766	0.001081748
0.8	0.278037301	0.279071421	0.280270520	0.00103412
0.9	0.236927759	0.237158321	0.239069733	0.000230562
1	0.205152843	0.206182831	0.207193943	0.001029988

Table 9. Comparisons of (Exact, MOHAM, and OHAM, $\alpha = 1$) at control u(t).

t_j	Exact u(t)	MOHAM u(t)	OHAM u(t)	Absolute Error
0.0	0.0000000	0.00000000	0.0000000	0.00000000
0.1	-0.178609739	-0.178300288	-0.178200169	0.000309451
0.2	-0.127572209	-0.127112234	-0.126901219	0.000459975

(Table 9) cont.....

0.3	-0.10322697	-0.102712624	-0.102402516	0.000514346
0.4	-8.72E-02	-8.68E-02	-8.63E-02	0.0004
0.5	-7.38E-02	-7.32E-02	-7.29E-02	0.0006
0.6	-6.06E-02	-5.99E-02	-5.97E-02	0.0007
0.7	-4.69E-02	-4.68E-02	-4.68E-02	1E-04
0.8	-3.22E-02	-3.21E-02	-3.21E-02	0.0001
0.9	-1.66E-02	-1.65E-02	-1.65E-02	1E-04
1	-1.70E-03	-1.69E-03	-1.69E-03	1E-05

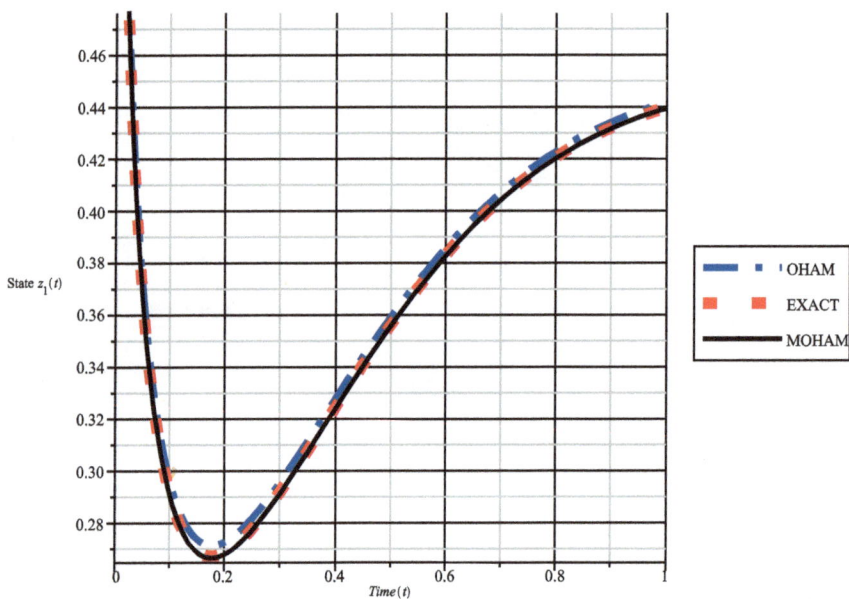

Fig. (3). For value $\alpha = 1$ (Exact=solid, MOHAM=dot, and OHAM=dash dot) at state $z_1(t)$.

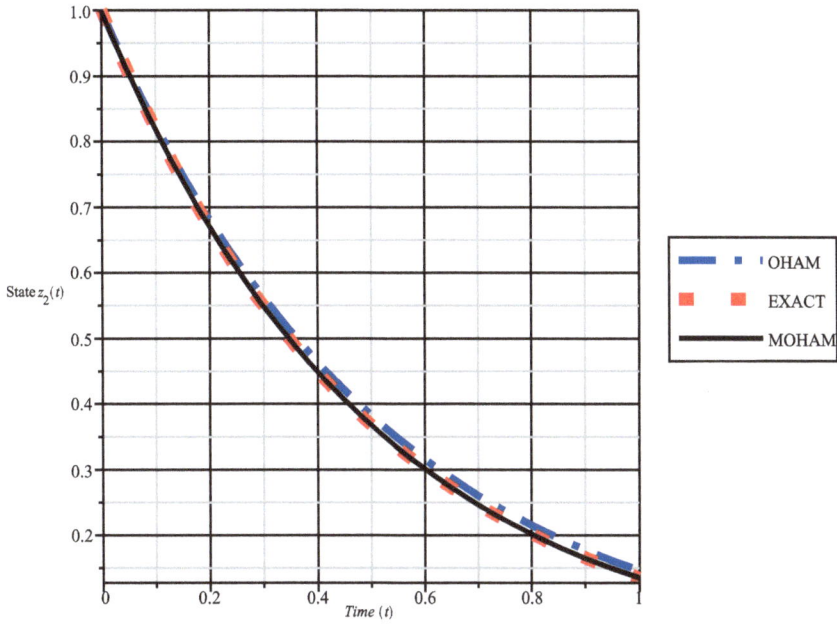

Fig. (4). For the value of $\alpha = 1$ (Exact=solid, MOHAM=dot, and OHAM=dash dot) at state $z_1(t)$.

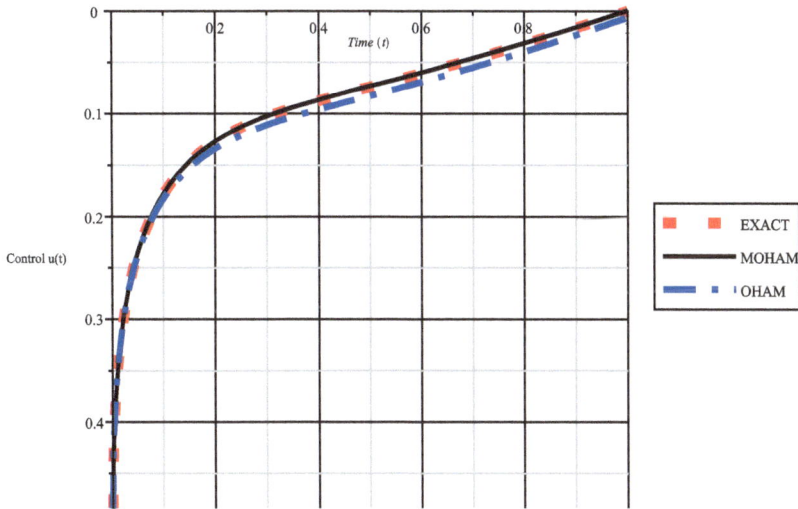

Fig. (5). For the value of $\alpha = 1$ (Exact=solid, MOHAM=dot, and OHAM=dash dot) at control u(t).

Table 10. Optimal value of J at different choices of α for example 2.

α	MOHAM	OHAM
1.0	0.4319087	0.4319088
0.9	0.4030890	0.4030893
0.8	0.3760121	0.3760122

CONCLUSION

In this paper, approximate analytical methods of MOHAM have been utilized to solve FOCPs successfully. The results are shown in tables and plotted in figures. A comparison of numerical simulation shows that MOHAM is more efficient than OHAM, especially for the point further from the initial point. The fractional derivative is in the conformable derivative sense. MOHAM gave an accurate optimal value of the control values, which minimizes the objective function more than other approximate methods. The second-order approximate solution was used to calculate the solution, which gave MOHAM an edge over mentioned method in fast convergence and CPU times. We show that the Galerkin method helps to determine the optimal values of the control-convergence parameter. For $\alpha = 1$, it can be seen that the approximate analytical solutions corresponded with the integer-order solutions. Finally, the MOHAM technique is straightforward to implement and efficient for solving FOCPs. Thus, the present formulation essentially extends the classical control theory to fractional dynamic systems. Researchers can extend the non-integer order to enlarge the stability region of the dynamical systems of the optimal control problems for behaviour, simulation, and mathematical representation for future research.

CONSENT FOR PUBLICATION

Not applicable.

CONFLICT OF INTEREST

The authors declare no conflict of interest, financial or otherwise.

ACKNOWLEDGEMENTS

The authors are thankful to the anonymous reviewers and editors for their valuable suggestions and comments to improve the quality of the chapter.

REFERENCES

[1]   I. Podlubny, "Fractional differential equations," *Mathematics in science and engineering,* vol. 198, pp. 41-119, 1999.

[2]   I. Podlubny, I. Petras, B. M. Vinagre, Y. Chen, P. O'Leary, and L. Dorcak, "Realization of fractional order controllers," *Acta Montanistica Slovaca,* vol. 8, no. 4, pp. 233-235, 2003.

[3]   K. Diethelm, *The analysis of fractional differential equations: An application-oriented exposition using differential operators of Caputo type.* Springer Science & Business Media, 2010.

[4]   M. Rahimy, "Applications of fractional differential equations," *Applied Mathematical Sciences,* vol. 4, no. 50, pp. 2453-2461, 2010.

[5]   T. Škovránek, I. Podlubny, and I. Petráš, "Modeling of the national economies in state-space: A fractional calculus approach," *Economic Modelling,* vol. 29, no. 4, pp. 1322-1327, 2012.

[6]   C. D. Constantinescu, J. M. Ramirez, and W. R. Zhu, "An application of fractional differential equations to risk theory," *Finance and Stochastics,* vol. 23, no. 4, pp. 1001-1024, 2019.

[7]   N. Sene, "Mittag-Leffler input stability of fractional differential equations and its applications," *Discrete & Continuous Dynamical Systems-S,* vol. 13, no. 3, p. 867, 2020.

[8]   W. Fan, X. Jiang, and S. Chen, "Parameter estimation for the fractional fractal diffusion model based on its numerical solution," *Computers & Mathematics with Applications,* vol. 71, no. 2, pp. 642-651, 2016.

[9]   A. D. Obembe, M. E. Hossain, and S. A. Abu-Khamsin, "Variable-order derivative time fractional diffusion model for heterogeneous porous media," *Journal of Petroleum Science and Engineering,* vol. 152, pp. 391-405, 2017.

[10]   I. Podlubny, *Fractional differential equations: an introduction to fractional derivatives, fractional differential equations, to methods of their solution and some of their applications.* Elsevier, 1998.

[11]   P. Kumar and O. P. Agrawal, "An approximate method for numerical solution of fractional differential equations," *Signal processing,* vol. 86, no. 10, pp. 2602-2610, 2006.

[12]   M. Caputo and M. Fabrizio, "A new definition of fractional derivative without singular kernel," *Progr. Fract. Differ. Appl,* vol. 1, no. 2, pp. 1-13, 2015.

[13]   H. Sun, A. Chang, Y. Zhang, and W. Chen, "A review on variable-order fractional differential equations: mathematical foundations, physical models, numerical methods and applications," *Fractional Calculus and Applied Analysis,* vol. 22, no. 1, pp. 27-59, 2019.

[14]   R. Scherer, S. L. Kalla, Y. Tang, and J. Huang, "The Grünwald–Letnikov method for fractional differential equations," *Computers & Mathematics with Applications,* vol. 62, no. 3, pp. 902-917, 2011.

[15]   N. Sene and G. Srivastava, "Generalized Mittag-Leffler input stability of the fractional differential equations," *Symmetry,* vol. 11, no. 5, p. 608, 2019.

[16]   N. Sene, "Cascade of Fractional Differential Equations and Generalized Mittag-Leffler Stability," *International Journal of Mathematical Modelling & Computations,* vol. 10, no. 1 (WINTER), pp. 25-35, 2020.

[17]   D. Qian, C. Li, R. P. Agarwal, and P. J. Wong, "Stability analysis of fractional differential system with Riemann–Liouville derivative," *Mathematical and Computer Modelling,* vol. 52, no. 5-6, pp. 862-874, 2010.

[18]   Y. Khan, Q. Wu, N. Faraz, A. Yildirim, and M. Madani, "A new fractional analytical approach via a modified Riemann–Liouville derivative," *Applied Mathematics Letters,* vol. 25, no. 10, pp. 1340-1346, 2012.

[19]   A. Atangana and J. Gómez-Aguilar, "Numerical approximation of Riemann-Liouville definition of fractional derivative: from Riemann-Liouville to Atangana-Baleanu," *Numerical Methods for Partial Differential Equations,* vol. 34, no. 5, pp. 1502-1523, 2018.

[20]   J. Losada and J. J. Nieto, "Properties of a new fractional derivative without singular kernel," *Progr. Fract. Differ. Appl,* vol. 1, no. 2, pp. 87-92, 2015.

[21]   R. Almeida, "Caputo–Hadamard fractional derivatives of variable order," *Numerical Functional Analysis and Optimization,* vol. 38, no. 1, pp. 1-19, 2017.

[22]   N. Sene and K. Abdelmalek, "Analysis of the fractional diffusion equations described by Atangana-Baleanu-Caputo fractional derivative," *Chaos, Solitons & Fractals,* vol. 127, pp. 158-164, 2019.

[23]   N. H. Tuan, H. Mohammadi, and S. Rezapour, "A mathematical model for COVID-19 transmission by using the Caputo fractional derivative," *Chaos, Solitons & Fractals,* vol. 140, p. 110107, 2020.

[24]   M. Farman, A. Ahmad, A. Akgül, M. U. Saleem, M. Naeem, and D. Baleanu, "Epidemiological Analysis of the Coronavirus Disease Outbreak with Random Effects," *Computers, Materials and Continua,* vol. 67, no. 3, 2021.

[25]   M. Farooq, M. Mohsin, M. Naeem, M. Farman, A. Akgül, and M. U. Saleem, "Discretization of the method of generating an expanded family of distributions based upon truncated distributions," *Thermal Science,* no. 00, pp. 3-3, 2021.

[26]   M. Farman, M. Aslam, A. Akgül, and A. Ahmad, "Modeling of fractional-order COVID-19 epidemic model with quarantine and social distancing," *Mathematical Methods in the Applied Sciences,* 2021.

[27]   M. Farman, A. Akgül, A. Ahmad, D. Baleanu, and M. U. Saleem, "Dynamical transmission of coronavirus model with analysis and simulation," *CMES-Computer Modeling in Engineering and Sciences,* vol. 127, no. 2, 2021.

[28]   A. Yokus and M. Yavuz, "Novel comparison of numerical and analytical methods for fractional Burger–Fisher equation," *Discrete & Continuous Dynamical Systems-S,* vol. 14, no. 7, p. 2591, 2021.

[29]   P. Veeresha and D. G. Prakasha, "A novel technique for (2+ 1)-dimensional time-fractional coupled Burgers equations," *Mathematics and Computers in Simulation,* vol. 166, pp. 324-345, 2019.

[30]   P. Veeresha, D. Prakasha, and D. Baleanu, "An efficient technique for fractional coupled system arisen in magnetothermoelasticity with rotation using Mittag–Leffler kernel," *Journal of Computational and Nonlinear Dynamics,* vol. 16, no. 1, p. 011002, 2021.

[31]   D. G. Prakasha, P. Veeresha, and M. S. Rawashdeh, "Numerical solution for (2+ 1)-dimensional time-fractional coupled Burger equations using fractional natural decomposition method," *Mathematical Methods in the Applied Sciences,* vol. 42, no. 10, pp. 3409-3427, 2019.

[32]   P. Veeresha, D. Prakasha, and H. M. Baskonus, "Novel simulations to the time-fractional Fisher's equation," *Mathematical Sciences,* vol. 13, no. 1, pp. 33-42, 2019.

[33]   P. Liu, Z. Zeng, and J. Wang, "Multiple Mittag–Leffler stability of fractional-order recurrent neural networks," *IEEE Transactions on Systems, Man, and Cybernetics: Systems,* vol. 47, no. 8, pp. 2279-2288, 2017.

[34]   R. Gorenflo and F. Mainardi, "Fractional calculus," in *Fractals and fractional calculus in continuum mechanics*: Springer, 1997, pp. 223-276.

[35]   V. E. Tarasov, *Fractional dynamics: applications of fractional calculus to dynamics of particles, fields and media.* Springer Science & Business Media, 2011.

[36]   T. M. Atanackovic, S. Pilipovic, B. Stankovic, and D. Zorica, *Fractional calculus with applications in mechanics: vibrations and diffusion processes.* John Wiley & Sons, 2014.

[37]   D. Baleanu and A. M. Lopes, *Handbook of Fractional Calculus with Applications.* De Gruyter, 2019.

[38]   V. E. Tarasov, *Applications in Physics, Part B.* Walter de Gruyter GmbH & Co KG, 2019.

[39]   I. Petráš, *Applications in Control.* Walter de Gruyter GmbH & Co KG, 2019.

[40]   O. P. Agrawal, O. Defterli, and D. Baleanu, "Fractional optimal control problems with several state and control variables," *Journal of Vibration and Control,* vol. 16, no. 13, pp. 1967-1976, 2010.

[41]   Z. Li, L. Liu, S. Dehghan, Y. Chen, and D. Xue, "A review and evaluation of numerical tools for fractional calculus and fractional order controls," *International journal of control,* vol. 90, no. 6, pp. 1165-1181, 2017.

[42]   F. Mohammadi, L. Moradi, D. Baleanu, and A. Jajarmi, "A hybrid functions numerical scheme for fractional optimal control problems: application to nonanalytic dynamic systems," *Journal of Vibration and Control,* vol. 24, no. 21, pp. 5030-5043, 2018.

[43]   O. P. Agrawal, "A general formulation and solution scheme for fractional optimal control problems," *Nonlinear Dynamics,* vol. 38, no. 1-4, pp. 323-337, 2004.

[44]   S. I. Muslih and D. Baleanu, "Hamiltonian formulation of systems with linear velocities within Riemann–Liouville fractional derivatives," *Journal of Mathematical Analysis and Applications,* vol. 304, no. 2, pp. 599-606, 2005.

[45]   T. Avkar and D. Baleanu, "Fractional Euler-Lagrange Equations for Constrained Systems," in *AIP Conference Proceedings*, 2004, vol. 729, no. 1: American Institute of Physics, pp. 84-90.

[46]   A. Jajarmi and D. Baleanu, "On the fractional optimal control problems with a general derivative operator," *Asian Journal of Control,* vol. 23, no. 2, pp. 1062-1071, 2021.

[47]   A. Lotfi, M. Dehghan, and S. A. Yousefi, "A numerical technique for solving fractional optimal control problems," *Computers & Mathematics with Applications,* vol. 62, no. 3, pp. 1055-1067, 2011.

[48]   N. H. Sweilam, S. M. Al-Mekhlafi, T. Assiri, and A. Atangana, "Optimal control for cancer treatment mathematical model using Atangana–Baleanu–Caputo fractional derivative," *Advances in Difference Equations,* vol. 2020, no. 1, pp. 1-21, 2020.

[49]   O. P. Agrawal and D. Baleanu, "A Hamiltonian formulation and a direct numerical scheme for fractional optimal control problems," *Journal of Vibration and Control,* vol. 13, no. 9-10, pp. 1269-1281, 2007.

[50]   D. Baleanu, R. P. Agarwal, R. K. Parmar, M. Alqurashi, and S. Salahshour, "Extension of the fractional derivative operator of the Riemann-Liouville," 2017.

[51]   D. Baleanu, O. Defterli, and O. P. Agrawal, "A central difference numerical scheme for fractional optimal control problems," *Journal of Vibration and Control,* vol. 15, no. 4, pp. 583-597, 2009.

[52]   G. S. Frederico and D. F. Torres, "Fractional conservation laws in optimal control theory," *Nonlinear Dynamics,* vol. 53, no. 3, pp. 215-222, 2008.

[53]   M. A. Herzallah and D. Baleanu, "Fractional-order Euler–Lagrange equations and formulation of Hamiltonian equations," *Nonlinear Dynamics,* vol. 58, no. 1, pp. 385-391, 2009.

[54]   T. A. Yildiz, A. Jajarmi, B. Yildiz, and D. Baleanu, "New aspects of time fractional optimal control problems within operators with nonsingular kernel," 2020.

[55]   R. K. Biswas and S. Sen, "Fractional optimal control problems: a pseudo-state-space approach," *Journal of Vibration and Control,* vol. 17, no. 7, pp. 1034-1041, 2011.

[56] R. K. Biswas and S. Sen, "Free final time fractional optimal control problems," *Journal of the Franklin Institute*, vol. 351, no. 2, pp. 941-951, 2014.

[57] V. Mehandiratta, M. Mehra, and G. Leugering, "Fractional optimal control problems on a star graph: Optimality system and numerical solution," *Mathematical Control & Related Fields*, vol. 11, no. 1, p. 189, 2021.

[58] Y. Ding, Z. Wang, and H. Ye, "Optimal control of a fractional-order HIV-immune system with memory," *IEEE Transactions on Control Systems Technology*, vol. 20, no. 3, pp. 763-769, 2011.

[59] A. Alizadeh and S. Effati, "Modified Adomian decomposition method for solving fractional optimal control problems," *Transactions of the Institute of Measurement and Control*, vol. 40, no. 6, pp. 2054-2061, 2018.

[60] F. Evirgen, "Analyze the optimal solutions of optimization problems by means of fractional gradient based system using VIM," *An International Journal of Optimization and Control: Theories & Applications (IJOCTA)*, vol. 6, no. 2, pp. 75-83, 2016.

[61] B. B. İ. Eroğlu, D. Avcı, and N. Özdemir, "Optimal control problem for a conformable fractional heat conduction equation," 2017.

[62] L. Zhang and Z. Zhou, "Spectral Galerkin approximation of optimal control problem governed by Riesz fractional differential equation," *Applied Numerical Mathematics*, vol. 143, pp. 247-262, 2019.

[63] L. Zada *et al.*, "New approximate-analytical solutions to partial differential equations via auxiliary function method," *Partial Differential Equations in Applied Mathematics*, vol. 4, p. 100045, 2021.

[64] A. B. Salati, M. Shamsi, and D. F. Torres, "Direct transcription methods based on fractional integral approximation formulas for solving nonlinear fractional optimal control problems," *Communications in Nonlinear Science and Numerical Simulation*, vol. 67, pp. 334-350, 2019.

[65] S. Liao, "The proposed homotopy analysis technique of nonlinear problems," Ph. D. Dissertation, Shanghai Jiao Tong University, Shanghai, 1992.

[66] S. Liao, "Notes on the homotopy analysis method: some definitions and theorems," *Communications in Nonlinear Science and Numerical Simulation*, vol. 14, no. 4, pp. 983-997, 2009.

[67] C.-s. Liu, "The essence of the generalized Taylor theorem as the foundation of the homotopy analysis method," *Communications in Nonlinear Science and Numerical Simulation*, vol. 16, no. 3, pp. 1254-1262, 2011.

[68] S. Liao, *Beyond perturbation: introduction to the homotopy analysis method*. CRC press, 2003.

[69] H. Sun, Y. Zhang, D. Baleanu, W. Chen, and Y. Chen, "A new collection of real world applications of fractional calculus in science and engineering," *Communications in Nonlinear Science and Numerical Simulation*, vol. 64, pp. 213-231, 2018.

[70] D. Baleanu, B. Agheli, and R. Darzi, "An optimal method for approximating the delay differential equations of noninteger order," *Advances in Difference Equations*, vol. 2018, no. 1, p. 284, 2018.

[71] B. Pan, Y. Ma, and Y. Ni, "A new fractional homotopy method for solving nonlinear optimal control problems," *Acta Astronautica*, vol. 161, pp. 12-23, 2019.

[72] B. B. İ. Eroğlu, D. Avcı, and N. Özdemir, "Constrained optimal control of a fractionally damped elastic beam," *International Journal of Nonlinear Sciences and Numerical Simulation*, vol. 21, no. 3-4, pp. 389-395, 2020.

[73] J. Biazar and R. Montazeri, "Optimal Homotopy Asymptotic and Multistage Optimal Homotopy Asymptotic Methods for Solving System of Volterra Integral Equations of the Second Kind," *Journal of Applied Mathematics*, vol. 2019, 2019.

[74] J. E. S. Pérez, J. F. Gómez-Aguilar, D. Baleanu, and F. Tchier, "Chaotic attractors with fractional conformable derivatives in the Liouville–Caputo sense and its dynamical behaviors," *Entropy*, vol. 20, no. 5, p. 384, 2018.

[75]   M. Yavuz and N. Özdemir, "European vanilla option pricing model of fractional order without singular kernel," *Fractal and Fractional,* vol. 2, no. 1, p. 3, 2018.

[76]   M. Matar, M. Abbas, J. Alzabut, M. Kaabar, S. Etemad, and S. Rezapour, "Investigation of the p-Laplacian nonperiodic nonlinear boundary value problem via generalized Caputo fractional derivatives," *Advances in Difference Equations,* vol. 2021, no. 1, pp. 1-18, 2021.

[77]   M. Yavuz, "European option pricing models described by fractional operators with classical and generalized Mittag-Leffler kernels," *Numerical Methods for Partial Differential Equations,* 2020.

[78]   F. Jarad, E. Uğurlu, T. Abdeljawad, and D. Baleanu, "On a new class of fractional operators," *Advances in Difference Equations,* vol. 2017, no. 1, p. 247, 2017.

[79]   D. Kumar, A. R. Seadawy, and A. K. Joardar, "Modified Kudryashov method via new exact solutions for some conformable fractional differential equations arising in mathematical biology," *Chinese journal of physics,* vol. 56, no. 1, pp. 75-85, 2018.

[80]   A. Yokuş, "Comparison of Caputo and conformable derivatives for time-fractional Korteweg–de Vries equation via the finite difference method," *International Journal of Modern Physics B,* vol. 32, no. 29, p. 1850365, 2018.

[81]   Q. Kang, S. I. Butt, W. Nazeer, M. Nadeem, J. Nasir, and H. Yang, "New variant of Hermite–Jensen–Mercer inequalities via Riemann–Liouville fractional integral operators," *Journal of Mathematics,* vol. 2020, 2020.

[82]   R. Khalil, M. Al Horani, A. Yousef, and M. Sababheh, "A new definition of fractional derivative," *Journal of Computational and Applied Mathematics,* vol. 264, pp. 65-70, 2014.

[83]   D. Avcı, B. B. İ. Eroğlu, and N. Özdemir, "Conformable fractional wave-like equation on a radial symmetric plate," in *Theory and Applications of Non-integer Order Systems*: Springer, 2017, pp. 137-146.

[84]   B. Xin, W. Peng, Y. Kwon, and Y. Liu, "Modeling, discretization, and hyperchaos detection of conformable derivative approach to a financial system with market confidence and ethics risk," *Advances in Difference Equations,* vol. 2019, no. 1, p. 138, 2019.

[85]   W. A. M. Othman, O. O. Okundalaye, and N. Kumaresan, "Optimal Homotopy Asymptotic Method-Least Square for Solving Nonlinear Fractional-Order Gradient-Based Dynamic System from an Optimization Problem," *Advances in Mathematical Physics,* vol. 2020, pp. 1-15, 2020.

[86]   M. Kaabar, "Novel methods for solving the conformable wave equation," *Journal of New Theory,* no. 31, pp. 56-85, 2020.

[87]   F. Martínez, I. Martínez, M. K. Kaabar, and S. Paredes, "New results on complex conformable integral," *AIMS Mathematics,* vol. 5, no. 6, pp. 7695-7710, 2020.

[88]   F. Martínez, I. Martínez, M. K. Kaabar, R. Ortíz-Munuera, and S. Paredes, "Note on the conformable fractional derivatives and integrals of complex-valued functions of a real variable," *IAENG International Journal of Applied Mathematics,* vol. 50, no. 3, pp. 609-615, 2020.

[89]   T. Abdeljawad, "On conformable fractional calculus," *Journal of computational and Applied Mathematics,* vol. 279, pp. 57-66, 2015.

[90]   A. El-Ajou, "A modification to the conformable fractional calculus with some applications," *Alexandria Engineering Journal,* vol. 59, no. 4, pp. 2239-2249, 2020.

[91]   Y. Qi and X. Wang, "Asymptotical stability analysis of conformable fractional systems," *Journal of Taibah University for Science,* vol. 14, no. 1, pp. 44-49, 2020.

[92]   F. Martínez, I. Martínez, M. K. Kaabar, and S. Paredes, "Generalized conformable mean value theorems with applications to multivariable calculus," *Journal of Mathematics,* vol. 2021, 2021.

[93]   F. Jarad, T. Abdeljawad, and Q. M. Al-Mdallal, "Fractional logistic models in the frame of fractional operators generated by conformable derivatives," 2019.

Current Developments in Mathematical Sciences, 2022, Vol. 3, 61-84   **61**

CHAPTER 3

Complex Chaotic Fractional-order Finance System in Price Exponent with Control and Modeling

Muhammad Farman[1], Parvaiz Ahmad Naik[2,*], Aqeel Ahmad[1], Ali Akgul[3] and **Muhammad Umer Saleem[4]**

[1]*Department of Mathematics and Statistics, University of Lahore, Lahore 54590, Pakistan*

[2]*School of Mathematics and Statistics, Xi'an Jiaotong University, Xi'an, Shaanxi 710049, P.R. China*

[3]*Department of Mathematics, Art and Science Faculty, Siirt University, 56100 Siirt, Turkey*

[4]*Department of Mathematics, Division of Science and Technology, University of Education, Lahore, Punjab 54770, Pakistan*

Abstract: The present chapter proposes modeling of complex fractional-order chaotic ifnancial system with control. Here, we have added critical minimum interest rate 'd' as a new parameter to get a novel stable ifnancial model. The fractional derivatives. are taken in Caputo and Caputo-Fabrizio sense for the proposed ifnance system. Dynamical models in ifnancial system with complicated behavior provide a new. perspective as result of trends and actual behavior of internal structure of the ifnancial. system. A theoretical stabilization of the equilibria, as well as the numerical simulations, are obtained. Furthermore, with sensitivity analysis, a certain threshold estimation of the basic reproductive number has been made. Also, the stability analysis of the model, together with uniqueness of the special solutions is provided. The concept of controllability and observability for the linearized control model is used for feedback control. The solution of the proposed fractional-order model has been procured by employing different numerical techniques with comparison among the solutions. The convergence analysis is carried out for the accuracy of the applied scheme. Finally, some numerical simulations are given for three fractional-order chaotic systems to verify the efectiveness for the obtained results. The fractal, stochastic processes and prediction are used in particular mechanism of its application to the macro and micro processes.

Keywords: Complex chaotic system, Caputo derivative, Caputo-Fabrizio derivative, Dynamical control, Fixed point theorem, Fractional-order ifnance. system, Stability analysis.

---

*Corresponding author Parvaiz Ahmad Naik:** School of Mathematics and Statistics, Xi'an Jiaotong University, Xi'an, Shaanxi 710049, P.R. China; E-mail: naik.parvaiz@ xjtu.edu.cn

Mehmet Yavuz & Necati Özdemir (Eds.)

INTRODUCTION

The financial system has two classes which are called Microeconomics and Macroeconomics. Microe- conomics is an individual's study, business decision and Macroeconomics is an extensive study. Consequently, the financial systems works at particular, domestic and international level. In the financial system, money, credit and finance are used as a media of exchange. Financial and economic systems are getting harder to understand and economic growth varies from low to high financial markets. The process of eco- nomic development and growth is more complex on the basis of multiple variables. There are non-linear factors like interest rate, good's price, investment demands and stock. In Economics, mathematicians started to apply chaos theory [1, 2] during the last decades of the 20th century.

The mathematical behavior of the fractional order system is studied. If input on the system is influenced according to time, only then it is called static system. If the current and past input on the system is influenced, then it is called dynamical system. To gain the desired doutput and adjust system input, a controller is used which is considered a system named as a controlled system. The controller is said to be a closed-loop controller if the controlled output on system directly depends on controller inputs otherwise, it is said to be an open-loop controller [3].

The oldest mathematical tool which provides attractive research in all kinds of fields [4] is fractional calculus. Fractional calculus plays the main role and has many benefits as compared to integer calculus that narrates the memory and hereditary characteristics of different procedures and materials [5]. Financial variables are appropriate to use fractional modelling, which narrates the actual behavior of the financial system and possesses long memories (actual behavior of the financial system which possesses long memories; stock market prices, exchanges rates and interest rates are considered to possess long memories with very influential behavior to initial prices or values and some chaotic attractors (28)). Chen proposed a fractional order financial system, which narrates the actual dynamical behavior, studied the period-doubling and identified intermittency routes to chaos [6]. Financial system is studied by the chaos control method for slide mode and feedback control [7, 8]. The numerical techniques are used to solve such complex financial systems as exact solution cannot be found easily, the most used technique to solve fractional differential equations is the GABMM [9]. Some analytical methods to solve nonlinear differential equations are VIM, HPM, ADM and homotopy analysis method [10-12].

Since in recent years, fractional calculus has been an important gadget to describe the dynamical behaviour of different physical systems [11]. In recent years, researchers have taken interest and attention to fractional calculus in different aspects under consideration for research of the said subject [13, 14]. In the last decade, derivatives and integrals of fractional orders had notable development as revealed by several monographs dedicated to it, studied differential-difference equation of fractional order [13-15].

In this chapter, to develop the system of complex nonlinear differential equations, we apply fractional parameters using the Caputo and Caputo Fabrizio derivatives method. Dynamical models of the financial system for complicated behavior are checked from a new perspective as result of trends and actual behavior of internal structure of the financial system. The stabilization of equilibrium is obtained by both theoretical analysis and simulation results. The linearized systems of controllability and observability are designed for the close loop of automatic control.

MATERIALS AND METHOD

The subject of Mathematical research is fractional calculus which is the result of integral value exponents from the traditional definition of integral calculus and derivative operations as fractional exponents [16-23].

Definition 2.1. For a function $g: \Re^+ \rightarrow \Re$, then the fractional integral of order $\beta > 0$ is given by

$$I_t^\beta\big(g(t)\big) = \frac{1}{\gamma(\beta)} \int_0^t (t-z)^{\beta-1} g(z)\,dz,$$

where γ shows the Gamma function and β is the fractional order parameter.

Definition 2.2 For a function $g \in C^n$, then the Caputo derivative of order $\beta > 0$ is defined by [9]

$$_cD_t^\beta\big(g(t)\big) = I^{n-\beta} D^n g(t) \frac{1}{\gamma(n-\beta)} \int_0^t \frac{g^n(z)}{(t-z)^{\beta+n-1}}\,dz,$$

That is defined for the absolute continuous functions and $n - 1 < \beta < n \in N$. Obviously, $_cD_t^\beta(g(t))$ tends to $g'(t)$ as $\beta \to 1$.

Definition 2.3 Let $f \in H^1(a, b)$, with $b > a$, and $0 \leq \beta \leq 1$, then the Caputo Fabrizio can be written as

$$D_t^\beta(f(t)) = \frac{\kappa(\beta)}{1 - \beta} \int_a^t f'(x) exp\left[-\beta \frac{t - x}{1 - \beta}\right] dx,$$

The normalized function is shown by $\kappa(\beta)$ and it holds $\kappa(0) = \kappa(1) = 1$. Consider the case for which $f \notin H^1(a, b)$ then, we have the following:

$$D_t^\beta(f(t)) = \beta \frac{\kappa(\beta)}{1 - \beta} \int_a^t (f(t) - f(x)) exp\left[-\beta \frac{t - x}{1 - \beta}\right] dx$$

Fractional Order Complex Chaotic System

The authors [17] reported a model consisting of three differential equations in first order to describe how the financial system operates:

$$\dot{x}(t) = z + x(y - a),$$

$$\dot{y}(t) = 1 - by - x^2, \tag{1}$$

$$\dot{z}(t) = -x - cz,$$

where x, y, z are the state variables, which represent interest rate, investment demand and pride index, respectively. The saving parameter a, cost per investment b and c is the elasticity of market demand *i.e.*, $a = 3, b = 0.1$ *and* $c = 1$.

From the above model in Equation (1), we use the parmeter '*d*' to modify the model, where d represents critical minimum interest rate. Now the equation will be written as:

$$\dot{x}(t) = z + xy - ax,$$

$$\dot{y}(t) = 1 - by - x^2, \tag{2}$$

$$\dot{z}(t) = d - x - cz.$$

The fractional differential system of Equation (2) is given below in Equation (3):

$$D^\beta x(t) = z + xy - ax$$

$$D^\beta y(t) = 1 - by - x^2$$

$$D^\beta z(t) = d - x - cz$$

(3)

Correspond to initial conditions $x(0) = n_1 \geq 0, y(0) = n_2 \geq 0, z(0) = n_3 \geq 0$.

Qualitative Analysis

System equilibrium points for qualitative analysis of Equation (3) by replacing the left hand equal to zero, we get the point of equilibrium as $E = (x, y, z)$ is endemic equilibrium in the system of fractional order, where

$$x = 0.840853, y = 2.92966, z = 0.0591468.$$

Stability Analysis

Taking the determinant of $|J - \lambda I|$, we get,

$$-\lambda^3 - 1.17034\lambda^2 - 2.591441906\lambda - 1.521101906.$$

Putting $|J - \lambda I| = 0$, we have

$\lambda_1 = -0.673977, \lambda_2 = -0.248181 - 1.48166i, \lambda_3 = -0.248181 + 1.48166i$.

The real parts of the jacobian matrix's values are negative, therefore the system is locally asymptotyically stable with parameters $a = 3, b = 0.1, c = 1$ and $d = 0.9$.

Sensitivity Analysis

Reproductive number of the system can are there with next generation method

$$R_0 = \frac{-abc - 2cx^2 + bcy}{b}.$$

The sensitivity of R_0

$$\frac{\partial R_0}{\partial a} = -c < 0,$$

$$\frac{\partial R_0}{\partial b} = \frac{-0.70703376761c}{b^2} < 0,$$

$$\frac{\partial R_0}{\partial c} = \frac{2.92966b - ab - 0.70703376761}{b} < 0.$$

Here a, b and c are decreasing with change R_0, we can reduce these parameters according to control situations. Fig. (**1**) represents the sensitivity and stability analysis of the complete model with given parameters values.

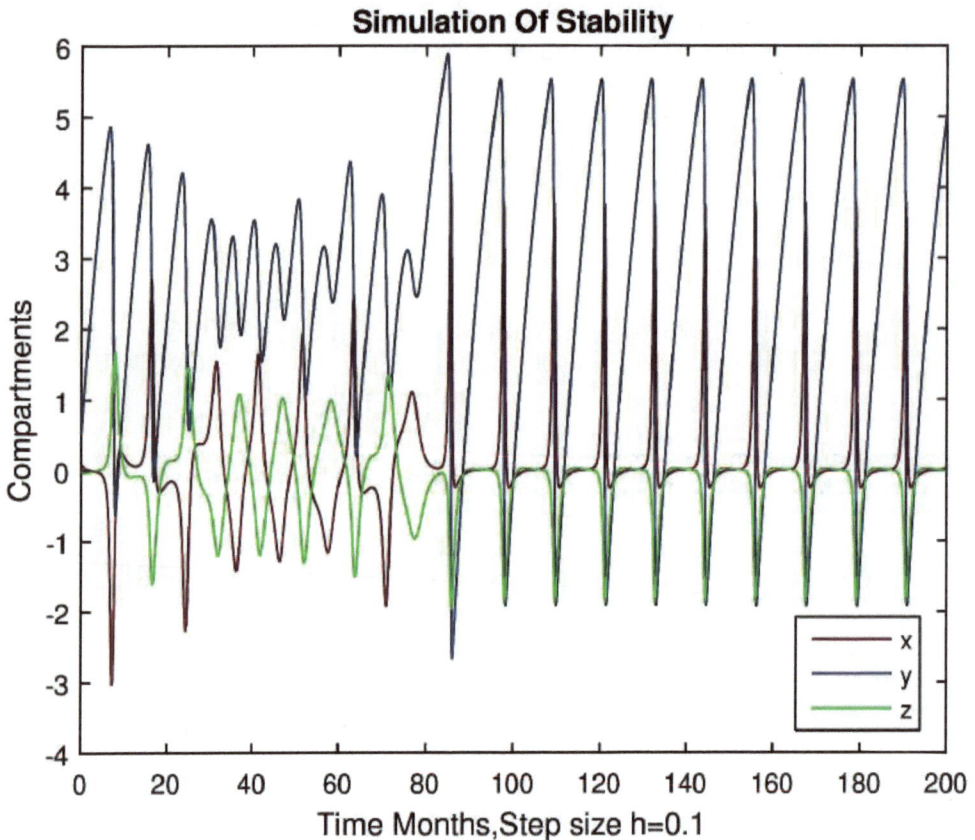

Fig. (1). Numerical Solutions for stability of complex fractional chaotic financial system in a time t (month) at different values.

Theorem.1 There is a unique solution for mathematical model given by (3), and the solution remains in $R^4, x \geq 0$.

Proof: Solution of (3), in $(0, \infty]$ is represented for uniqueness and existence, since

$$D^\beta x|_{x=0} = -c < 0,$$

$$D^\beta y|_{y=0} = \frac{-0.70703376761c}{b^2} < 0,$$

$$D^\beta z|_{z=0} = \frac{2.92966b - ab - 0.70703376761}{b} < 0.$$

Complex Financial System with Caputo Sense

By using the Laplace transform, we have

$$L\{D^\beta x(t)\} = L\{z\} + L\{x.y\} - aL\{x\},$$

$$L\{D^\beta y(t)\} = L\{1\} - bL\{y\} - L\{x.x\},$$

$$L\{D^\beta x(t)\} = dL\{1\} - L\{x\} - cL\{z\},$$

$$L\{x(t)\} = \frac{S^{\beta-1}x(0)}{S^\beta} + \frac{L\{x\}}{S^\beta} + \frac{L\{x.y\}}{S^\beta} - \frac{aL\{x\}}{S^\beta},$$

$$L\{y(t)\} = \frac{S^{\beta-1}y(0)}{S^\beta} + \frac{L\{1\}}{S^\beta} - b\frac{L\{y\}}{S^\beta} - \frac{L\{x.x\}}{S^\beta},$$

$$L\{z(t)\} = \frac{S^{\beta-1}z(0)}{S^\beta} + d\frac{L\{1\}}{S^\beta} - \frac{L\{x\}}{S^\beta} - c\frac{L\{z\}}{S^\beta}.$$

Correspond to initial conditions $x(0) = n_1 = 0.1, \ y(0) = n_2 = 4, z(0) = n_3 = 0.5$, the nonlinearity xy and xx can be written as

$$xy = \sum_{k=1}^{\infty} A_k, xx = \sum_{k=1}^{\infty} B_k,$$

where A_k *and* B_k can be written as

$$A_k = \frac{1}{k!}\frac{d^k}{d\lambda_k}\left|\sum_{j=0}^{k}\lambda^j x_j \sum_{j=0}^{k}\lambda^j y_j\right||_{\lambda=0}$$

$$B_k = \frac{1}{k!}\frac{d^k}{d\lambda_k}\left|\sum_{j=0}^{k}\lambda^j x_j \sum_{j=0}^{k}\lambda^j x_j\right||_{\lambda=0}$$

$$L\{x_{k+1}(t)\} = \frac{0.1}{S} + \frac{L\{x_k\}}{S^\beta} + \frac{L\{A_k\}}{S^\beta} - \frac{aL\{x_k\}}{S^\beta}$$

$$L\{y_{k+1}(t)\} = \frac{4}{S} + \frac{1}{S^{\beta+1}} - b\frac{L\{y_k\}}{S^\beta} - \frac{L\{B_k\}}{S^\beta}$$

$$L\{z_{k+1}(t)\} = \frac{0.5}{S} + \frac{d}{S^{\beta+1}} - \frac{L\{x_k\}}{S^\beta} - c\frac{L\{z_k\}}{S^\beta}$$

The technique gives analytical solution as:

$$x = \sum_{k=1}^{\infty} x_k, y = \sum_{k=1}^{\infty} y_k, \qquad z = \sum_{k=1}^{\infty} z_k$$

Complex Finance System with Caputo-Fabrizio Sense

Applying the transformation of Sumudu to both sides of Equation (3) which is constructed by the Caputo-Fabrizo derivative with fractional order $\beta \in [0,1]$, we have

$$M(\beta)\frac{S(x(t)-x(0))}{1-\beta+\beta s} = S[z + x(y - a)],$$

$$M(\beta)\frac{S(y(t)-y(0))}{1-\beta+\beta s} = S[1 - by - x^2],$$

$$M(\beta)\frac{S(z(t)-z(0))}{1-\beta+\beta s} = S[d - x - cz].$$

We obtain the rearrangement and application of inverse Sumudu transformation

$$x_{(n+1)}(t) = x_n(0) + S^{-1}\frac{1-\beta+\beta s}{M(\beta)}S[z + x(y - a)],$$

$$y_{(n+1)}(t) = y_n(0) + S^{-1}\frac{1-\beta+\beta s}{M(\beta)} S[1-by-x^2],$$

$$z_{n+1}(t) = z_n(0) + S^{-1}\frac{1-\beta+\beta s}{M(\beta)} S[d-x-cz].$$

Then the solution is provided by

$$x(t) = \lim_{n\to\infty} x_n(t), y(t) = \lim_{n\to\infty} y_n(t), z(t) = \lim_{n\to\infty} z_n(t).$$

Stability and Fixed-Point Theorem Analysis as an Application

Let us suppose $(X_1, ||.\,||)$ as a Banach space and P as a self-map of X_1. Let $y_{n+1} = g(P, y_n)$ be particular recursive procedure. Our aim is to prove that P is picard P-stable.

Theorem 2: Let $(X_1, ||.\,||)$ as a Banach space and P be a self-map of X_1 satisfying $||Px - Py|| \leq C||x - px|| + c||x - y||$ For all $x, y \in X_1$ where $0 \leq c, 0 \leq c < 1$. Suppose that P is a picard P-stable. Suppose the following recursive formula

$$x_{(n+1)}(t) = x_n(0) + S^{-1}\left[\frac{1-\beta+\beta s}{M(\beta)} S[z_n + x_n(y_n - a)]\right]$$

$$y_{(n+1)}(t) = y_n(0) + S^{-1}\left[\frac{1-\beta+\beta s}{M(\beta)} S[1 - by_n - x_n^2]\right]$$

$$z_{n+1}(t) = z_n(0) + S^{-1}\left[\frac{1-\beta+\beta s}{M(\beta)} S[d - x_n - cz_n]\right]$$

where $\frac{1-\beta+\beta s}{G(v)}$ is the fractional Lagrange multiplier.

Theorem 3 Suppose self-map P is defined as

$$x_{(n+1)}(t) = x_n(0) + S^{-1}\left[\frac{1-\beta+\beta s}{M(\beta)} S[z_n + x_n(y_n - a)]\right]$$

$$y_{(n+1)}(t) = y_n(0) + S^{-1}\left[\frac{1-\beta+\beta s}{M(\beta)} S[1 - by_n - x_n^2]\right]$$

$$z_{n+1}(t) = z_n(0) + S^{-1}\left[\frac{1-\beta+\beta s}{M(\beta)} S[d - x_n - c z_n]\right]$$

We have specified a self-map as 'p' and in $L_1(a,b)$ we are P-stable if

$$\{1 + f(q) - ag(q) + M\,h(q) + LI(q)\} < 1,$$

$$\{1 - bf(q) - (M + M1)g1(q)\} < 1,$$

$$\{1 - f2(q) - cg2(q)\} < 1.$$

Proof: We will demonstrate that P has fixed point in the first phase, for which we will assess the following for all $(m, n) \in N$

$$P(x_n \ (t)) - P(x_m \ (t)) = x_n \ (t) - x_m \ (t) + S^{-1}\left[\frac{1-\beta+\beta s}{G(v)} S[z_n + x_n(y_n - a)]\right] - S^{-1}\left[\frac{1-\beta+\beta s}{G(v)} S[z_m + x_m(y_m - a)]\right]$$

$$P(y_n(t)) - P(y_m \ (t)) = y_n \ (t) - y_m \ (t) + S^{-1}\left[\frac{1-\beta+\beta s}{G(v)} S[1 - by_n - x_n{}^2]\right] - S^{-1}\left[\frac{1-\beta+\beta s}{G(v)} S[1 - by_m - x_m{}^2]\right] \tag{4}$$

$$P(z_n \ (t)) - P(z_m \ (t)) = z_n \ (t) - z_m \ (t) + S^{-1}\left[\frac{1-\beta+\beta s}{G(v)} S[d - x_n - c z_n]\right] - S^{-1}\left[\frac{1-\beta+\beta s}{G(v)} S[d - x_m - c z_m]\right].$$

Taking norm on both sides of Equation (4) above 1st equation, we get,

$$\|P(x_n \ (t)) - P(x_m \ (t))\| = \left\|x_n \ (t) - x_m \ (t) + S^{-1}\left[\frac{1-\beta+\beta s}{G(v)} S[z_n + x_n(y_n - a)]\right] - S^{-1}\left[\frac{1-\beta+\beta s}{G(v)} S[z_m + x_m(y_m - a)]\right]\right\|$$

$$\|P(y_n \ (t)) - P(y_m \ (t))\| = \left\|y_n \ (t) - y_m \ (t) + S^{-1}\left[\frac{1-\beta+\beta s}{G(v)} S[1 - by_n - x_n{}^2]\right] - S^{-1}\left[\frac{1-\beta+\beta s}{G(v)} S[1 - by_m - x_m{}^2]\right]\right\|$$

$$\|P(z_n \ (t)) - P(z_m \ (t))\| = \left\|z_n \ (t) - z_m \ (t) + S^{-1}\left[\frac{1-\beta+\beta s}{G(v)} S[d - x_n - c z_n]\right] - S^{-1}\left[\frac{1-\beta+\beta s}{G(v)} S[d - x_m - c z_m]\right]\right\|, \tag{5}$$

by making use of triangular inequality, Equation (5) implies

$$\|P(x_n(t)) - P(x_m(t))\| = \|x_n(t) - x_m(t)\| + \left\|S^{-1}\left[\frac{1-\beta+\beta s}{G(v)}S[z_n + x_n(y_n - a)]\right] - S^{-1}\left[\frac{1-\beta+\beta s}{G(v)}S[z_m + x_m(y_m - a)]\right]\right\|$$

$$\|P(y_n(t)) - P(y_m(t))\| = \|y_n(t) - y_m(t)\| + \left\|S^{-1}\left[\frac{1-\beta+\beta s}{G(v)}S[1 - by_n - x_n^2]\right] - S^{-1}\left[\frac{1-\beta+\beta s}{G(v)}S[1 - by_m - x_m^2]\right]\right\|$$

$$\|P(z_n(t)) - P(z_m(t))\| = \|z_n(t) - z_m(t)\| + \left\|S^{-1}\left[\frac{1-\beta+\beta s}{G(v)}S[d - x_n - cz_n]\right] - S^{-1}\left[\frac{1-\beta+\beta s}{G(v)}S[d - x_m - cz_m]\right]\right\|$$

Further simlification,

$$\|P(x_n(t)) - P(x_m(t))\| \leq \|x_n(t) - x_m(t)\| + S^{-1}\left[\frac{1-\beta+\beta s}{G(v)}[\|z_n - z_m\| - \|a(x_n - x_m)\| + \|x_n(y_n - y_m)\| + \|y_m(x_n - x_m)\|]\right] \tag{6}$$

Replacing this Equation (6), we obtain

$$\|P(x_n(t)) - P(x_m(t))\| \leq \|x_n(t) - x_m(t)\| + S^{-1}\left[\frac{1-\beta+\beta s}{G(v)}[\|x_n - x_m\| - \|a(x_n - x_m)\| + \|x_n(y_n - y_m)\| + \|y_m(x_n - x_m)\|]\right]. \tag{7}$$

Since x_n, x_m, y_m are bounded and they are convergent, so we get M, M_1 and L for all t like

$$\|x_n\| < M, \|x_m\| < M_1, \|y_m\| < L \tag{8}$$

Consider now Equation (7) with Equation (8), we have

$$\|P(x_n(t)) - P(x_m(t))\| \leq 1 + f(q) - ag(q) + M\,h(q) + LI(q)\|x_n - x_m\|, \tag{9}$$

where f, g, h and I are functions from $S^{-1}[\frac{1-\beta+\beta s}{G(v)}]$ In the same path we obtain

$$\|P(y_n(t)) - P(y_m(t))\| \leq 1 - bf(q) - (M + M1)g_1(q)\|y_n - y_m\| \tag{10}$$
$$\|P(z_n(t)) - P(z_m(t))\| \leq d - f_2(q) - cg_2(q)\|z_n - z_m\| \tag{11}$$

where

$$1 + f(q) - ag(q) + M\,h(q) + LI(q) \leq 1$$

$$1 - bf(q) - (M + M1)g_1(q) \leq 1$$

$$1 - f_2(q) - cg_2(q) \leq 1$$

Thus, the self-mapping of non-linear p has a fixed point. Next, we demonstrate that p meets the requirement. Let Equations (9) - (11) hold and using

$C = (0,0,0)$

$$1 + f(q) - ag(q) + M\,h(q) + LI(q)$$

$$C = 1 - bf(q) - (M + M_1)g_1(q)$$

$$d - f_2(q) - cg_2(q)$$

It shows that the given condition satisfied the Theorem (2) and Theorem (3) for the nonlinear mapping P, so P is picard P-stable.

Uniqueness of the Special Solution

Here the Hilbert space $H = L(a,b) \times (0,T) \to R$, such that $\iint < \infty$. The operator is used as

$$z - x(y - a)$$

$$P(x,y,z) = 1 - by - x^2$$

$$d - x - cz$$

The objective of this part is to demonstrate the inner product as

$$P((X_{11} - X_{12}, X_{21} - X_{22}, X_{31} - X_{32})(w_1, w_2, w_3))$$

$$= [(X_{31} - X_{32}) + (X_{11} - X_{12})[(X_{11} - X_{12}) - a], w_1] \qquad \textbf{(12)}$$

$$[-b(X_{21} - X_{22}) - (X_{11} - X_{12})^2, w_2][-(X_{11} - X_{12}) - c(X_{31} - X_{32}), w_3]$$

We shall evaluate the first equation in the system without loss of generality

$$[(X_{31} - X_{32}) + (X_{11} - X_{12})[(X_{11} - X_{12}) - a], w_1]$$

$$\cong [(X_{11} - X_{12}), w_1] + [-a(X_{11} - X_{12}) + (X_{11} - X_{12})(X_{21} - X_{22}), w_1] \qquad (13)$$

Assume that $(X_{11} - X_{12}) \cong (X_{21} - X_{22}) \cong (X_{31} - X_{32})$, Equation (13) becomes

$$[(X_{11} - X_{12}) + [-a(X_{11} - X_{12}) + (X_{11} - X_{12})^2], w_1].$$

We obtain the following based on the relation between the norm and the inner product and repeat the same fusion from the system's 2nd and 3rd formula (12).

$$[-b(X_{21} - X_{22}) - (X_{11} - X_{12})^2, w_2] \leq (-b - \overline{w}_2)\| X_{21} - X_{22}\|\|w_2\|$$

$$[-(X_{11} - X_{12}) - c(X_{31} - X_{32}), w_3] \leq (- (\overline{w}_3) - c(\overline{w}_3))\| X_{31} - X_{32}\|\|w_3\|,$$

we have,

$$P((X_{11} - X_{12}, X_{21} - X_{22}, X_{31} - X_{32})(w_1, w_2, w_3))$$

$$\leq (1 - a + (\overline{w}_1))\| X_{11} - X_{12}\|\|w_1\|$$

$$\leq (-b - (\overline{w}_2))\| X_{21} - X_{22}\|\|w_2\| \qquad (14)$$

$$\leq (-(\overline{w}_3) - c \leq (w_3))\| X_{31} - X_{32}\|\|w_3\|.$$

Large values of M with $i = 1,2,3$. Both results converge to same solution using the topological principle and there are three small non-negative parameters of lm1, lm2, lm3, such that

$$\|x - X_{11}\|, \|x - X_2\| < \frac{lm_1}{z(1 - a + (\overline{w}1_1))\| X_{11} - X_{12}\|\|w1\|}$$

$$\|y - X_{21}\|, \|y - X_{22}\| < \frac{lm_2}{z(-b - (\overline{w}_2))\| X_{21} - X_{22}\|\|w2\|}$$

$$\|z - X_{31}\|, \|z - X_{32}\| < \frac{lm_3}{z(-(\overline{w}_3) - c \leq (w3))\| X_{31} - X_{32}\|\|w3\|}$$

Connecting the exact solution to the right side of the Equation (14) and introduce the triangular inequality

$$(1 - a + (\overline{w}_1))\| X_{11} - X_{12}\|\|w_1\| < l$$

$$(-b- (\overline{w}_2))\| X_{21} - X_{22}\|\|w_2\|< l$$

$$(-(\overline{w}_3)-c\leq(w_3))\| X_{31} - X_{32}\|\|w_3\|< l$$

As we know l is a small positive parameter, we obtain

$$(1-a+ (\overline{w}_1))\| X_{11} - X_{12}\|\|w_1\|< 0,$$

$$(-b- (\overline{w}_2))\| X_{21} - X_{22}\|\|w_2\|< 0,$$

$$(-(\overline{w}_3)-c\leq(w_3))\| X_{31} - X_{32}\|\|w_3\|< 0.$$

But it is obvious that

$$(1-a+ (\overline{w}_1)) \neq 0$$

$$(-b- (\overline{w}_2))\neq 0$$

$$(-(\overline{w}_3)-c\leq(w_3))\neq 0$$

So, we get,

$$\| X_{11} - X_{12}\|= 0,$$

$$\| X_{21} - X_{22}\|= 0,$$

$$\| X_{31} - X_{32}\|= 0.$$

These results show that

$$X_{11}=X_{12},$$

$$X_{21}=X_{22},$$

$$X_{31}=X_{32.}$$

In this way, we complete the proof.

Numerical Results and Discussion

The analysis has been made using Caputo and Caputo-Fabrizio fractional technique for complex chaotic financial systems. By using Caputo-Fabrizio and Caputo fractional derivatives, the numerical outcomes of interest rate, investment demand and price exponent for different fractional values of β are obtained. Figs. (**2-4**) represent the graphical solution of fiance system with Caputo derivative and we observed that the interest rate, investment demand and the price exponent have more degree of freedom as compared to ordinary derivatives. From Figs. (**5-7**), Caputo-Fabrizio fractional order derivative is used for finance system, we easily observed that all compartments provide better estimation at fractional values as compared to ordinary derivatives. Comparison of Caputo and Caputo-Fabrizio fractional derivative are shown in Figs. (**8-10**), we easily got the result with Caputo-Fabrizio derivative result is more reliable than Caputo fractional derivative. Relationship between these variables is increasing or decreasing factor with by Caputo-Fabrizio and Caputo fractional derivatives. From Figs. (**2-11**) by taking non-integer fractional parameter values, remarkable responses are acquired from the compartments. Numerical results indicate that system preserves the chaotic motion for β. Interest rate starts increasing while investment demand and price exponent start decreasing which shows the actual macroeconomic behavior of the proposed system. This scheme should remain stable and consistent for complex chaotic financial system.

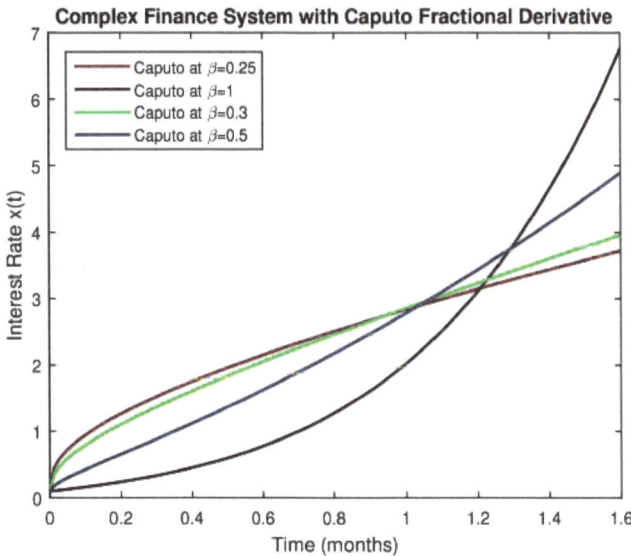

Fig. (2). Numerical solutions for Interest rate $x(t)$ in a time t (month) at different values of β with Caputo fractional derivatives.

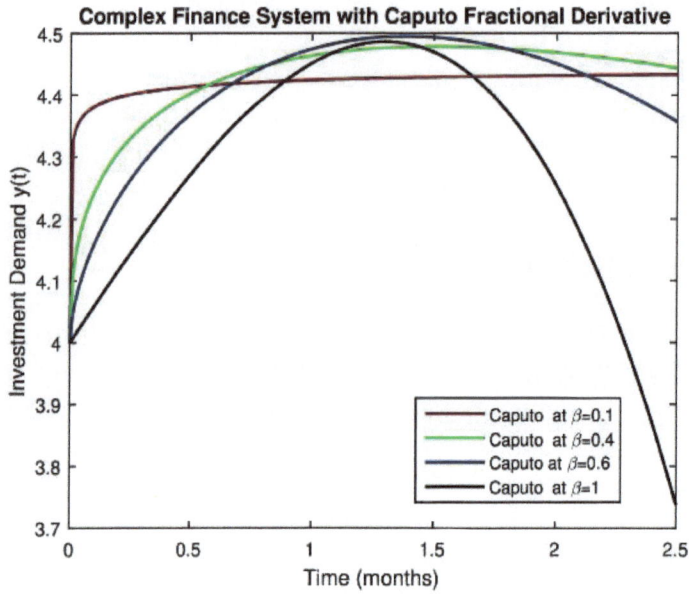

Fig. (3). Numerical solutions for Investment demand $y(t)$ in a time t (month) at different values of β with Caputo fractional derivatives.

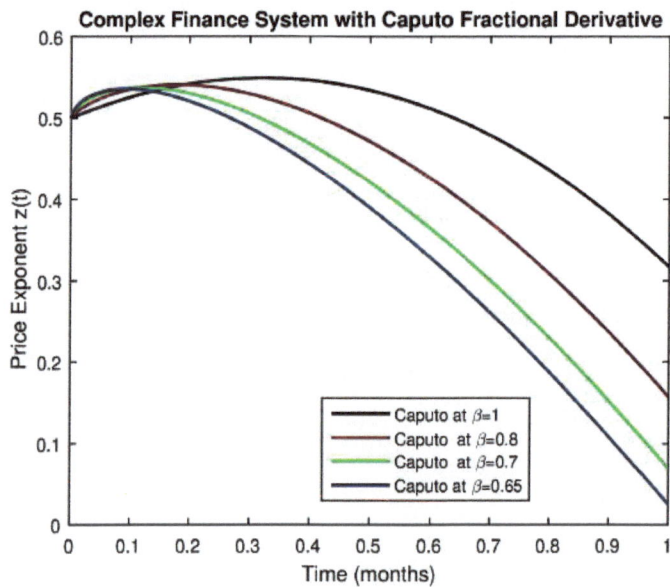

Fig. (4). Numerical solutions for Price Exponent $z(t)$ in a time t (month) at different values of β with Caputo fractional derivatives.

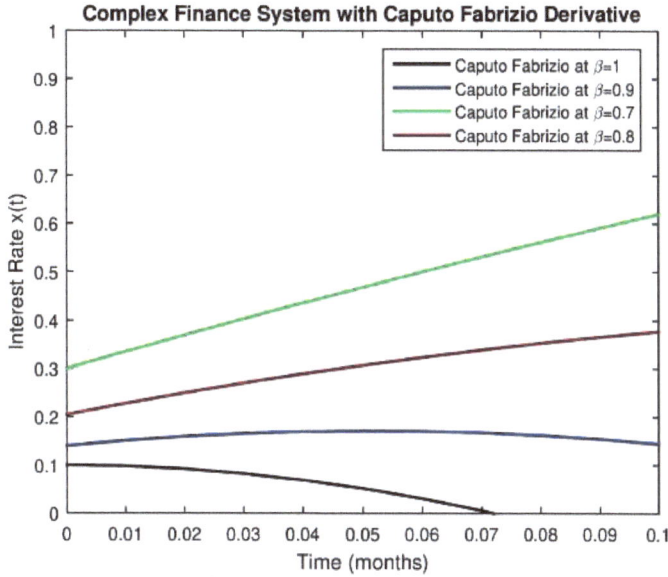

Fig. (5). Numerical Solutions for Interest Rate $x(t)$ in a time t (month) at different values of β with caputo-fabrizio derivatives.

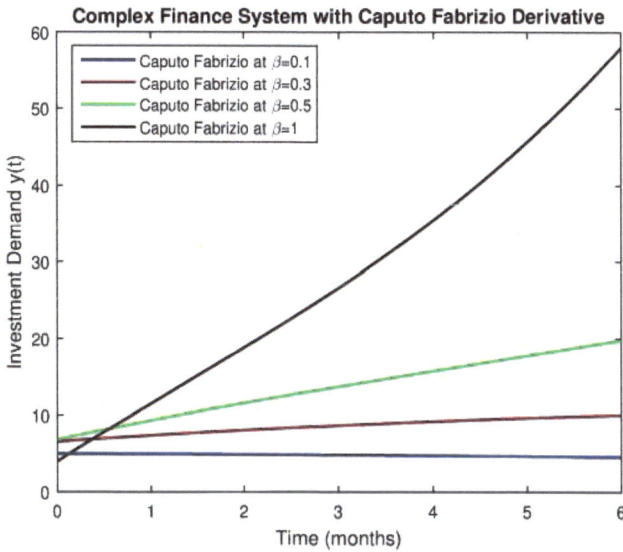

Fig. (6). Numerical Solutions for Investment Demond $y(t)$ in a time t (month) at different values of β with caputo-fabrizio derivatives.

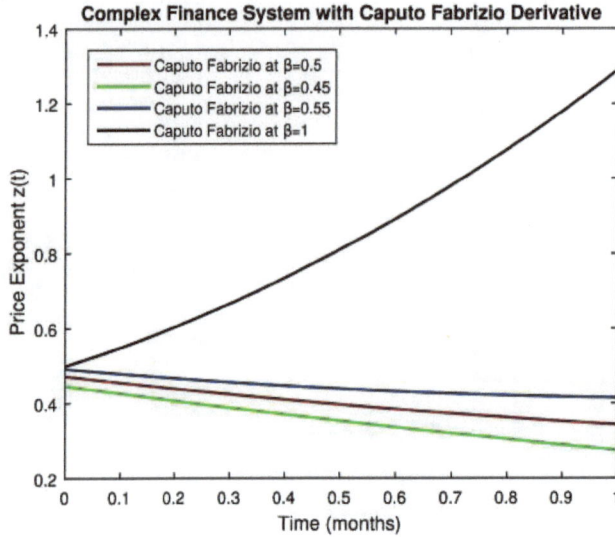

Fig. (7). Numerical Solutions for Price Exponent $z(t)$ in a time t (month) at different values of β with caputo-fabrizio derivatives.

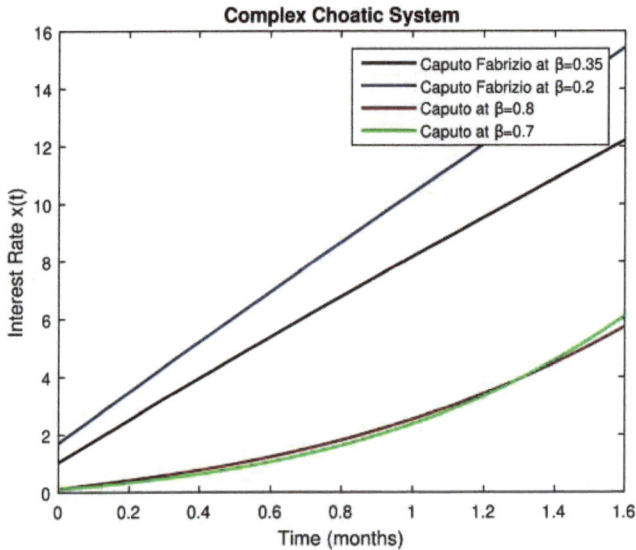

Fig. (8). Numerical Solutions for Interest Rate $x(t)$ in a time t (month) different values of β with CF and caputo derivatives.

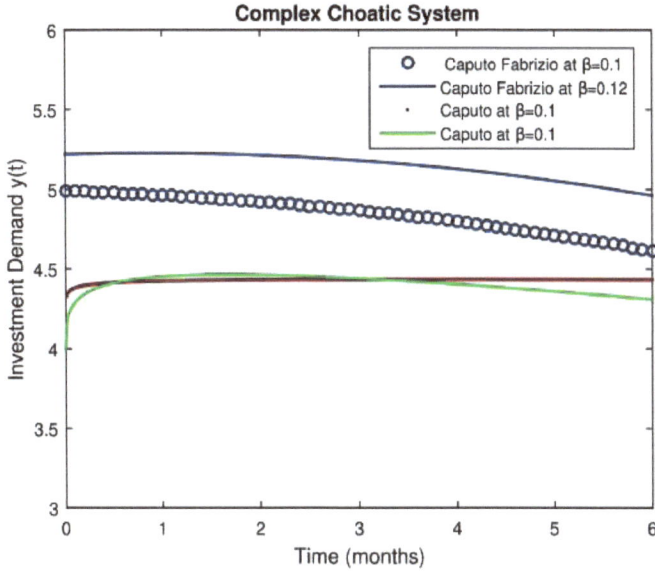

Fig. (9). Numerical Solutions for Investment Demond $y(t)$ in a time t (month) different values of β with CF and caputo derivatives.

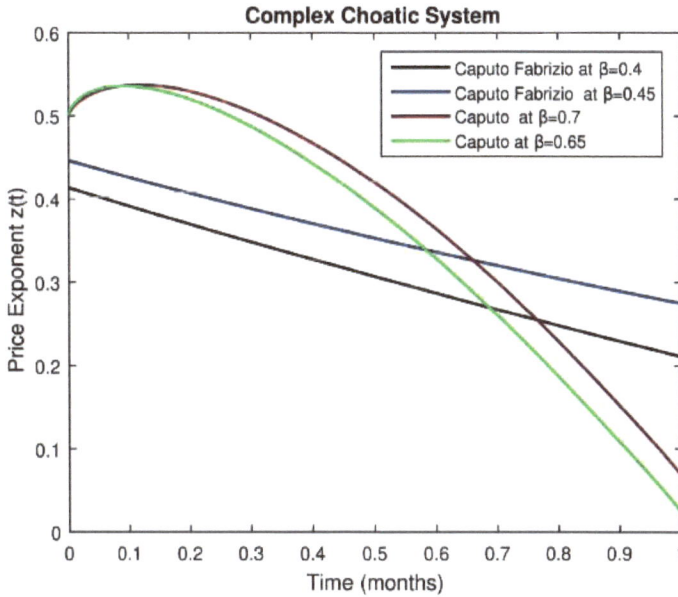

Fig. (10). Numerical Solutions for Price Exponent $z(t)$ in a time t (month) different values of β with CF and caputo derivatives.

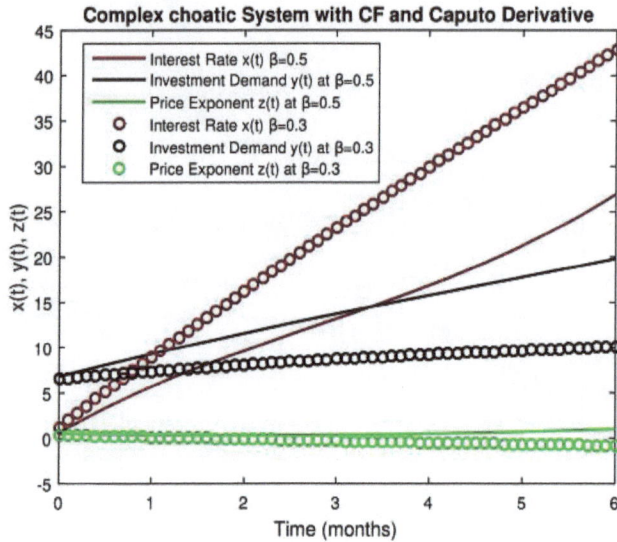

Fig. (11). Complex choatic system with CF and Caputo derivatives.

Input and Output Stability

Stability is a main anxiety in feedback control design for automatic control systems because a feedback controls law can stabilize and also can destabilize a system. We use Lyapunov's indirect method [19, 20] to examine the system stability (2.1 – 2.3). The equilibrium points of the system depend on the steady state of interest rate, investment demand, and exponent of price. The equilibrium point as $(0.840853, 2.92966, 0.0591468)$. Suppose the linear system

$$\dot{x}(t) \; = \; Gx + Hu, \quad t \in I, \tag{15}$$

$$y(t) \; = \; Jx, \quad t \in I, \tag{16}$$

where $x = [x, y, z]T$, Here $H = [\,0\,1\,1\,]T$ and $J = [\,1\,0\,0\,]$. We find the following eigenvalues of system: -0.673977, $-0.248181 - 1.48166i$, $-0.248181 + 1.48166i$. Since the eigenvalues of the system are negative real parts, it satisfies the input output stability theorem.

The following is a mathematically linear control system:

$$\dot{x}(t) \; = \; Gx + Hu, \quad t \in I, \tag{17}$$

$$y(t) = Jx, \quad t \in I, \tag{18}$$

where the matrices g, h and j are described in i and the dimensions are right (*i.e.*, g is $n \times n$, h is $n \times p$ and j is a matrix of $k \times n$). *I* is a closed interval, $i = [t_0, t_e]$, $t_0 < t_e < \infty$, respectively, the matrix of controllability $i = [t_0, \infty)$ is $n \times n\,p$ specified by $R = [H, GH, G^2H, G^3H, \ G^{n-1}H]^T$. The rank (*i.e. Rank(R) = n*), so it is said that the system can be controlled. Then $k \times n$ the observability matrix is specified by $O = [J, JG, JG^2, JG^3, JG^{n-1}]^T$. The rank (*i.e. Rank(R) = n*), hence it is said that the system is observable [16, 17]. We consider the two cases according to the input and output in complex chaotic financial system. For case 1, if consider the price index as inputs and interest rate as the only output than $x = [x, y, z]^T, H = [0\,0\,1]^T$ and $J = [1\,0\,0]$. The controllability matrix is given as R= $[H\ GH\ G^2H]$ and the rank of matrix for controllability is 3. The matrix of observability is $O = [J; JG; JG^2]$ and the rank of matrix for observability is 2. The system can therefore be controlled and observed. Figs. (**12 & 13**) represent the control observability which represents the close loop design half system. It supported the result. For case 2, if consider the price index and investment demand as input and interest rate being the only output than $x = [x, y, z]^T, H = [0\,1\,1]^T$ and $J = [1\,0\,0]$. The level of controllability and observability matrix is 3. The system can therefore be controlled and observed. Figs. (**12-14**) represent the controllable and observable state with different initial conditions of the state variables, when interest rate is an input, which form a closed loop to stabilize the finance system with price exponent.

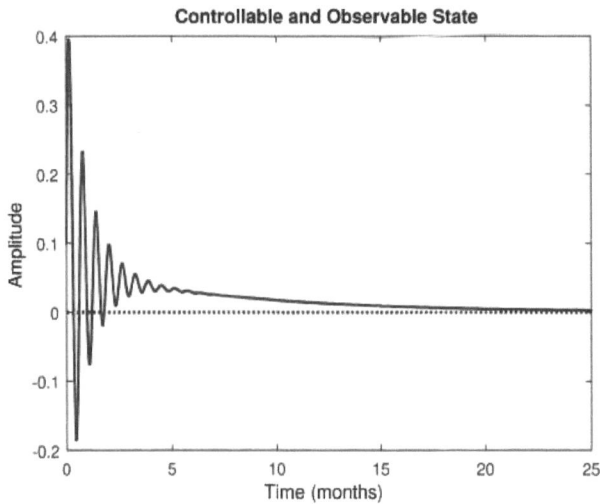

Fig. (12). Controllable and observable state of model with different values.

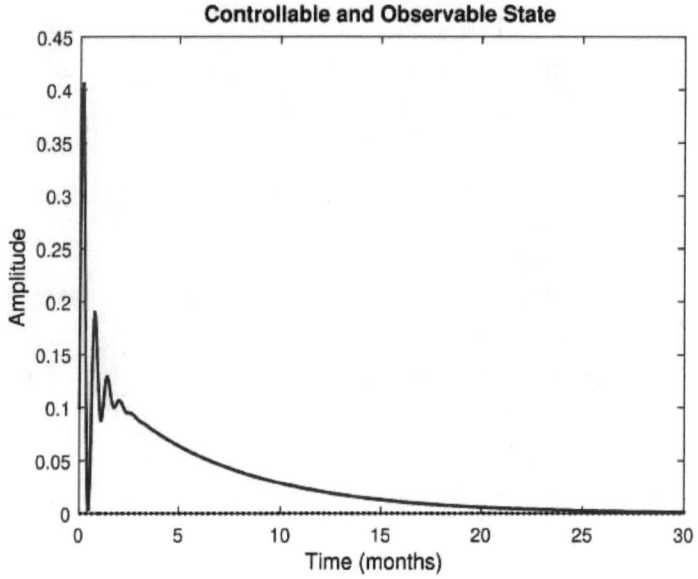

Fig. (13). Controllable and observable state of model with different values.

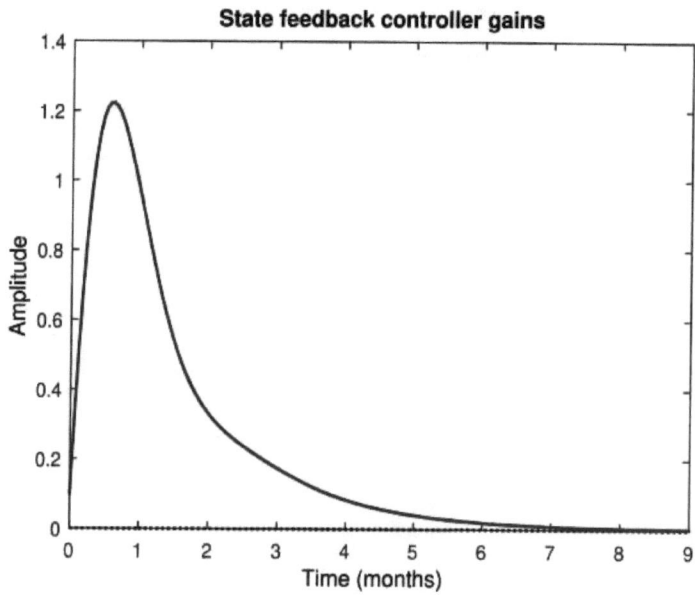

Fig. (14). Controllable and observable state of model with different values.

CONCLUSION

Fractional-order model using Caputo and Caputo-Fabrizio derivatives is studied for complex chaotic finance system. Dynamical models of the financial system for complicated behavior provide a new perspective as a result of trends and actual behavior of internal structure of the financial system. The efficiency of the proposed scheme is provided by performing convergence analysis. The fractional parameter is affected by our alternatives as displayed in the Figures. It is essential to note that significant modifications and memory impacts in fractional derivatives can be seen in comparison with ordinary derivatives. The fractional financial system's control is tested by the condition of the financial system's controllability and observability for close design of the model. It is proved that the financial system should meet the quadratic form that can be interpreted as the preservation of financial instruments. In addition, a system of financial state feedback is intended. Fixed point theory and Picard Lindelof approach are used for uniqueness and stability analysis for the proposed fractional order model. The effect of the different orders is analysed using numerical techniques. We can examine $\beta = 1$ by numerical simulation for the Caputo-Fabrizio non-integer order derivative, which reveals appropriate absorbing characteristics. Graphical representation demonstrates the impacts of fractional parameters on complex chaotic system, which has important outcomes for finance system practitioners and its related complications. The current study will be helpful for further understanding and to control the economic system.

CONSENT FOR PUBLICATION

Not applicable.

CONFLICT OF INTEREST

The authors declare no conflict of interest, financial or otherwise.

ACKNOWLEDGEMENTS

The authors are thankful to the anonymous reviewers and editors for their valuable suggestions and comments to improve the quality of the chapter.

REFERENCES

[1]   G. Gandolfo and D. Federici, "International Finance and Open-economy Macroeconomics", Springer, 2nd ed., 2016.

[2]   M. Lines, "Nonlinear Dynamical Systems in Economics", Springer-Verlag Wien, 2005.

[3]   J. M. Coron, "Control and Nonlinearity", American Mathematical Society, vol. 136, pp. 426, 2007.

[4]   Y. Zhou and F. Jiao, "Existence of mild solutions for fractional neutral evolution equations", Comput. Math. Appl., vol. 59, pp. 1063-1077, 2010.

[5]   R. Hilfer, "Applications of Fractional Calculus in Physics", World Scientific, Hackensack, NJ, pp. 472, 2000.

[6]   W. C. Chen, "Nonlinear dynamics and chaos in a fractional-order financial system", Chaos Soliton. Fract., vol. 36, pp. 1305-1314, 2008.

[7]   M. Salah, N. Hamri, and J. Wang, "Chaos control of a fractional-order financial system", Math. Prob. Eng., vol. 2010, pp. 270646, 2010.

[8]   W. Zhen, H. Xia, and S. Guodong, "Analysis of nonlinear dynamics and chaos in a fractional order financial system with time delay", Comput. Math. Appl., vol. 62, pp. 1531-1539, 2011.

[9]   K. Diethelm, N. J. Ford, and A. D. Freed, "A predictor-corrector approach for the numerical solution of fractional differential equations", Nonlinear Dyn., vol. 29, pp. 3-22, 2002.

[10]   S. Momani and A. Yildirim, "Analytical approximate solutions of the fractional convection-diffusion equation with nonlinear source term by He's homotopy perturbation method", Int. J. Comput. Math., vol. 87, pp. 1057-1065, 2010.

[11]   S. M. Goh, M. S. M. Noorani, and I. Hashim, "On solving the chaotic Chen system: a new timemarching design for the variational iteration method using Adomian's polynomial", Numer. Algorithms, vol. 54, pp. 245-260, 2010.

[12]   P. A. Naik, J. Zu, and M. Ghoreishi, "Stability analysis and approximate solution of SIR epidemic model with crowley-martin type functional response and holling type-ii treatment rate by using homotopy analysis method", J. Appl. Anal. Comput., vol. 10, no. 4, pp. 1482-1515, 2020.

[13]   M. Ghoreishi, A. I. B. M. Ismail, and A. K. Alomari, "Application of the homotopy analysis method for solving a model for HIV infection of CD4+ T-cells", Math. Comput. Model., vol. 61, pp. 2528-2534, 2011.

[14]   M. U. Saleem, M. Farman, M. O. Ahmad, and M. Rizwan,"Control of an artificial human pancreas", Chin. J. Phys., vol. 55, pp. 2273-2282, 2017.

[15]   H. Yu, G. Cai, and Y. Li, "Dynamic analysis and control of a new hyperbolic finance system", Nonlinear Dyn., vol. 67, pp. 21-71, 2012.

[16]   J. J. Losada and J. J. Nieto, "Properties of the new fractional derivative without singular kernel", Prog. Fract. Differ. Appl., vol. 1, pp. 87-92, 2015.

[17]   M. Caputo and M. Fabrizio, "A new definition of fractional derivative without singular kernel", Prog. Fract. Differ. Appl., vol. 1, pp. 73-85., 2015.

[18]   P. A. Naik, J. Zu, and K. M. Owolabi, "Modeling the mechanics of viral kinetics under immune control during primary infection of HIV-1 with treatment in fractional order", Physica A, vol. 545, pp. 123816, 2020.

[19]   J. H. Ma and Y. S. Chen, "Study for the bifurcation topological structure and the global complicated character of a kind of non-linear finance system", Appl. Math. Mech., vol. 22(11), pp. 1240-1250, 2001.

[20]   P. A. Naik, M. Yavuz, and J. Zu, "The role of prostitution on HIV transmission with memory: A modeling approach", Alexandria Eng. J., vol. 59, no. 4, pp. 2513-2531, 2020.

[21]   P. Veeresha, "A numerical approach to the coupled atmospheric ocean model using a fractional operator", Math. Model. Numer. Simul. Appl., vol. 1, no. 1, pp. 1-10, 2021.

[22]   Z. Hammouch, M. Yavuz, and N. Özdemir, "Numerical solutions and synchronization of a variable-order fractional chaotic system", Math. Model. Numer. Simul. Appl., vol. 1, no. 1, pp. 11-23, 2021.

[23]   Yokuş, "Construction of different types of traveling wave solutions of the relativistic wave equation associated with the Schrödinger equation", Math. Model. Numer. Simul. Appl., vol. 1, no. 1, pp. 24-31, 2021.

CHAPTER 4

The Duhamel Method in Transient Heat Conduction: A Rendezvous of Classics and Modern Fractional Calculus

Jordan Hristov[1]*

[1]*Department of Chemical Engineering, University of Chemical Technology and Metallurgy, Sofia, Bulgaria*

Abstract: This chapter presents an attempt to demonstrate that the Duhamel theorem applicable for time-dependent boundary conditions (or time-dependent source terms) of heat conduction in a finite domain and the use of the Fourier method of separation of variable (superposition version) naturally lead to appearance of the Caputo-Fabrizio operators in the solution. The fractional orders of the emerging series of Caputo-Fabrizio operators are directly related to the eigenvalues determined by the Fourier's method. The general expression of the solution in terms of Caputo-Fabrizio operators has been developed followed by four examples.

Keywords: Caputo-Fabrizio derivative, Duhamel theorem, Heat conduction.

INTRODUCTION

The chapter is especially devoted to the idea of demonstrating how the classical methods in analytical heat transfer (diffusion) meet the modern fractional operators. The target is the well-known Duhamel's method allowing the transient heat diffusion problem with time-dependent boundary conditions to be presented as a convolution integral (see in the sequel). The general form of the Caputo-type fractional operator can be expressed as

$$D_t^\alpha f(t) = \frac{M(\alpha)}{N(\alpha)} \int_0^t R(\alpha, t - \tau) \frac{df(\tau)}{d\tau} d\tau,$$

where $M(\alpha)/N(\alpha)$ is a normalization function and $R(\alpha, t - r)$ is the relaxation function (memory kernel).

***Corresponding author Jordan Hristov:** Department of Chemical Engineering, University of Chemical Technology and Metallurgy, Sofia, Bulgaria; E-mails: jordan.hristov@mail.bg and jyh@uctm.edu

Mehmet Yavuz & Necati Özdemir (Eds.)

From this general definition, we may briefly present two general members of this group.

Caputo derivative [1]

$$^{C}D_t^\alpha f(t) = \frac{1}{\Gamma(1-\alpha)} \int_0^t \frac{1}{(t-\tau)^\alpha} \frac{df(\tau)}{d\tau} d\tau, \quad 0 < \alpha \le 1,$$

with a singular (power-law) memory kernel, $M = 1$ and $N(\alpha) = \Gamma(1-\alpha)$

and *Caputo-Fabrizio fractional operator* [2]

$$^{CF}D_t^\alpha f(t) = \frac{M(\alpha)}{1-\alpha} \int_0^t exp\left[-\frac{\alpha}{1-\alpha}(t-\tau)\right] \frac{df(\tau)}{d\tau} d\tau, \quad 0 < \alpha \le 1$$

with a non-singular (regular) memory kernel where $N(\alpha) = 1 - \alpha$, while the function $M(\alpha)$ satisfies the conditions $M(0) = M(1) = 1$.

From this position, we try to show that many solutions to transient problems developed by Duhamel's method can be presented in terms of the Caputo-Fabrizio fractional operator by applying classical solution methods, such as the Fourier's separation of variables.

The Main Focus of this Chapter

The modern treatments in fractional calculus try to make the fractional operators useful and versatile tools for solving real-world problems. Nowadays, there are too many controversial opinions about the adequate applications of fractional operators with different memory kernels. This requires the development of examples where it is possible clearly and on the basis of well-known methods of solutions to show that the availability of the new fractional operators, especially that of Caputo-Fabrizio, is natural. The main focus of this chapter is on the basis of classical solutions available in many textbooks on heat transfer to demonstrate that by applying the Duhamel's method and the method of separation of variables, the Caputo-Fabrizio operators appears naturally: precisely, solutions as series of operators with fractional parameters depending on the eigenvalues of the auxiliary transient problem.

TRANSIENT HEAT CONDUCTION AS A PRINCIPLE PROBLEM

Let us consider the classical heat conduction problem with constant heat diffusivity

$$\frac{\partial T}{\partial t} = a\frac{\partial^2 T}{\partial x^2} \tag{1}$$

formulated in a finite domain (1-D slab), $0 \leq x \leq L$ under homogenous initial and time-dependent boundary condition at $x = 0$, namely

$$T(x,0) = 0, \quad T(0,t) = Q(t), \quad T(L,t) = 0. \tag{2}$$

The solution of this problem needs application of the Duhamel method briefly presented next.

Duhamel's Method

In one dimensional case, with zero initial conditions, which are the mandatory conditions of the Duhamel theorem [3], in a finite domain and Cartesian coordinates, the models is (1) with boundary and initial conditions (2).

The application of the Duhamel method needs a solution of an auxiliary problem where the model (1) has to be solved with a unit step change at the $x = 0$ (Dirichlet problem) $((t) = U(t), U(t) = 0$ for $t < 0$ and $U(t) = 1$ for $t \geq 0$), namely

$$T(0,t) = U(t), \quad T(L,t) = 0, \quad T(x,0) = 0$$

Therefore, we may formulate the auxiliary problem as

$$\frac{\partial R}{\partial t} = a\frac{\partial^2 R}{\partial x^2}, \quad 0 < x < L \tag{3}$$

with with boundary and initial conditions

$$R(0,t) = 1, \quad R(L,t) = 0, \quad R(x,0) = 0 \tag{4}$$

In order to solve the original problem (1)-(2) the solution can be presented in the form, within a limited time interval $0 \leq t \leq r_1$, by applying the solution of the auxiliary problem (3)-(4), in a general form

$$T(x,t) \equiv Q(0)R(x,t)$$

In accordance with the Duhamel theorem, the solution to the main problem can be presented as a superposition with a memory (convolution) integral

$$T(x,t) = Q(0)R(x,t) + \int_0^t R(x,t-\tau)\frac{dQ(\tau)}{d\tau}d\tau \tag{5}$$

Taking into account the properties of the convolution integrals [1], we may present (5) in an alternative form

$$T(x,t) = Q(0)R(x,t) + \int_0^t Q(\tau)\frac{dR(x,t-\tau)}{dt}d\tau \tag{6}$$

where taking into account that $R(x,0) = 0$, we have that $\frac{\partial R(x,t-\tau)}{\partial \tau} = \frac{\partial R(x,t-\tau)}{\partial t}$.

Duhamel's Method And Separations of Variables: the General Case

Time-Dependent Boundary Condition

However, in the beginning, we have to find the solution to the auxiliary problem when $(t) = 1$ (Dirichlet problem) by assuming the solution as a sum (a linear superposition) of the stationary problem profile $u(x)$ and the transient profile (x,t), namely

$$R(x,t) = u(x) + P(x,t)$$

with respect to the steady-state solution $u(x)$, the problem can be defined as

$$\frac{\partial^2 u}{\partial x^2} = u(L) = 0, \quad u(0) = 1.$$

The problem with respect to (x,t) should be resolved with respect to the time t and the dimensionless coordinate $0 \le x/L \le 1$. Hence, the problem is defined as

$$\frac{\partial P}{\partial t} = a\frac{\partial^2 P}{\partial x^2}, \quad P(x,0) = -\frac{x}{L}, \quad P(0,t) = P(L,t) = 0.$$

The solution about (x,t) by the method of separation of variables (the Fourier method) is

$$P(x,t) = \sum_{n=1}^{\infty} C_n e^{-a\lambda_n t}\sin\left(\pi n\frac{x}{L}\right), \quad \lambda_n = \left(\frac{\pi^2}{L^2}\right)n^2 \tag{7}$$

where λ_n are eigenvalues and $\sin\left(\pi n \frac{x}{L}\right)$ are the corresponding eigen functions.

The coefficients C_n are solutions of the initial condition

$$-\frac{x}{L} = \sum_{n=1}^{\infty} C_n \sin\left(\pi n \frac{x}{L}\right)$$

and defined as $C_n = \frac{2}{\pi} \frac{(-1)^n}{n}$.

Finally, we may write

$$R(x,t) = \frac{x}{L} + \frac{2}{\pi} \sum_{n=1}^{\infty} \frac{(-1)^n}{n} \sin\left(\pi n \frac{x}{L}\right) \exp\left(-a\lambda_n t\right) \tag{8}$$

Therefore, the solution to the main problem is

$$T(x,t) = \int_0^t Q(\tau) \frac{d}{dt}\left[R(x,t-\tau)\right] d\tau$$

or alternatively as

$$T(x,t) = \int_0^t Q(\tau) \frac{d}{dt}\left[R(x,t-\tau)\right] d\tau + \int_0^t Q(\tau)\left[R(x,t-\tau)\delta(t-\tau)\right] d\tau =$$

$$= \int_0^t Q(\tau) \frac{d}{dt}\left[R(x,t-\tau)\right] d\tau + Q(t) R(x,0) \tag{9}$$

where $R(x,0) = 0$ for $x \in [0,L]$ and for $x = L$ we have $R(x,L) = 0$.

The solutions (6) and (9) (the last expression) can be easily presented in an equivalent convolution form

$$T(x,t) = \int_0^t R(x,t-\tau) \frac{dQ(\tau)}{d\tau} d\tau$$

Time-Dependent Internal Heat Generation

Consider a homogeneous boundary problem in Cartesian coordinates with homogeneously distributed internal and discontinuous heat generation (t), $[W/m^3]$ [3]

$$\frac{\partial T\left(x,t\right)}{\partial t} = a\frac{\partial^2 T\left(x,t\right)}{\partial x^2} + \frac{a}{k_T}g\left(t\right), \quad 0 < x < L, \quad t > 0$$

with boundary and initial conditions

$$T\left(0,t\right) = T\left(L,t\right) = 1, \quad T\left(x,0\right) = 0$$

Now, auxiliary problem is defined with unity heat generation source is [3]

$$\frac{\partial T\left(x,t\right)}{\partial t} = a\frac{\partial^2 T\left(x,t\right)}{\partial x^2} + \frac{a}{k_T}, \quad 0 < x < L, \quad t > 0$$

with $R(0,t) = R(L,t) = 1$, $R(x,0) = 0$.

From the Duhamela's theorem we have

$$T\left(x,t\right) = \int_0^t g\left(\tau\right)\frac{d}{d\tau}R\left(t-\tau\right)d\tau \tag{10}$$

Therefore, the Laplace transform of the convolution (10) is

$$\mathcal{L}\left[T\left(x,t\right)\right] = g\left(p\right) * pR\left(p\right) = R\left(p\right) * pg\left(p\right) \tag{11}$$

which allows (10) to be presented as

$$T\left(x,t\right) = \int_0^t R\left(t-\tau\right)\frac{dg\left(\tau\right)}{d\tau}d\tau$$

This result will be used in the examples (4.3 and 4.4) developed further in this chapter.

CAPUTO-FABRIZIO TRANSFORM OF THE CLASSICAL SOLUTION

Caputo-Fabrizio Operator: Some Basic Properties

In the seminal article of Caputo and Fabrizio [2], a new time-fractional operator with exponential kernel was conceived

$$^{CF}D_t^\alpha f(t) = \frac{M(\alpha)}{1-\alpha}\int_0^t exp\left[-\frac{\alpha(t-s)}{1-\alpha}\right]\frac{df(s)}{ds}ds, \tag{12}$$

where the normalizing function $M(\alpha)$ should obey the conditions $M(0) = M(1) = 1$.

In the context of the preceding discussion in this article, it is obvious that this operator is a convolution of the time derivative of function (t) and the memory kernel [2, 4]

$$K = exp\left[-\frac{\alpha(t-s)}{1-\alpha}\right].$$

From the definition (12) it follows that if $(t) = C = const.$, then $^{CF}D^\alpha f(t) = 0$, an expected results as in the classical Caputo derivative. Further, the Laplace transform of $^{CF}D_t^\alpha(t)$ is [2]

$$\mathcal{L}\left[^{CF}D_t^\alpha f(t)\right] = \frac{p\mathcal{L}\left[f(t)-f(0)\right]}{p+\alpha(1-p)}. \tag{13}$$

Since in the constitution of the Caputo-Fabrizio operator the stretched time is multiplied by a dimensional factor $\alpha/(1-\alpha)$ which should have a dimension of inverse time (t^{-1}) we have a principle problem to be resolved. Actually α is a dimensionless parameter and the answer avoiding ambiguities was developed in [4]. In fact, the exponential relaxation kernel is defined by (see earlier about the exponential relaxation function) $ex(-t/r)$ and as memory $exp[(t-s)/r]$. Thus, a nondimensalization of the exponential function, by help of the characteristic time of the relaxation process t_0, yields (see analyzes of this problem in [4, 5])

$$exp\left(\frac{t-s}{\tau}\right) = exp\left(\frac{t/t_0 - s/t_0}{\tau/t_0}\right) = exp\left(\frac{\bar{t}-\bar{s}}{\bar{\tau}}\right).$$

The nondimensalization preserves the meaning of the exponential memory function and avoids doubts about the definition of the fractional order as [4, 9] (see also comments next).

$$\frac{1-\alpha}{\alpha} = \bar{\tau} = \frac{\tau}{t_0} \Rightarrow \alpha = \frac{1}{1+\tau/t_0}. \tag{14}$$

Examples how experimental data about the relaxation times (by Prony decomposition) converted into fractional orders of Caputo-Fabrizio operators were published recently in [5, 6].

Now, we address the nondimensalization of the convolution integral, as it was done in [6]. Let the Caputo-Fabrizio operator of the function (t) is defined as

$$^{CF}D_t^\alpha U(t) = \frac{1}{1-\alpha}\int_0^t exp\left[-\frac{\alpha}{1-\alpha}(t-s)\right]\frac{d}{ds}U(s)ds.$$

Changing the variables as $Y = U/U_0$ and $\bar{t} = t/t_0$, where U_0 and t_0 are characteristic scales of the process (and they are constants), so that $0 < Y < 1$ and $0 < \bar{t} < 1$.

It is easy to demonstrate that if we may present $\frac{\alpha}{1-\alpha}(t-s)$ as $(\frac{t-s}{c})$, where r is the relaxation time of the process (see also (14)), then, after non-dimensalization we get

$$\left(\frac{\frac{t}{T}-\frac{s}{t_0}}{\frac{\tau}{t_0}}\right) = \left(\frac{\bar{t}-\bar{s}}{\bar{\tau}}\right) \rightarrow \bar{\tau} = \frac{\tau}{t_0} = \frac{1-\alpha}{\alpha} \Rightarrow \alpha = \frac{1}{1+(\tau/t_0)}. \qquad (15)$$

Then, the dimensionless form of $^{CF}D_{\bar{t}}^{\alpha}(t)$ is

$$^{CF}D_t^{\alpha}U(t) \Rightarrow \frac{1}{1-\alpha} \int_0^{\bar{t}} \exp\left[-\frac{\alpha}{1-\alpha}(\bar{t}-\bar{s})\right] U_0 \frac{dY(\bar{s})}{d\bar{s}} d\bar{s}$$

or equivalently

$$^{CF}D_t^{\alpha}U(t) \Rightarrow U_0 \left\{ \frac{1}{1-\alpha} \int_0^{\bar{t}} \exp\left[-\frac{\alpha}{1-\alpha}(\bar{t}-\bar{s})\right] \frac{dY(\bar{s})}{d\bar{s}} d\bar{s} \right\} \Rightarrow$$

$$\Rightarrow {}^{CF}D_t^{\alpha}U(t) = U_0 \left[{}^{CF}D_t^{\alpha}Y(\bar{t}) \right]$$

In addition, the Laplace transform of the scaled (dimensionless) operator is (see (13))

$$\mathcal{L}\left[{}_{cf}^c D_t^{\alpha} f(t) \right] = U_0 \mathcal{L} \left\{ \left[{}_{cf}^c D_t^{\alpha} U(t) \right] \right\}$$

Besides, if the non-dimensalization is only with respect to the time by $\bar{t} = t/t_0$ we get

$$^{CF}D_t^{\alpha}U(t) \Rightarrow \frac{1}{1-\alpha} \int_0^{\bar{t}} \exp\left[-\frac{\alpha}{1-\alpha}(\bar{t}-\bar{s})\right] \frac{dU(\bar{s})}{d\bar{s}} d\bar{s} \Rightarrow {}^{CF}D_t^{\alpha}U(t) = \left[{}^{CF}D_t^{\alpha}U(\bar{t}) \right] \qquad (16)$$

and, again, no additional prefactors emerge in front of the convolution integral unlike the case of the Caputo-derivative with a power-law kernel (see the analyzes in [6]). The result (16) can be attributed to the invariant properties of the exponential function used as memory kernel.

Caputo-Fabrizio Derivatives of Some Elementary Functions

Power-Law Function

In general, the Caputo-Fabrizio derivative of a power-law function $(x) = bt^n$ is defined by the integral

$$^{CF}D_x^\alpha [f(t)] = \frac{1}{1-\alpha} \int_0^t e^{-A(t-s)} \frac{df(s)}{ds} ds = bn \int_0^t e^{-A(t-s)} s^{n-1} ds,$$

where

$$A = \frac{\alpha}{1-\alpha}, \quad \alpha = \frac{1}{1+A}, \quad 1-\alpha = \frac{A}{1+A} \Rightarrow \frac{1}{1-\alpha} = \frac{1+A}{A}.$$

The general expression obtained by Maple is follows

$$^{CF}D_x^\alpha [bt^n] = \frac{(1+A)}{A} \left[(-1)^n bn \frac{\Gamma(n)}{A^{n+1}} e^{-At} + \frac{b}{A} t^n + (-1)^n bnt^n \frac{\Gamma(n,-At)e^{-At}}{A^{n+1}} \right]. \quad \textbf{(17)}$$

Obviously, it is hard to use this expression for particular calculations due to problems emerging with Gamma function, and because of that some cases will be outlined next.

Linear Ramp

With a liner ramp function we have

$$f(t) = bt \Rightarrow \frac{df(t)}{dt} = b,$$

then

$$^{CF}D_x^\alpha [bt] = \frac{1}{1-\alpha} \int_0^t e^{-A(t-s)} \frac{df(s)}{ds} ds = \frac{b}{1-\beta} \int_0^t e^{-A(t-s)} ds = \frac{b}{\alpha} \left(1 - e^{-\frac{\alpha}{1-\alpha}t}\right) \quad \textbf{(18)}$$

That is, for long times we get $\lim_{t\to\infty} \{^{CF}D_x^\alpha [bt]\} = \frac{b}{\alpha}$, a damped integer-order derivative (because $\frac{b}{\alpha} > b$ and $\frac{b}{\alpha} = b$ only when $\alpha = 1$). When the time is taken into account, the damping in time is due to the second (exponential) term. Obviously, for $\alpha \to 1$ we recover the integer-order derivative $\frac{d(bt)}{dt} = b$.

Quadratic Ramp

Now, let us calculate Caputo-Fabrizio derivative when $(t) = bt^2$, then

$$^{CF}D_x^\alpha \left[bt^2 \right] = (2bt)\frac{(1+A)\left[(At-1)^2 - 2e^{-At}\right]}{A^5 t}.$$

For $\alpha \to 1$ we have $A \to \infty$ and the second term approaches (the limit is) unity, that is we get the integer-order derivative.

Exponential Ramp

With $(t) = b\exp(kt)$ we have

$$^{CF}D_x^\alpha \left[b\exp(kt) \right] = \frac{bk}{\alpha + k(1-\alpha)}e^{kt} - \frac{bk}{\alpha + k(1-\alpha)}e^{-\frac{\alpha}{1-\alpha}t} \tag{19}$$

For $\alpha = 1$ we get $^{CF}D_x^\alpha \left[b\exp(kt) \right] = \frac{d}{dt}\left[b\exp(kt) \right] = bk\exp(kt)$. In (19) the

second term decays in time and for long times its nominator goes to unity and we get the integer order derivative damped by a factor $1/[\alpha + k(1-\alpha)]$, namely

$$\lim_{t\to\infty}\left\{ ^{CF}D_x^\alpha \left[b\exp(kt) \right] \right\} \to \frac{bk}{\alpha + k(1-\alpha)}e^{kt}.$$

The compact form of (19) is

$$^{CF}D_x^\alpha \left[b\exp(kt) \right] = bke^{kt}\left[\frac{1 - e^{-[\alpha - k(1-\alpha)]t}}{\alpha + k(1-\alpha)} \right]. \tag{20}$$

For $\alpha \to 1$, we have $A \to \infty$ and the exponentially decaying term goes to unity, that is we get the integer order derivative.

Exponential Decay

With $(t) = b\exp(-kt)$ we have

$$^{CF}D_x^\alpha \left[b\exp(-kt) \right] = -bke^{-kt}\left[\frac{1 - e^{-(A+k)t}}{A-k} \right] = -bke^{-kt}\left[\frac{1 - e^{-\left[\frac{\alpha + k(1-\alpha)}{1-\alpha}\right]t}}{\alpha - k(1-\alpha)} \right]. \tag{21}$$

For $\alpha \to 1$, we have $A \to \infty$ and exponentially decaying term goes to unity, that is we get the integer order derivative. Moreover, for large times the nominator of this term goes to unity and we get the integer order derivative damped by a factor $1/[\alpha - k(1-\alpha)]$.

Power-Law Decay

With $f(t) = bt^{-\beta}$, where $0 < \beta < 1$, which commonly can appear as a function of decaying homogeneously distributed energy (heat source), the derivative $^{CF}D_t^\alpha [bt^{-\beta}]$ is hard to be expressed analytically in a closed form (see (17)). We will demonstrate two examples which support this standpoint and then will try to find a solution how to overcome the emerging problems.

Case with $\beta = 1$,

Then for $^{CF}D_x^\alpha [bt^{-1}]$ one obtain

$$^{CF}D_x^\alpha [bt^{-1}] = \frac{1+A}{A^2} bt.$$

Case with $\beta = 2$,

$$^{CF}D_x^\alpha [bt^{-2}] = \frac{1+A}{A^4} [A^2t^2 - 2At + 2 - \exp(-At)]. \tag{22}$$

It is obvious that it is hard to find well-behaving expression $^{CF}D_t^\alpha [bt^{-\beta}]$, so we refer to an approximation of $bt^{-\beta}$ as a series of functions where $^{CF}D_x^\alpha [f(t)]$ is easily available. Here we suggest the power-law function $(t) = bt^{-\beta}$, where $0 < \beta < 1$ to be approximated as a sum of decaying exponential functions, that is, an approximation by Prony's series, namely

$$f(t) = bt^{-\beta} \approx \sum_{j=1}^{N} a_j e^{-b_j t} \tag{23}$$

The approximation of the power-law function $f(t) = bt^{-\beta}$ by sum of exponents (Prony' series) is a problem studied in many studies and skipping citations here we refer to the work of McLean [7] as well as to [8] where references sources thoroughly analyzing the problem can be found. If we accept the approximation (23), then $^{CF}D_t^\alpha [bt^{-\beta}]$ can be approximated as a sum of finite number of derivatives from exponential functions, namely

$$^{CF}D_t^\alpha [bt^{-\beta}] \approx {}^{CF}D_t^\alpha \left[\sum_{j=1}^{N} a_j e^{-b_j t} \right] = \sum_{j=1}^{N} {}^{CF}D_t^\alpha [a_j e^{-b_j t}].$$

Taking into account the result (21) we get

$$^{CF}D_t^\alpha \left[bt^{-\beta} \right] \approx \sum_{j=1}^{N} {}^{CF}D_t^\alpha \left[b_j e^{-k_j t} \right] = \sum_{j=1}^{N} -b_j k_j e^{-k_j t} \left[\frac{1 - e^{-\left[\frac{\alpha + k_j (1-\alpha)}{1-\alpha} \right] t}}{\alpha - k_j (1 - \alpha)} \right] \quad \textbf{(24)}$$

At the end of this section we have to summarize the experience in obtaining derivatives from elementary functions in following note

Note 1. It is easy to determine a fractional derivative from a given function $f(t)$, *if both the kernel and the function are of one the same type.* Precisely, it is a well-known fact the Caputo derivatives (with power-law kernel) from a power-law functions $f(t) = bt^\beta$ and $f(t) = bt^{-\beta}$ are

$$D_t^\alpha \left[bt^\beta \right] = \frac{\Gamma(\beta+1)}{\Gamma(\beta+1-\alpha)} t^{\beta-\alpha}, \quad D_t^\alpha \left[bt^{-\beta} \right] = \frac{\Gamma(\beta+1)}{\Gamma(\beta)} t^{-(\beta+\alpha)}.$$

Comparable cases, with exponential memory, are the above-mentioned examples with Caputo-Fabrizio derivatives from exponential functions, that is (20) and (21).

Unfortunately, when the kernel and the function are not of one the same type we got problems as it was demonstrated above. Since all models are approximations, then the function which has to be differentiated can be approximated by Prony's series (sum of exponentials), that is to get the approximation (24), and this will allow to resolve the problem with the differentiation approximately. Prony's approximations are available for many useful functions (monotonically decaying) applicable to real-world models (see more information in [8]).

Time-Scales

Relations to the nth Fractional Order

The construction of Caputo-Fabrizio operator by definition, implicitly, contains a dimensionless kernel since it comes from the classical $(exp - t/r)$, which is dimensionless. Additional scaling inside the kernels such as $(t/t_0)/(r/t_0)$ does not change the ratio $(exp - t/r)$, but yields the product $(1/\bar{r})[(-t/t_0)]$. Now, $(1/\bar{r}) \in [0, \infty)$ can be to be replaced by the ratio $[\alpha/(1 - \alpha)] \in [0, \infty)$, where $\alpha \in [0,1]$. The time scale t_0 should be defined as the maximum time of the process (or the observation time), as in the viscoelastic experiments [5]. Otherwise, in diffusion processes such as the heat conduction in a finite domain, for instance, we have a *diffusion time scale* $t_0 = L^2/a$, where L is characteristic length scale, while $a = \lambda/\rho C_p$ $[m^2/\ ]$ is the thermal diffusivity, λ is the heat conductivity $[W/mK]$. Such a case was discussed in [4] and [9].

The nth exponential term in the solution (7) can be presented as

$$\exp\left(-a\lambda_n t\right) = \exp\left(-\frac{t}{\tau_n}\right), \quad \tau_n = \frac{1}{a\lambda_n} = \frac{1}{\frac{a}{L^2}\left(\pi^2 n^2\right)} \tag{25}$$

That is, the nth relaxation time scale is $r_n = 1/a\lambda$. The non-dimensalization of the exponential terms, as it was done in the preceding section, yields

$$\frac{\tau_n}{t_0} = \frac{1}{\left(\frac{\alpha}{1-\alpha}\right)}.$$

Now, using the time scale $t_0 = L^2/a$ we get

$$\frac{\tau_n}{t_0} = \frac{1}{a\lambda_n}\frac{1}{\frac{L^2}{a}} = \frac{1}{\pi^2 n^2} \tag{26}$$

Therefore, the nth dimensionless fractional factor in the nth exponential terms is

$$\frac{1-\alpha_n}{\alpha_n} = \frac{\tau_n}{t_0} = \frac{1}{\pi^2 n^2} \Rightarrow \frac{\alpha_n}{1-\alpha_n} = \pi^2 n^2, \quad 0 < \alpha_n < 1.$$

Then, the fractional order α_n can be simply determined by the relationship

$$\frac{\alpha_n}{1-\alpha_n} = \pi^2 n^2 \Rightarrow \alpha_n = \frac{1}{1 + 1/(\pi^2 n^2)}, \quad 1 \le n < \infty, \quad 0 < \alpha_n < 1 \tag{27}$$

Therefore, each exponential term in the general solution (8) can be expressed as

$$cxp(-a\lambda_n t) = exp\left(-\frac{\alpha_n}{1-\alpha_n}t\right) \tag{28}$$

because as it was defined above $r_n = 1/a\lambda_n$ and taking into account the relations (25)-(26).

Fourier's Numbers

Alternatively, from the classical solution it follows that

$$\exp\left(-a\lambda_n t\right) = \exp\left(-\frac{at}{L^2}\pi^2 n^2\right) = \exp\left(-Fo\pi^2 n^2\right),$$

where $Fo = at/L^2 = t/t_0$ is the Fourier number (*i.e.* dimensionless time).

Then,

$$\exp[-a\lambda_n(t-s)] = \exp[-(Fo - Fos)\pi^2 n^2], \quad Fos = \frac{s}{t_0} = \frac{as}{L^2}.$$

Here Fos is a Fourier number based on the time delay s. Since $s \ll t$ it follows that $Fos \ll Fo$. Hence, the exponential term $\exp[-a\lambda_n(t-s)]$ can be presented as

$$\exp[-a\lambda_n(t-s)] = \exp\left[-\frac{a_n}{1-a_n}(Fo - Fos)\right].$$

Therefore, we related the present results to the classical presentation through the Fourier number.

Solution in Terms of Caputo-Fabrizio Operator: Basic Case

From (8) we may present $(x, t - r)$ as

$$R(x, t - \tau) = \frac{x}{L} + \frac{2}{\pi}\sum_{n=1}^{\infty} \frac{(-1)^n}{n} \sin\left(\pi n \frac{x}{L}\right) \exp\left[-\beta_n(t - \tau)\right], \quad \beta_n = a\lambda_n \qquad (29)$$

The rate constants $\beta_n = \frac{a_n}{1-a_n}$, where $0 \le \alpha_n \le 1$, should depend only on the eigenvalues λ_n through the relation

$$\beta_n = \frac{\alpha_n}{1 - \alpha_n} \Rightarrow \alpha_n = \frac{\beta_n}{1 + \beta_n},$$

where α_n is defined by (27) Then,

$$R(x, t - \tau) = \frac{x}{L} + \frac{2}{\pi}\sum_{n=1}^{\infty} \frac{(-1)^n}{n} \sin\left(\pi n \frac{x}{L}\right) \frac{(1 - \alpha_n)}{M(\alpha)}\left\{\frac{M(\alpha)}{1 - \alpha_n} \exp\left[-\frac{\alpha_n}{1 - \alpha_n}(t - \tau)\right]\right\} (30)$$

or in a compact form as

$$R(x, t - \tau) = \frac{x}{L} + \frac{2}{\pi}\left\{\sum_{n=1}^{\infty} \frac{M_n(\alpha_n)}{1 - \alpha_n} \exp\left[-\frac{\alpha_n}{1 - \alpha_n}(t - \tau)\right]\right\} \qquad (31)$$

$$M_n(\alpha_n) = \frac{(-1)^n}{n} \sin\left(\pi n \frac{x}{L}\right) \frac{(1 - \alpha_n)}{M(\alpha)}.$$

In terms of Losada-Nieto definition [10], where $(\alpha) = 1$ the simplified form of (30) is

$$R\left(x,t-\tau\right) = \frac{x}{L} + \frac{2}{\pi}\sum_{n=1}^{\infty}\frac{(-1)^n}{n}\sin\left(\pi n\frac{x}{L}\right)(1-\alpha_n)\left\{\frac{1}{1-\alpha_n}\exp\left[-\frac{\alpha_n}{1-\alpha_n}(t-\tau)\right]\right\} \quad \textbf{(32)}$$

Taking into account the properties of the convolution we may present the solution (32) (for the sake of simplicity in terms of Losada-Nieto definition of the operator) as

$$T\left(x,t\right) = \int_0^t R\left(x,t-\tau\right)\frac{d}{d\tau}Q\left(\tau\right)d\tau + Q\left(t\right)R\left(x,0\right),$$

or with more details

$$T\left(x,t\right) = \int_0^t \frac{x}{L}\frac{d}{d\tau}Q\left(\tau\right)d\tau +$$

$$\int_0^t \frac{2}{\pi}\sum_{n=1}^{\infty}N_n\left(\frac{x}{L},n,\alpha_n\right)\left\{\frac{1}{1-\alpha_n}\exp\left[-\frac{\alpha_n}{1-\alpha_n}(t-\tau)\right]\right\}\frac{d}{d\tau}Q\left(\tau\right)d\tau +$$

$$Q\left(t\right)R\left(x,0\right),$$

where

$$N_n\left(\frac{x}{L},n,\alpha_n\right) = \frac{(-1)^n}{n}\sin\left(\pi n\frac{x}{L}\right)(1-\alpha_n),$$

or

$$T\left(x,t\right) = \int_0^t \frac{x}{L}\frac{d}{d\tau}Q\left(\tau\right)d\tau + \int_0^t \frac{2}{\pi}\sum_{n=1}^{\infty}N_n\left(\frac{x}{L},n,\alpha_n\right)\left\{{}^{CF}D_t^{\alpha_n}\left[Q\left(t\right)\right]\right\} + Q\left(t\right)R\left(x,0\right),$$

or for $R(x,0) = 0$ we simply get

$$T\left(x,t\right) = \int_0^t \frac{x}{L}\frac{d}{dt}Q\left(\tau\right)d\tau + \int_0^t \frac{2}{\pi}\sum_{n=1}^{\infty}N_n\left(\frac{x}{L},n,\alpha_n\right)\left\{{}^{CF}D_t^{\alpha_n}\left[Q\left(t\right)\right]\right\}.$$

EXAMPLES

Example 1. Linear Ramp as Boundary Condition

A slab of thickness L with zero initial conditions and a time-dependent boundary condition (t) at $x = L$, namely

$$f\left(t\right) = \begin{cases} bt, & 0 < t < t_1, \\ \\ 0, & t > t_1. \end{cases}$$

Applying the superposition method the solution for $0 < t < t_1$ the solution is

$$T(x,t) = f(0) R(x,t) + \int_0^t R(x, t - \tau) \frac{df(\tau)}{d\tau}.$$

That is with $f(0) = 0$, assuming fort the sake of simplicity $M(\alpha) = 1$, and on the basis of the general solution (30) (or (31)), we have

$$T(x,t) = \int_0^t \frac{x}{L} \left\{ \frac{2}{L} \sum_{n=1}^{\infty} \frac{(-1)^n}{\beta_n} \sin(\beta_n, x) \exp\left[-\frac{\beta_n^2}{a}(t - \tau)\right] \right\} \frac{df(\tau)}{d\tau},$$

where $\beta_n = \sqrt{\lambda_n} = \frac{\pi n}{L}$.

With $\frac{df(t)}{dt} = b$ and the result (18) we get for the time range $0 < t < t_1$

$$T(x,t) = bt\left(\frac{x}{L}\right) + \frac{2b}{L} \left\{ \sum_{n=1}^{\infty} \frac{(-1)^n}{\beta_n} \sin(\beta_n, x) \int_0^t \exp\left[-\frac{\beta_n^2}{a}(t - \tau)\right] d\tau \right\}.$$

In terms of the fractional order α_n of the Caputo-Fabrizio operator we have

$$T(x,t) = bt\left(\frac{x}{L}\right) + \frac{2b}{L} \left\{ \sum_{n=1}^{\infty} \frac{(-1)^n}{\beta_n} \sin(\beta_n, x) \int_0^t \exp\left[-\frac{\beta_n^2}{a}(t - \tau)\right] d\tau \right\} \qquad (33)$$

That is, (33) matches the classical solution [3].

Note 2 The solution for $t > t_1$ is not presented here since it contains the same structure (its 2nd term, see [3]) which can be transformed in terms of Caputo-Fabrizio derivatives.

Example 2. Exponential Ramp as Boundary Condition

If the boundary condition is exponential function $f_{\exp}(x = 0, t) = b\exp(kt)$, with $f_{\exp}(x = 0,0) = 0$ (no exponential rise at $t = 0$), $\frac{df_{\exp}(\tau)}{d\tau} = bke^{k\tau}$, and applying (20) we get

$$T(x,t) = bk\frac{x}{L}e^{kt} + bk\frac{2}{L} \sum_{n=1}^{\infty} \frac{(-1)^n}{\beta_n} \sin(\beta_n, x) e^{kt} \left[\frac{1 - e^{-[\alpha_n - k(1-\alpha_n)]t}}{\alpha_n + k(1 - \alpha_n)}\right] \qquad (34)$$

The classical solution is [3]

$$T(x,t) = k\frac{x}{L}e^{kt} + k\frac{2}{L} \sum_{n=1}^{\infty} \frac{(-1)^n}{\beta_n} \sin(\beta_n, x) \exp\left(-\frac{\beta_n^2}{a}t\right) \int_0^t \left(\exp\left(k + \frac{\beta_n^2}{a}\tau\right)\right) d\tau \qquad (35)$$

and it be easy transformed in the form of (34).

Note 3. In the classical solution [3] the convolution integral is transformed as

$$\int_0^t e^{k(t-\tau)} \frac{df(\tau)}{d\tau} d\tau = e^{kt} \int_0^t e^{-k\tau} \frac{df(\tau)}{d\tau} d\tau \tag{36}$$

Then, with

$$\int_0^t \exp - \left(\frac{\beta_n^2}{a}\tau\right) d\tau = -\frac{a}{\beta_n^2}\left[\exp - \left(\frac{\beta_n^2}{a}t\right) - 1\right] = \frac{a}{\beta_n^2}\left[1 - \exp - \left(\frac{\beta_n^2}{a}t\right)\right].$$

Hence, the solution (35) is straightforwardly transformable to (34) taking into account that $\beta_n^2/a = \alpha_n/(1 - \alpha_n)$. It is noteworthy to mark that when non-locality is not at issue, such approach (i.e. the transform (36)) is commonly encounter in the literature. However, if we consider appearance of a non-local operator, such as the Caputo-Fabrizio derivative, this approach is unacceptable because it violates the memory principle imposed by the convolution integral. In the simple case used here the step with (36) can be applied only to test if the new approach provides the same result as the classical one.

Example 3. Exponential Source Term

Consider a boundary problem in Cartesian coordinates with homogeneously distributed internal and discontinuous heat generation $g(t)$, $[W/m^3]$

$$\frac{\partial T(x,t)}{\partial t} = a\frac{\partial^2 T(x,t)}{\partial x^2} + \frac{a}{k}g(t), \quad 0 < x < L, \quad t > 0,$$

with boundary and initial conditions

$$T(0,t) = T(L,t) = 1, \quad T(x,0) = 0.$$

Now, auxiliary problem is defined with unity heat generation source, namely

$$\frac{\partial T(x,t)}{\partial t} = a\frac{\partial^2 T(x,t)}{\partial x^2} + \frac{a}{k}, \quad 0 < x < L, \quad t > 0,$$

$$R(0,t) = R(L,t) = 1, \quad R(x,0) = 0. \tag{37}$$

From the Duhamela's theorem, we have

$$T(x,t) = \int_0^t g(\tau)\frac{d}{d\tau}R(t-\tau)\,d\tau \tag{38}$$

The Laplace transform of the convolution (38) is

$$\mathcal{L}[T(x,t)] = g(p) * pR(p) = R(p) * pg(p) \tag{39}$$

which allows (38) to be presented as

$$T(x,t) = \int_0^t R(t-\tau)\frac{dg(\tau)}{d\tau}\,d\tau \tag{40}$$

This result will be used in the next examples. Here we will demonstrate the main idea of this chapter with an example taken from [3] and two heat generation terms:

• Exponentially decaying heat generation term which may model heat of extinguishing exothermic chemical reaction.

• Exponentially growing heat generation term which may model heat of an exothermic chemical reaction.

Consider one dimensional cylinder with radius $0 < r < B$, initially at zero, with internal heat generation $g(t)\ [W/m^3]$ per unit volume for $t > 0$ modelled by

$$\frac{\partial^2 T(r,t)}{\partial r^2} + \frac{1}{r}\frac{\partial T(r,t)}{\partial r} + \frac{1(t)}{k_T} = \frac{1}{a}\frac{\partial T(r,t)}{\partial t}, \quad 0 \leq r \leq B, \quad t \geq 0,$$

with boundary and initial conditions

$$T(B,t) = 0, \quad T(r \to 0, t) \to T_0, \quad T(r,0) = 0,$$

where T_0 has a finite value.

The auxiliary problem with unit step generation is [3]

$$\frac{\partial^2 R(r,t)}{\partial r^2} + \frac{1}{r}\frac{\partial R(r,t)}{\partial r} + \frac{1}{k_T} = \frac{1}{a}\frac{\partial R(r,t)}{\partial t}, \quad 0 \leq r \leq B, \quad t \geq 0,$$

$$R(B,t) = 0, \quad R(r \to 0, t) \to R_0, \quad R(r,0) = 0.$$

The final solution is

$$T(x,t) = \int\limits_0^t \frac{2}{Bk_T} \sum_{n=1}^{\infty} \underbrace{\frac{J_0(\beta_n r)}{\beta_n^3 J_1(\beta_n B)}}_{time-independent} e^{-a\beta_n^2(t-s)} \frac{dg(s)}{ds} ds \tag{41}$$

where t he eigenvalues β_n are positive roots of $J_0(\beta_n B) = 0$.

Now, using the formulation (40) and taking into account that in (41) the integration is with respect to s only, we have

$$T(x,t) = \frac{2}{Bk_T} \sum_{n=1}^{\infty} \frac{J_0(\beta_n r)}{\beta_n^3 J_1(\beta_n B)} \int_0^t e^{-a\beta_n^2(t-s)} \frac{dg(s)}{ds} ds \tag{42}$$

The term

$$\int_0^t e^{-a\beta_n^2(t-s)} \frac{dg(s)}{ds} ds$$

has the same construction as the Caputo-Fabrizio operator.

In this context with $r_n = 1/a\beta_n^2$ we can define the fractional order α_n as

$$\frac{\tau_n}{t_0} = \frac{1}{a\beta_n^2 \frac{L^2}{a}} = \frac{1}{\beta_n^2 L^2} = \frac{1-\alpha_n}{\alpha_n} \Rightarrow \alpha_n = \frac{1}{1+\tau_n/t_0} = \frac{1}{1+\beta_n^2 L^2} \tag{43}$$

bearing in mind that $\beta_n^2 L^2$ is dimensionless while $a\beta_n^2 L^2$ has a dimension of inverse time $[s^{-1}]$. Now, in terms of Caputo-Fabrizio operator we may express (42) as

$$T(x,t) = \frac{2}{Bk_T} \sum_{n=1}^{\infty} \frac{J_0(\beta_n r)}{\beta_n^3 J_1(\beta_n B)} (1-\alpha_n) \left[\frac{1}{(1-\alpha_n)} \int_0^t e^{-\frac{\alpha_n}{1-\alpha_n}(t-s)} \frac{dg(s)}{ds} ds \right],$$

or in compact form as

$$T(x,t) = \frac{2}{Bk_T} \sum_{n=1}^{\infty} \frac{J_0(\beta_n r)}{\beta_n^3 J_1(\beta_n B)} \left[(1-\alpha_n) \, {}^{CF} D_t^{\alpha_n} g(t) \right].$$

Exponentially Decaying Heat Generation

Now, with an exponentially decaying heat generation term and the expression (21), we have

$$T(x,t) = \frac{2}{Bk_T} \sum_{n=1}^{\infty} \frac{J_0(\beta_n r)}{\beta_n^3 J_1(\beta_n B)} \left\{ (1-\alpha_n) \, bke^{-kt} \left[1 - \frac{e^{-[\alpha_n - k(1-\alpha_n)t]}}{\alpha_n - k(1-\alpha_n)} \right] \right\}.$$

Exponentially Growing Heat Generation

Similarly, with exponentially growing heat sources (see section 3.2.4) and applying (39) with (20) we get

$$T(x,t) = \frac{2}{Bk_T} \sum_{n=1}^{\infty} \frac{J_0(\beta_n r)}{\beta_n^3 J_1(\beta_n B)} \left\{ (1-\alpha_n) \, bke^{kt} \left[1 - \frac{e^{-[\alpha_n - k(1-\alpha_n)t]}}{\alpha_n + k(1-\alpha_n)} \right] \right\}. \quad \textbf{(44)}$$

Example 4. Power-Law Decay as a Source Term

From the solution (44) we may use the Prony's series approximation (24) of $t^{-\beta}$, $0 < \beta < 1$ as

$$^{CF}D_t^{\alpha} \left[bt^{-\beta} \right] \approx \sum_{j=1}^{N} -b_j k_j e^{-k_j t} \left[\frac{1 - e^{-\left[\frac{\alpha + k_j(1-\alpha)}{1-\alpha} \right] t}}{\alpha - k_j(1-\alpha)} \right]$$

to obtain

$$T(x,t) = \frac{2}{Bk_T} \sum_{n=1}^{\infty} \frac{J_0(\beta_n r)}{\beta_n^3 J_1(\beta_n B)} \left\{ (1-\alpha_n) \sum_{j=1}^{N} -b_j k_j e^{-k_j t} \left[\frac{1 - e^{-\left[\frac{\alpha_n + k_j(1-\alpha_n)}{1-\alpha} \right] t}}{\alpha_n - k_j(1-\alpha_n)} \right] \right\}.$$

EMERGING QUESTIONS ON THE FOURIER SERIES TRUNCATION

The relationship (27) defining α_n through the eigenvalues is equivalent to the relationship (15) which is more general.

In general, the relaxation time should be within the range $r_n \in (0, \infty)$ but the condition that $n \in [1, \infty)$ in the Fourier's method imposes a lower limit $n = 1$ and this leads to:

$$r_1/t_0 = 1/(\pi^2) \approx 0.3183$$

and therefore we get $\alpha_1 \approx 0.758$.

Hence, with increase in the number of terms when $1/(\pi^2 n^2) \ll 1$ the fractional order α_n will approach unity as it shown in Fig. (**1**).

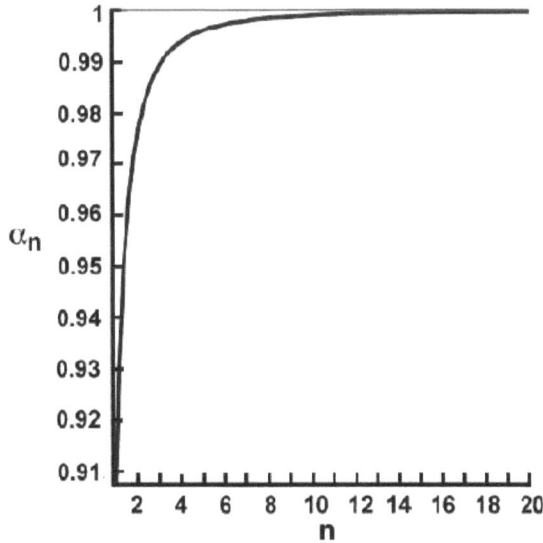

Fig. (1). Functional relationship $\alpha_n = (n)$ defining the fractional order as a function of the eigenvalues.

In other words, with increase in the number of terms the relaxation time r_n (see (25)) becomes large and the decay of the nth exponential term becomes very fast. This immediately formulates two principle questions:

- Question 1: How many terms of the infinite series (29) can be practically used, *i.e.* how to truncate the Fourier series?
- Question2 : How many terms of exponential terms (Caputo-Fabrizio fractional operators-see 3.4) will be involved in the solution?

With the time scale approach we are interested to see when $\alpha_n \to 1$ since in such a case the prefactors $(1 - \alpha_n)$ in the final solution go to zero. Here, no unique solution can be suggested and the problem may be related to the desired accuracy with acceptable number of terms and looking for acceptable convergence with increase in n. This can be easily controlled by the (27) where, for example for $n = 5$ we get $\alpha_n \approx 0.996$ and $1 - \alpha_5 \approx 3.6 \times 10^{-3}$, as well as $-\alpha_5/(1 - \alpha_5) \approx 276.239$ (which makes the exponential terms conditionally negligible).

CONCLUSION

Actually, the main task in this study is neither new solutions of the transient heat conduction with time-dependent boundary conditions or source terms, not accuracies and related convergences of the Fourier' series. The classic Fourier's method of separation of variables, precisely its superposition version (the product version is not considered here) allows clearly demonstrating that by applying the Duhamel theorem, the resulting solution automatically can be presented through a convolution integral of sums of exponentially decaying terms (or as a sum of convolution integrals with exponential kernels). The rate constants of the exponential terms are controlled by the eigenvalues, determined by the Fourier method. Moreover, as demonstrated here, these sums can be easily converted into sums of Caputo-Fabrizio operators (derivatives) acting on the time-dependent boundary conditions or source terms.

To a greater extent, this study is qualitative, demonstrating that the Caputo-Fabrizio operator naturally appears in classical solutions. The expressions developed raise questions about the number of terms and convergence of the series involved but they, for now, only draw open problems to be resolved. It is important that the principle job demonstrating the appearance of the Caputo-Fabrizio derivatives in the classical solutions was done.

CONSENT FOR PUBLICATION

Not applicable.

CONFLICT OF INTEREST

The author declares no conflict of interest, financial or otherwise.

ACKNOWLEDGEMENT

Declared none.

REFERENCES

[1]   Podlubny I.,Fractional Differential Equations, New York,Academic Press, 1999.

[2]   Caputo, M.C., Fabrizio, M., A new definition of fractional derivative withoutsingular kernel, Progr. Fract. Differ. Appl.,vol.1, no.1, pp.73-85,2015.

[3]   Hahn, D. W., Ozisik, N.M., Heat conduction, 3rd ed., Wiley, New York, 2012.

[4]   Hristov J. Derivatives with non-singular kernels: From the Caputo-Fabrizio definition and beyond: Appraising analysis with emphasis on diffusion models. In: Frontiers in Fractional Calculus, 2017, S. Bhalekar, Ed., 269–342,Bentham Science Publishers, Sharjah, 2017.

[5]   Hristov J, Response functions in linear viscoelastic constitutive equations and related fractional operators Math. Model. Nat. Phenom., vol.14, 305, 2019, doi: 10.1051/$mmnp$/2018067.

[6]   Hristov J, Linear viscoelastic responses and constitutive equations in terms of fractional operators with non-singular kernels: Pragmatic approach, memory kernel correspondence requirement and analyses,Eur.Phys. J. Plus, vol.134, 283, 2019, doi: 10.1140/$epjp$/i2019 − 12697 − 7.

[7]   McLean, W., Exponential sum approximations for $t^{-\beta}$, In: J. Dick *et al.* (Eds.), Contemporary computational mathematics-A celebration of the 80th birthday of Ian Sloan, Springer. Int. Publ., 2018.

[8]   Hristov, J., Prony's series and modern fractional calculus, In: Y. Karaca,, Dumitru Baleanu, Yu-Dong Zhang, Osvaldo Gervasi, Majaz Moonis, (Eds.),Multi-Chaos, Fractal, and Multi-Fractional Artificial Intelligence of DifferentComplex Systems, Elsevier Inc., 2021, in press.

[9]   Hristov, J., Derivation of fractional Dodson's equation and beyond:Transient mass diffusion with a non-singular memory and exponentially fading-out diffusivity, Progr. Fract. Differ. Appl.,vol.3, no.4, pp. 255-270, 2017 , doi: 10.18576/$pfda$/030402

[10]   Losada, J., Nieto, J.J., Properties of a New Fractional Derivative withoutSingular Kernel, Progr. Fract. Differ. Appl., vol.1,no.2, pp. 87–92, 2015.

CHAPTER 5

Oscillatory Heat Transfer Due to the Cattaneo-Hristov Model on the Real Line

Derya Avcı[1]* and **Beyza Billur İskender Eroğlu[1]**

[1]*Department of Mathematics, Faculty of Arts and Sciences, Balıkesir University, Balıkesir, Turkey*

Abstract: This chapter aims to discuss the analytical solutions for heat waves observed in Cattaneo-Hristov heat conduction modelled with Caputo-Fabrizio fractional derivative. This operator includes a non-singular exponential kernel and also requires physically interpretable initial conditions for its Laplace transform property. These provide significant advantages to obtain analytical solutions. Two different types of harmonic heat sources are assumed to elicit heat waves. The analytical solutions are obtained by applying Laplace transform with respect to the time variable and the exponential Fourier transform with respect to spatial coordinate. The temperature curves for varying values of the fractional parameter, angular frequency, and the velocity of the moving heat source are drawn using MATLAB.

Keywords: Caputo-Fabrizio fractional derivative, Cattaneo-Hristov heat diffusion model, Exponential fading memory, Fourier transform, Harmonic source effect, Laplace transform, Oscillatory heat transfer.

INTRODUCTION

The diffusion phenomenon describes the movement of various materials in nature, such as molecules, heat, liquids and atoms. Physically, this movement occurs from a higher concentration region to a lower concentration region. Also, it is based on the relationship between flux and concentration gradient.

The chance in this relation determines the type of diffusion model. In the classical theory of diffusion, Fick's diffusion model is a parabolic-type partial differential equation [1-4]:

$$\frac{\partial T}{\partial t} = \kappa \frac{\partial^2 T}{\partial x^2},$$

*Corresponding author Derya Avcı: Department of Mathematics, Faculty of Arts and Sciences, Balıkesir University, Balıkesir, Turkey; E-mail: dkaradeniz@balikesir.edu.tr

Mehmet Yavuz & Necati Özdemir (Eds.)

where κ denotes the diffusion (or thermal conductivity) coefficient. This model corresponds to the heat conduction equation constructed on the Fourier's law. Analytical solution of the diffusion (or heat) model is a Gaussian function which implies the infinite speeds of diffused particles:

$$T(x,t) = \frac{1}{(4\pi\kappa t)^{1/2}} \exp\left(-\frac{x^2}{4\kappa t}\right).$$

This is called as casuality problem and is, unfortunately an unphysical concept for many diffusion processes on different scales [5]. To overcome this weakness, generalizations of the constitutive laws for diffusion processes and the resulting partial differential equations have been studied [6].

In the generalized theory of heat diffusion, Cattaneo's law between heat flux and temperature gradient is in the following form [7,8]:

$$q + \tau_0 \frac{\partial q}{\partial t} = -k \operatorname{grad} T,$$

where τ_0 is the relaxation time. Its corresponding heat diffusion model for rigid heat conductors given by

$$\tau_0 \frac{\partial^2 T}{\partial t^2} + \frac{\partial T}{\partial t} = a\Delta T,$$

in which $\sqrt{a/\tau_0}$ represents the propagation speed of heat waves, is one of the generalized models frequently studied in thermal sciences. The heat waves arising from this model were called *"second sound"* [9]. The emergence of this concept has revealed that wave phenomenon is not only specific to the hyperbolic type partial differential equations. This awareness has aroused great interest among researchers, and it has led to the discovery of unnoticeable waves in parabolic-type diffusion processes [10-13].

All generalized diffusion models in classical theory are defined by integer-order derivatives based on the local description. However, it has been observed that locally generalized models are insufficient to describe the anomalous diffusion phenomenon noticed in many processes in nature. Fractional operators gain importance in eliminating this inadequacy. These operators very realistically describe the properties of memory and inheritance in many processes, thanks to their non-local definitions.

Heat conduction is one of the physical processes in which fractional operators are used most effectively. For example, the time non-local relation between heat flux and temperature gradient was first offered in the following concept by the long-tail power kernel [14-16]:

$$q(t) = \begin{cases} -\dfrac{k}{\Gamma(\alpha)}\dfrac{\partial}{\partial t}\displaystyle\int_0^t (t-\tau)^{\alpha-1}\,\mathrm{grad}T(\tau)\,d\tau,\; 0 < \alpha \le 1, \\[2ex] -\dfrac{k}{\Gamma(\alpha-1)}\displaystyle\int_0^t (t-\tau)^{\alpha-2}\,\mathrm{grad}T(\tau)\,d\tau,\; 1 < \alpha \le 2, \end{cases}$$

where Γ is the Euler's gamma function. This relation reveals the time-fractional heat conduction equation as follows:

$$\frac{\partial^\alpha T}{\partial t^\alpha} = a\Delta T,\; 0 < \alpha \le 2, \qquad\qquad (1)$$

where $\dfrac{\partial^\alpha}{\partial t^\alpha}$ denotes the Caputo fractional derivative. Eq.(1) is also known as "*diffusion-wave equation*" depending the variation of fractional order α. Similarly, the anomalous behaviors in particle jumps reveal the space or space-time fractional diffusion equations modeled with Riesz, Weyl, or fractional Laplacian operators. These models have been studied many times, both mathematically and physically [17,18].

Another non-local relation incorporating the heat flux with its history was considered by [7] as

$$q(x,t) = -\int_{-\infty}^t R(x,t)\nabla T(x,t-\tau)\,d\tau.$$

Assuming $R(x,t)$ as the Jeffrey's time-dependent kernel function $R(t) = \exp\left[-(t-s)/\tau\right]$, where τ is the relaxation time, yielded an integro-differential heat equation [7]:

$$\frac{\partial T(x,t)}{\partial t} = \frac{a_2}{\tau} \int_0^t e^{-\left(\frac{t-s}{\tau}\right)} \frac{\partial^2 T(x,t)}{\partial x^2} ds, \; a_2 = \frac{k_2}{\rho C_p},$$

in which k_2, C_p, ρ represent the elastic conductivity, specified heat capacity and density of heated particles, respectively. Joseph and Preziosi [19] modified this model by considering both of the elastic and thermal conductivities:

$$\frac{\partial T(x,t)}{\partial t} = a_1 \frac{\partial^2 T(x,t)}{\partial x^2} + \frac{a_2}{\tau} \int_0^t e^{-\left(\frac{t-s}{\tau}\right)} \frac{\partial^2 T(x,s)}{\partial x^2} ds,$$

$$a_2 = \frac{k_1}{\rho C_p}, \; a_2 = \frac{k_2}{\rho C_p},$$

(2)

where k_1 is the thermal conductivity. Recently, Hristov has reconstructed Eq.(2) by considering the Caputo-Fabrizio (CF) fractional derivative [20]:

$$\frac{\partial T(x,t)}{\partial t} = a_1 \frac{\partial^2 T(x,t)}{\partial x^2} + a_2 (1-\alpha) \; {}^{CF}D_t^\alpha \left(\frac{\partial^2 T(x,t)}{\partial x^2} \right), \; \alpha \in (0,1).$$

This model, called the Cattaneo-Hristov (CH) heat equation, is a heat diffusion model with non-singular exponential fading memory that emerges from physical reality, not by displacement of integer-order derivative and fractional derivative. In addition, it demonstrates how other constitutive equations could be modified with non-singular fading memories. Analytical and numerical solutions of the model (10) have been studied [21-25].

In classical models, heat waves can be researched in two ways [26, 27]: In the first way, the harmonic effect is assumed at the boundary of the domain. Thus, Dirichlet or Neumann boundary-value problems are formulated. In the second way, a heat source function is thought to give the harmonic effect and a source problem is analyzed. Heat waves occuring in fractional heat diffusion models with singular or non-singular kernels also depend on the variation of the fractional parameter as well as initial or boundary conditions. For the time-fractional heat equation in terms of Caputo fractional derivative, Povstenko deeply studied the heat waves in different coordinate systems [28-34].

This chapter aims to clarify the heat waves in the CH heat diffusion occurring on the real axis. Heat waves are studied under the assumption of non-moving and time-moving harmonic heat sources. For future work, it is remarkable to consider this problem on the half-real line or in curvilinear coordinate systems.

In the Section 2, the basic concepts used in problem formulation are reminded. In Section 3, the problem is solved applying the Laplace and exponential Fourier integral transforms in two sub-sections. Also in this section, solutions are physically interpreted by 2 and 3-dimensional graphics drawn using MATLAB software. Finally, the obtained results are commented in Section 4.

MATHEMATICAL PRELIMINARIES

The CH heat diffusion considered in the present chapter is described by CF fractional derivative. Therefore, we can briefly give some basic definitions related to this operator as follows:

Definition 1: [35] Let $(a,b) \subset R$ and then the Sobolev space of order 1 is defined by

$$H^1(a,b) = \{f \in L^2(a,b): f' \in L^2(a,b)\},$$

where f' is the weak derivative of f.

Definition 2: [35, 36] Let $f \in H^1(a,b)$ and $\alpha \in (0,1)$. The CF fractional derivative of order α is defined as follows:

$$\left({}^{CF}D_{a+}^{\alpha}f\right)(t) = \frac{N(\alpha)}{1-\alpha}\int_a^t f'(\tau)\exp\left(-\frac{\alpha}{1-\alpha}(t-\tau)\right)d\tau,$$

in which $N(\alpha)$ denotes the normalization function satisfying $N(0) = N(1) = 1$.

This definition can also be called the left CF fractional derivative. The commonly used examples of normalization functions that supply computational simplicity are $N(\alpha) = 1$ and $N(\alpha) = 1 - \alpha + \dfrac{\alpha}{\Gamma(\alpha)}$.

The right-side definition of this derivative was stated by Abdeljawad and Baleanu [37]:

$$\left({}^{CF}D_{b-}^{\alpha}f \right)(t) = -\frac{N(\alpha)}{1-\alpha} \int_{t}^{b} f'(\tau)\exp\left(-\frac{\alpha}{1-\alpha}(\tau-t) \right)d\tau,$$

which is particularly needed to determine the optimality conditions for optimal control problems. The Laplace transform of CF fractional derivative is given by [36]

$$\mathcal{L}\left[\left({}^{CF}D_{t}^{(\alpha+n)}f \right)(t) \right] = \frac{s^{n+1}f^{*}(s) - s^{n}f(0) - s^{n-1}f'(0) - \ldots - f^{(n)}(0)}{s+\alpha(1-s)},$$

where $n \geq 0$ and $\mathcal{L}\left[f(t) \right] = f^{*}(s)$.

Note that this property requires integer-order physically interpretable initial conditions. This is an important advantage for analytically solvable application problems modelled with the CF fractional derivative.

Since the present problem is considered on the real axis, we apply the exponential Fourier transform [17, 38]:

$$\mathcal{F}\{f(x)\} = \tilde{f}(\xi) = \frac{1}{\sqrt{2\pi}} \int_{-\infty}^{\infty} f(x)\exp(ix\xi)dx,$$

with its inverse tranform:

$$\mathcal{F}^{-1}\{\tilde{f}(\xi)\} = f(x) = \frac{1}{\sqrt{2\pi}} \int_{-\infty}^{\infty} \tilde{f}(\xi)\exp(-ix\xi)d\xi.$$

The exponential Fourier transform converts the differentiation of order 2 into the following form:

$$\mathcal{F}\left\{ \frac{d^{2}f(x)}{dx^{2}} \right\} = -\xi^{2}\tilde{f}(\xi).$$

The conventional integral transform techniques reduce partial differential equations to algebraic equations. Thus, they allow the initial-boundary value problems of differential equations to be solved analytically by removing the computational complexity.

PROBLEM FORMULATION

We first discuss the effect of a non-moving harmonic heat source on the CH model:

The *non-moving* Harmonic Heat Source Effects

Consider the following source problem:

$$\frac{\partial T(x,t)}{\partial t} = a_1 \frac{\partial^2 T(x,t)}{\partial x^2} + a_2(1-\alpha)\ {}^{CF}D_t^{\alpha}\left(\frac{\partial^2 T(x,t)}{\partial x^2}\right) + \delta(x)\exp(i\omega t),$$

$$-\infty < x < \infty,\ 0 < t < \infty,\ \alpha \in (0,1),$$

under the homogeneous initial condition:

$$t = 0:\quad T(x,0) = 0,\ -\infty < x < \infty,$$

where $\delta(x)$ is the well-known Dirac delta function, and ω denotes the angular frequency also known as radial or circular frequency, which measures angular displacement per unit time.

Applying the Laplace transform to the time variable t and also the exponential Fourier transform to spatial coordinate x give the solution into the transformed domain:

$$\tilde{T}^*(\xi,s) = \operatorname{Re}\tilde{T}^*(\xi,s) + i\operatorname{Im}\tilde{T}^*(\xi,s)$$

$$= \frac{1}{\sqrt{2\pi}}\frac{\left[(1-\alpha)s^2 + \alpha s\right] + i\left[(1-\alpha)s + \alpha\right]\omega}{\left(s^2+\omega^2\right)\left[(1-\alpha)s^2 + (a_1+a_2)(1-\alpha)\xi^2 s + a_1\alpha\xi^2\right]}.$$

Inverting the Laplace transform reveals

$$\mathrm{Re}\,\tilde{T}\left(\xi,t\right)=\frac{1}{\sqrt{2\pi}}\frac{1}{P_1\left(\xi\right)}\left\{\frac{1}{2}\left(\frac{P_2\left(\xi\right)}{B(\xi)}-P_4\left(\xi\right)\right)e^{\frac{B(\xi)-A(\xi)}{2\beta}t}-\frac{1}{2}\left(\frac{P_2\left(\xi\right)}{B(\xi)}+P_4\left(\xi\right)\right)e^{\frac{-B(\xi)-A(\xi)}{2\beta}t}\right.$$
$$\left.+P_3\left(\xi\right)\sin\left(\omega t\right)+P_4\left(\xi\right)\cos\left(\omega t\right)\right\},$$

$$\mathrm{Im}\,\tilde{T}\left(\xi,t\right)=\frac{1}{\sqrt{2\pi}}\frac{1}{P_1\left(\xi\right)}\left\{\frac{1}{2}\left(\frac{P_5\left(\xi\right)}{B(\xi)}+P_3\left(\xi\right)\right)e^{\frac{B(\xi)-A(\xi)}{2\beta}t}-\frac{1}{2}\left(\frac{P_5\left(\xi\right)}{B(\xi)}-P_3\left(\xi\right)\right)e^{\frac{-B(\xi)-A(\xi)}{2\beta}t}\right.$$
$$\left.+P_4\left(\xi\right)\sin\left(\omega t\right)-P_3\left(\xi\right)\cos\left(\omega t\right)\right\},$$

where $\beta=1-\alpha$, $a=a_1+a_2$, $P_m\left(\xi\right)$ $(m=1,...,5)$ are the polynomials given explicitly in Appendix and

$$A(\xi)=a\beta\xi^2+\alpha,\; B(\xi)=\sqrt{a^2\beta^2\xi^4+2\left(a_2-a_1\right)\alpha\beta\xi^2+\alpha^2}.$$

Note that a, β, $A(\xi)$, and $B(\xi)$ are also valid for the next problem formulation.

In this study, since we will only draw the real part of the temperature solution, we focus on the real part solutions. Based on this, inverse Fourier transform gives

$$\mathrm{Re}\,T\left(x,t\right)=\frac{1}{\sqrt{2\pi}}\int_{-\infty}^{\infty}\left[\mathrm{Re}\,\tilde{T}\left(\xi,t\right)\cos\left(x\xi\right)+\mathrm{Im}\,\tilde{T}\left(\xi,t\right)\sin\left(x\xi\right)\right]d\xi.$$

Fig. (**1**) shows the dependence of the temperature function on the fractional parameter α. For increasing values of α, some increase in amplitude of waves is observed. However, the physical behavior of the temperature curves is similar.

In Fig. (**2**), the change in the behavior of temperature curves according to the angular frequency parameter ω is observed for the arbitrary choice of $\alpha=0.6$. As the angular frequency value increases from 0 to π, the wavelength gets smaller.

Fig. (**3**) shows the behavior of the temperature surface. This graph, unlike the others, is given also to indicate the behavior of the temperature according to the $x-$ spatial coordinate.

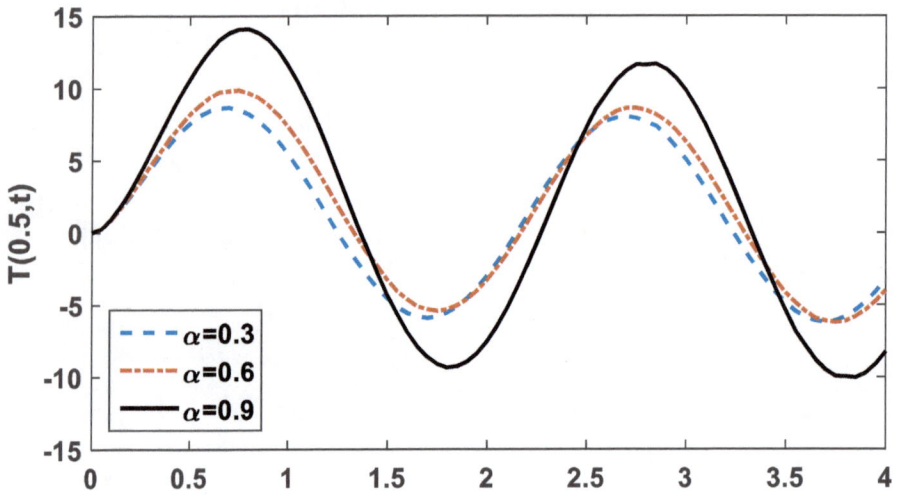

Fig. (1). Dependence of temperature distribution under the effect of non-moving heat source to the variation of fractional order α for $\omega = \pi$.

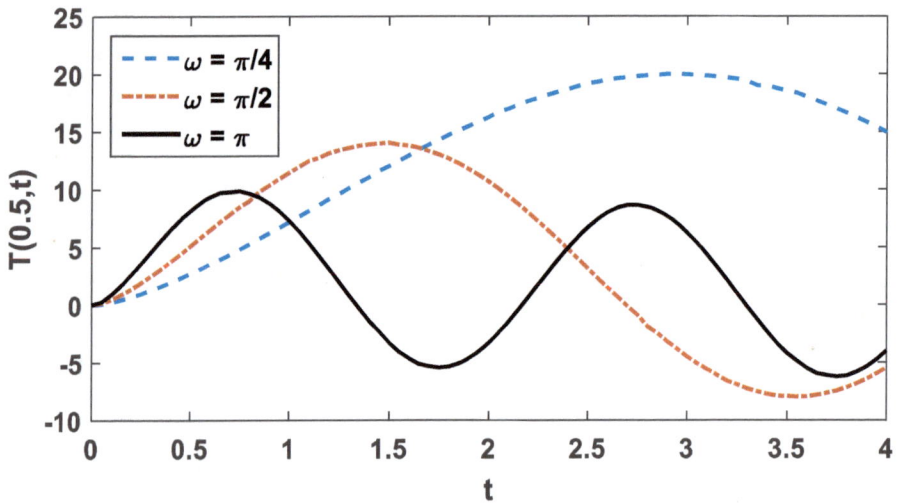

Fig. (2). Dependence of temperature distribution under the effect of a non-moving heat source to the variation of angular frequency ω for $\alpha = 0.6$.

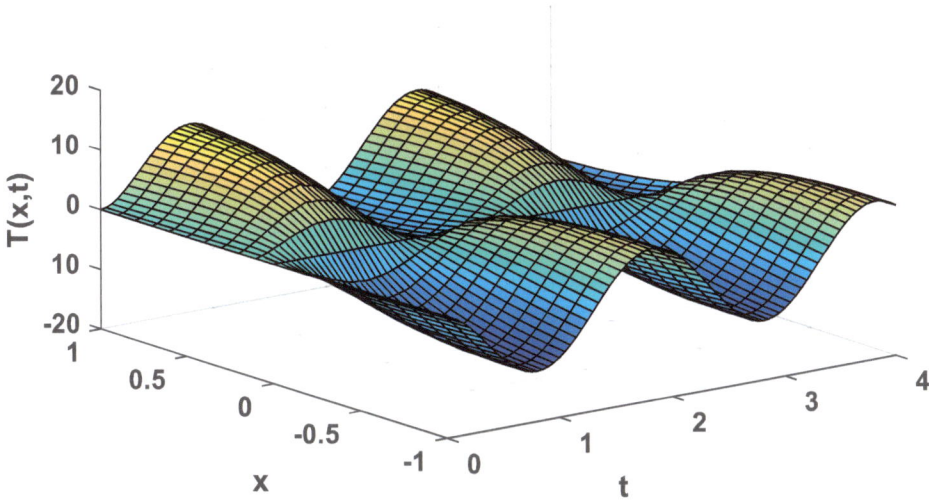

Fig. (3). Temperature surface under the effect of a non-moving heat source for $\alpha = 0.6$ and $\omega = \pi$.

The *time-moving* Harmonic Heat Source Effect

In this case, we consider the harmonic effect changing with respect to the time variable. The problem is formulated as follows:

$$\frac{\partial T(x,t)}{\partial t} = a_1 \frac{\partial^2 T(x,t)}{\partial x^2} + a_2(1-\alpha) \; {}^{CF}D_t^\alpha \left(\frac{\partial^2 T(x,t)}{\partial x^2} \right) + \delta(x-\upsilon t)\exp(i\omega t),$$

$$-\infty < x < \infty, \; 0 < t < \infty, \; \alpha \in (0,1),$$

$$t = 0: \quad T(x,0) = 0, \quad -\infty < x < \infty,$$

where υ is the velocity of moving heat source.

Applying the Laplace and Fourier integral transforms gives:

$$\tilde{T}^*(\xi,s) = \mathrm{Re}\,\tilde{T}^*(\xi,s) + i\,\mathrm{Im}\,\tilde{T}^*(\xi,s)$$

$$= \frac{1}{\sqrt{2\pi}} \frac{\left[(1-\alpha)s^2 + \alpha s\right] + i\left[(1-\alpha)s + \alpha\right](\omega + \xi\upsilon)}{\left(s^2 + (\omega+\xi\upsilon)^2\right)\left[(1-\alpha)s^2 + \left[(1-\alpha)(a_1+a_2)\xi^2 + \alpha\right]s + a_1\xi^2\alpha\right]}.$$

Then, inverting Laplace transform leads to

$$\operatorname{Re}\tilde{T}(\xi,t)=\frac{1}{\sqrt{2\pi}}\frac{1}{Q_1(\xi)}\left\{\frac{1}{2}\left(\frac{Q_2(\xi)}{B(\xi)}-Q_4(\xi)\right)e^{\frac{B(\xi)-A(\xi)}{2\beta}t}-\frac{1}{2}\left(\frac{Q_2(\xi)}{B(\xi)}+Q_4(\xi)\right)e^{\frac{-B(\xi)-A(\xi)}{2\beta}t}\right.$$

$$\left.+Q_3(\xi)\sin\left((\upsilon\xi+\omega)t\right)+Q_4(\xi)\cos\left((\upsilon\xi+\omega)t\right)\right\},$$

$$\operatorname{Im}\tilde{T}(\xi,t)=\frac{1}{\sqrt{2\pi}}\frac{1}{Q_1(\xi)}\left\{\frac{1}{2}\left(\frac{Q_5(\xi)}{B(\xi)}+Q_3(\xi)\right)e^{\frac{B(\xi)-A(\xi)}{2\beta}t}-\frac{1}{2}\left(\frac{Q_5(\xi)}{B(\xi)}-Q_3(\xi)\right)e^{\frac{-B(\xi)-A(\xi)}{2\beta}t}\right.$$

$$\left.+Q_4(\xi)\sin\left((\upsilon\xi+\omega)t\right)-Q_3(\xi)\cos\left((\upsilon\xi+\omega)t\right)\right\},$$

where $Q_m(\xi)$ $(m=1,...,8)$ represents the polinomials which are exactly given in Appendix.

Again, we only need to calculate the real part of the temperature solution. This solution is arrived at as follows:

$$\operatorname{Re}T(x,t)=\frac{1}{\sqrt{2\pi}}\int_{-\infty}^{\infty}\left[\operatorname{Re}\tilde{T}(\xi,t)\cos(x\xi)+\operatorname{Im}\tilde{T}(\xi,t)\cos(x\xi)\right]d\xi.$$

As seen in Fig. (4), the change of α does not make a significant change in the wave amplitude, unlike the non-moving heat source effect. Moreover, unlike (Fig. 1), it is observed that the heat waves weaken over time.

In Fig. (5), the dependence on the angular frequency parameter has a similar character as in Fig. (2).

Unlike the non-moving heat source case, the velocity parameter for the moving heat source is also considered here. Fig. (6) is given for this purpose. As the velocity parameter gets closer to 1, a damping occurs in the heat wave.

Finally, the temperature surface is presented under the effect of a time-moving heat source in Fig. (7). In this graph, weakening the effect of heat waves over time is observed.

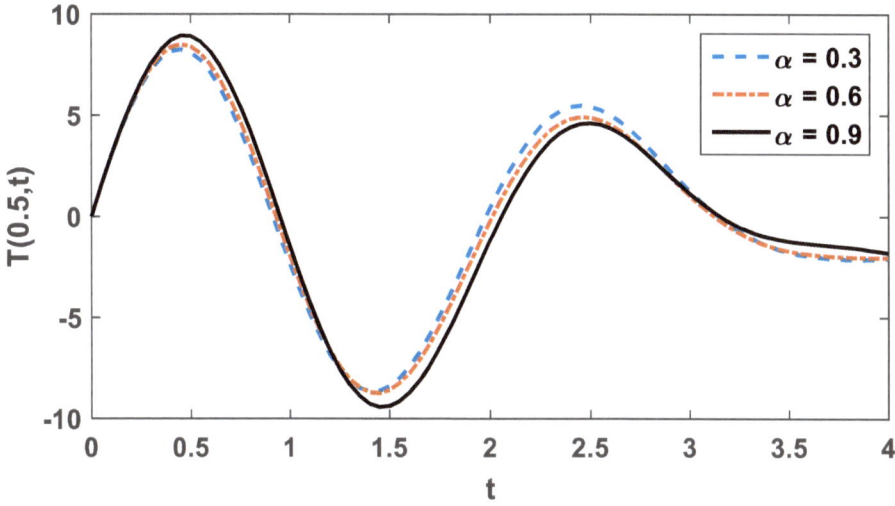

Fig. (4). Dependence of temperature distribution under the effect of a time-moving heat source to the variation of fractional order α for $\upsilon = 1$, $\omega = \pi$.

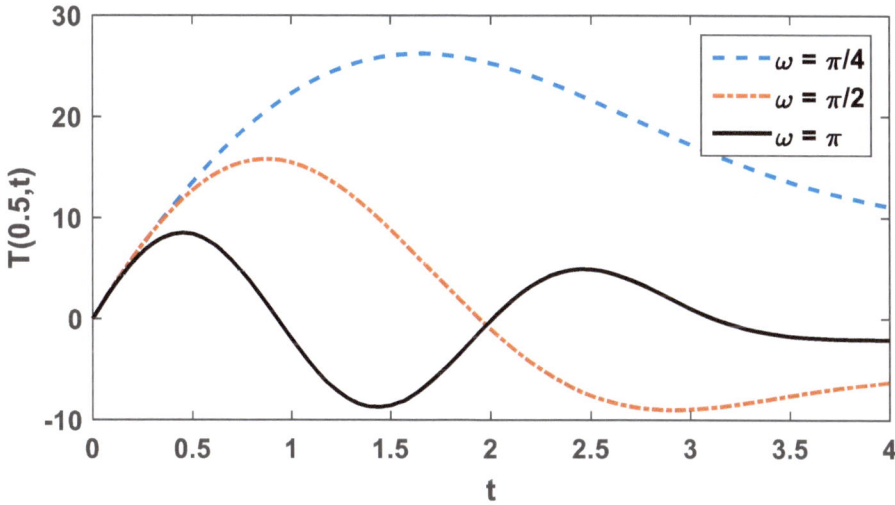

Fig. (5). Dependence of temperature distribution under the effect of a time-moving heat source to the angular frequency ω for $\alpha = 0.6$, $\upsilon = 1$.

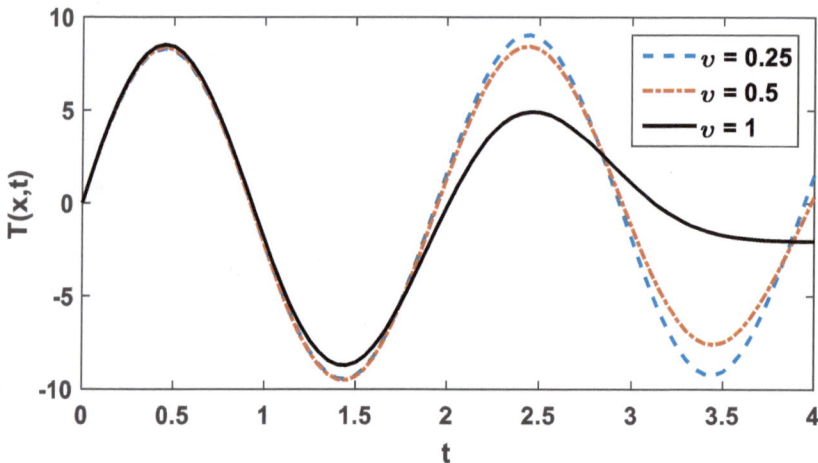

Fig. (6). Dependence of temperature distribution on the velocity of a moving source υ for $\alpha = 0.6$ and $\omega = \pi$.

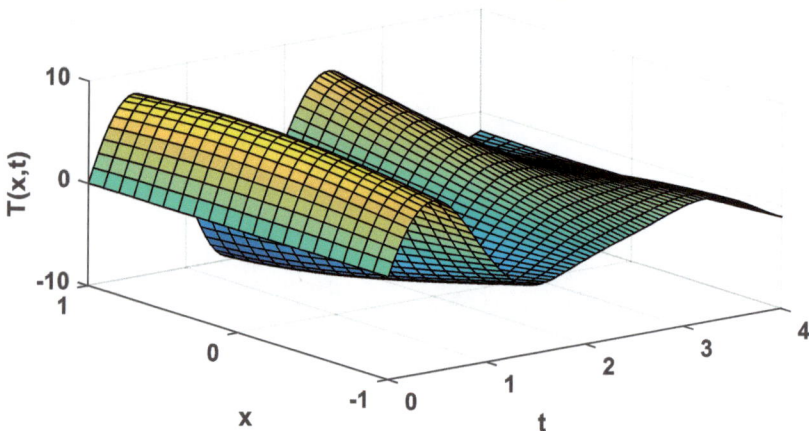

Fig. (7). Temperature surface under the effect of a time-moving heat source for $\alpha = 0.6$, $\omega = \pi$ and $\upsilon = 1$.

CONCLUDING REMARKS

The existence of the wave fronts in both classical and fractional diffusion processes has attracted considerable attention since Ångström [39]. Thus, the idea that the wave phenomenon arises only from hyperbolic partial differential models has weakened and gained a more realistic perspective. The occurrence of the wave phenomenon in both classical and fractional diffusion processes can be examined by considering a harmonic boundary condition or a harmonic source function. Motivated by these, this chapter examines the Cattaneo-Hristov heat diffusion model under the influence of two types of harmonic non-moving and time-moving heat sources. The problem is considered on the real axis. The Laplace and the exponential Fourier integral transforms have given the analytical solutions of wave-like temperature functions. The numerical results obtained for both sub-problems have been graphed using MATLAB according to the velocity of moving source, angular frequency, and fractional order parameters.

APPENDIX

$$P_1(\xi) = \left(a^2\beta^2\omega^2 + a_1^2\alpha^2\right)\xi^4 + 2a_2\alpha\beta\omega^2\xi^2 + \beta^2\omega^4 + \alpha^2\omega^2$$

$$P_2(\xi) = \left(a^2\beta^3\omega^2 - aa_1\alpha^2\beta + 2a_1^2\alpha^2\beta\right)\xi^4 + \left((a_2 - a_1)\alpha\beta^2\omega^2 - a_1\alpha^3\right)\xi^2$$

$$P_3(\xi) = \omega\left(a_2\alpha\beta\xi^2 + \beta^2\omega^2 + \alpha^2\right)$$

$$P_4(\xi) = \left(a\beta^2\omega^2 + a_1\alpha^2\right)\xi^2$$

$$P_5(\xi) = \omega\left(\left(a^2\alpha\beta^2 - aa_1\alpha\beta^2\right)\xi^4 + \left((2a_2 - a_1)\alpha^2\beta - a\beta^3\omega^2\right)\xi^2 + \alpha\beta^2\omega^2 + \alpha^3\right)$$

$$\begin{aligned}Q_1(\xi) &= a^2\beta^2\upsilon^2\xi^6 + 2a^2\beta^2\upsilon\omega\xi^5 + \left(a^2\beta^2\omega^2 + \beta^2\upsilon^4 + 2a_2\alpha\beta\upsilon^2 + a_1^2\alpha^2\right)\xi^4 \\ &\quad + \left(4\beta^2\upsilon^3\omega + 4a_2\alpha\beta\upsilon\omega\right)\xi^3 + \left(6\beta^2\upsilon^2\omega^2 + 2a_2\alpha\beta\omega^2 + \alpha^2\upsilon^2\right)\xi^2 \\ &\quad + \left(4\beta^2\upsilon\omega^3 + 2\alpha^2\upsilon\omega\right)\xi + \beta^2\omega^4 + \alpha^2\omega^2\end{aligned}$$

$$\begin{aligned}Q_2(\xi) &= a^2\beta^3\upsilon^2\xi^6 + 2a^2\beta^3\upsilon\omega\xi^5 + \left(a^2\beta^3\omega^2 + a\alpha\beta^2\upsilon^2 - 2a_1\alpha\beta^2\upsilon^2 - aa_1\alpha^2\beta + 2a_1^2\alpha^2\beta\right)\xi^4 \\ &\quad + 2(a_2 - a_1)\alpha\beta^2\upsilon\omega\xi^3 + \left((a_2 - a_1)\alpha\beta^2\omega^2 - a_1\alpha^3\right)\xi^2\end{aligned}$$

$$Q_3(\xi) = (\upsilon\xi + \omega)\left(\left(\beta^2\upsilon^2 + a_2\alpha\beta\right)\xi^2 + 2\beta^2\upsilon\omega\xi + \beta^2\omega^2 + \alpha^2\right)$$

$$Q_4(\xi) = a\beta^2\upsilon^2\xi^4 + 2a\beta^2\upsilon\omega\xi^3 + \left(a\beta^2\omega^2 + a_1\alpha^2\right)\xi^2$$

$$\begin{aligned}Q_5(\xi) &= (\upsilon\xi + \omega)\left(\left(a^2\alpha\beta^2 - a\beta^3\upsilon^2 - aa_1\alpha\beta^2\right)\xi^4 + \left(\alpha\beta^2\upsilon^2 - a\beta^3\omega^2 + (2a_2 - a_1)\alpha^2\beta\right)\xi^2 \\ &\quad + 2\alpha\beta^2\upsilon\omega\xi + \alpha\beta^2\omega^2 + \alpha^3\right)\end{aligned}$$

CONSENT FOR PUBLICATION

Not applicable.

CONFLICT OF INTEREST

The authors declare no conflict of interest, financial or otherwise.

ACKNOWLEDGEMENT

Declared none.

REFERENCES

[1]   J. Crank, The Mathematics of Diffusion. Oxford: Oxford University Press, 1956.

[2]   M. N. Özışık, Heat Conduction. New York: John Wiley, 1980.

[3]   H. U. Fucks, The dynamics of heat, New York: Springer, 1996.

[4]   L. Boltzmann, "Zur Integration der Diffusionsgleichung bei variabeln Diffusions- coefficienten", Annalen der Physik und Chemie, vol. 289(13), pp. 959-964, 1894.

[5]   G. Aranovich, M.D. Donohue, "Limitations and generalizations of the classical phenomenological model for diffusion in fluids", Molecular physics, vol. 105(8), pp. 1085-1093, 2007.

[6]   J. Masoliver, G. H. Weiss, "Finite-velocity diffusion", Eur. J. Phys., vol 17(4), pp. 190-196, 1996.

[7]   C. Cattaneo, "On the Conduction of Heat (in Italian) ", Atti Sem. Mat. Fis. Universit´a Modena, vol. 3(1), pp. 83-101, 1948.

[8]   C. Cattaneo, "On a form of heat equation which eliminates the paradox of instantaneous propagation", CR Acad. Sci, vol. 247, pp. 431-433, July 1958.

[9]   Bargmann, S.: Second sound waves in solids. In: Hetnarski, R.B. (ed.) Encyclopedia of Thermal Stresses, pp. 4273-4275 Springer, Dordrecht (2014)

[10]   J. Gembarovic, V. Majernik, "Non-Fourier propagation of heat pulses in finite medium", Int. J. Heat Mass Tran., vol. 31(5), 1073-1080, May 1988.

[11]   D.W. Tang, N. Araki, "Analytical solution of non-Fourier temperature response in a finite medium under laser-pulse heating", Heat Mass Transf. vol. 31, pp. 359-363, June 1996.

[12]   M. Lewandowska, L. Malinowski, "An analytical solution of the hyperbolic heat conduction equation for the case of a finite medium symmetrically heated on both sides", Int. Commun. Heat Mass Transf. vol. 33(1), pp. 61-69, January 2006.

[13]   G. Atefi, M.R. Talaee, "Non-Fourier temperature field in a solid homogeneous finite hollow cylinder", Arch. Appl. Mech., vol. 81(5), pp. 569-583 (2011).

[14]   Y.Z. Povstenko, "Fractional heat conduction equation and associated thermal stress". J. Therm. Stresses, vol. 28(1), pp. 83-102, 2005.

[15]   Y.Z. Povstenko, "Fractional heat conduction equation and associated thermal stresses in an infinite solid with spherical cavity". Q. Jl Mech. Appl. Math, vol. 61(4), pp. 523-547, July 2008.

[16]   Y.Z. Povstenko, Fractional Thermoelasticity. New York: Birkhäuser, 2015.

[17]   F. Mainardi, "Fractional relaxation-oscillation and fractional diffusion-wave phenomena", Chaos Soliton Fract., vol. 7(9), pp. 1461-1477, September 1996.

[18]   Y.Z. Povstenko, Linear Fractional Diffusion-Wave Equation for Scientists and Engineers. New York: Birkhäuser, 2015.

[19]   D.D. Joseph, L. Preciozi, "Heat Waves", Rev. Mod. Phys., vol. 61(1), pp. 41-73, January 1989.

[20]   J. Hristov, "Transient heat diffusion with a non-singular fading memory: from the Cattaneo constitutive equation with Jeffrey's kernel to the Caputo-Fabrizio time-fractional derivative", Therm. Sci., vol. 20, 757-762, 2016.

[21]   B.S.T. Alkahtani, A. Atangana, "A note on Cattaneo-Hristov model with non-singular fading memory", Therm. Sci., vol. 21 (1 Part A), pp. 1-7, 2017.

[22]   I. Koca, A. Atangana, "Solutions of Cattaneo-Hristov model of elastic heat diffusion with Caputo-Fabrizio and Atangana-Baleanu fractional derivatives", Therm. Sci., vol. 21 (6 Part A), pp. 2299-2305, April 2017.

[23]   N. Sene, "Solutions of fractional diffusion equations and Cattaneo-Hristov diffusion model", International Journal of Analysis and Applications, vol. 17 (2), pp. 191-207, March 2019.

[24]   Y. Singh, D. Kumar, K. Modi, V. Gill, "A new approach to solve Cattaneo-Hristov diffusion model and fractional diffusion equations with Hilfer-Prabhakar derivative", AIMS Math., vol. 5, pp. 843-855, December 2020.

[25]   B.B. İskender Eroğlu, D. Avcı , "Separable solutions of Cattaneo-Hristov heat diffusion equation in a line segment: Cauchy and source problems", Alex. Eng. J., vol. 60, 2347-2353, April 2021.

[26]   W. Nowacki, Thermoelasticity, 2nd edn. Oxford: PWN-Polish Scientfic Publishers, Warsaw and Pergamon Press, 1986.

[27]   H.S. Carslaw, J.C. Jaeger, Conduction of Heat in Solids, 2nd edn. Oxford: Oxford University Press, 1959.

[28]   Y. Povstenko, "Harmonic impact on the surface of a half-plane in the framework of Time-Fractional Heat Conduction", Scientific Issues of Jan Dlugosz University in Czestochowa. Mathematics, vol. 21, pp. 85-21, July 2016.

[29]   Y. Povstenko, "Time-fractional heat conduction in a half-line domain due to boundary value of temperature varying harmonically in time", Math. Probl. Eng., vol. 2016(Article ID 8605056), 7 pages, November 2016.

[30]   Y. Povstenko, T. Kyrylych, "Time-fractional diffusion with mass absorption in a half-line domain due to boundary value of concentration varying harmonically in time". Entropy, 20(5), pp. 346, May 2018.

[31]   Y. Povstenko, T. Kyrylych, "Time-fractional diffusion with mass absorption under harmonic impact", Fract. Calc. Appl. Anal., vol. 21(1), pp. 118-133, March 2018.

[32]   B. Datsko, I. Podlubny, and Y. Povstenko, "Time-fractional diffusion-wave equation with mass absorption in a sphere under harmonic impact", Mathematics, vol. 7(5), pp. 433, May 2019.

[33]   Y. Povstenko, "Fractional thermoelasticity problem for an infinite solid with a cylindrical hole under harmonic heat flux boundary condition". Acta Mech., vol. 230(6), pp. 2137-2144, March 2019.

[34]   Y. Povstenko, M. Ostoja-Starzewski, "Doppler effect described by the solutions of the Cattaneo telegraph equation", Acta Mech., vol. 232(2), pp. 725-740, November 2021.

[35]   N.R.O. Bastos, "Calculus of variations involving Caputo-Fabrizio fractional differentiation", Stat., Optim. Inf. Comput., vol. 6, pp. 12-21, March 2018.

[36]   M. Caputo, M. Fabrizio, "A new definition of fractional derivative without singular kernel", Prog. Fract. Differ. Appl., vol. 1(2), pp. 73-85, April 2015.

[37]   T. Abdeljawad, D. Baleanu, "On fractional derivatives with exponential kernel and their discrete versions", Rep. Math. Phys., vol. 80(1), pp. 11-27, August 2017.

[38]   P.L. Butzer, R.J. Nessel, Fourier Analysis and Approximation Volume 1. (Pure and applied mathematics; a series of monographs and textbooks). New York: Academic Press, 1971.

[39]   A.J. Ångström, "Neue Methode, das Wärmeleitungsvermögen der Körper zu bestimmen", Ann. Phys. Chem., vol. 144, pp. 513–530, 1861.

Optimal Homotopy Analysis of a Nonlinear Fractional-order Model for HTLV-1 Infection of CD4+ T-Cells

Mohammad Ghoreishi[1], Parvaiz Ahmad Naik[2],* and Mehmet Yavuz[3]

[1]*School of Mathematical Sciences, Universiti Sains Malaysia, 11800 Gelugor, Penang, Malaysia*

[2]*School of Mathematics and Statistics, Xi'an Jiaotong University, Xi'an, Shaanxi 710049, P. R. China*

[3]*Department of Mathematics and Computer Sciences, Necmettin Erbakan University, Konya 42090, Turkey*

Abstract: In this chapter, a series solution of a nonlinear fractional-order mathematical model of human T-cells lymphotropic virus-1 (HTLV-1) infection of CD4+ T-cells is obtained by using a strong and capable technique so-called Homotopy Analysis Method (HAM). The proposed model is a system of nonlinear ordinary differential equations that divides CD4+ T-cells into four components: uninfected cells, latently infected cells, actively infected cells and leukemia cells. The fractional model is more general than the classical one, as in the fractional model, the next state depends not only upon its current state but also upon all of its historical states. The homotopy analysis method (HAM) is applied for a strongly nonlinear fractional-order system as it utilizes a simple method to adjust and control the convergence region of the infinite series solution by using an auxiliary parameter and allows to obtain a one-parametric family of explicit series solutions. By using the homotopy series solutions, firstly, several β-curves are plotted to demonstrate the regions of convergence, then the square residual errors are obtained for different values of these regions. Secondly, the numerical solutions are presented to show the accuracy of the applied homotopy analysis method. In this chapter, a detailed proof of the convergence of this method for nonlinear fractional-order model of HTLV-1 infection of CD4+ T-cells is also given. The results indicate that the HAM is accurate and capable to obtain an accurate approximate analytical solution for HTLV-1 infection of CD4+ T-cells.

Keywords: Human T-cells lymphotropic virus-1 (HTLV-1) infection of CD4+ T-cells, Homotopy analysis method, $\hbar$-curve, β-curves, Convergence-control parameter, Least square.

---

* Corresponding author Parvaiz Ahmad Naik: School of Mathematics and Statistics, Xi'an Jiaotong University, Xi'an, Shaanxi 710049, P. R. China; E-mail: naik.parvaiz@xjtu.edu.cn

INTRODUCTION

Human T-cells lymphotropic virus-I (HTLV-1) infection of $CD4^+$ T-cells (HLTV-1) was first discovered in 1980. The first human retrovirus, many epidemiologists, mathematicians and biologists are interested in investigating this virus due to several biological characteristics [1-2]:

1. This retrovirus shows relationship between viruses and cancer,

2. The association of HTLV-1 with a disease similar to multiple sclerosis (MS) created an opportunity to study the mechanisms that lead to the disease,

3. Its identification facilitated the discovery and isolation of the human immunodeficiency virus (HIV) [3-6], which caused a global epidemic of acquired immune deficiency syndrome (AIDS) [2].

According to the study [7], HTLV-1 is a single-stranded RNA retrovirus with reverse transcriptase activity that leads to a DNA copy of the viral genome. The viral DNA copy is then integrated into the DNA of the host genome. After integration, the viral DNA can latently persist within a T-cell for a long time. The latent infected T-cells contain the viral DNA but are not producing it, and they cannot cause new infections of susceptible cells. Stimulation of the latent infected $CD4^+$ T-cells by antigen can initiate activation of the infected cells. Actively infected T-cells can produce virus and can cause new infections of susceptible T-cells. Actively infected T-cells may then convert to adult T-cells leukemia (ATL) through certain mechanisms which are not yet known. Like HIV, HTLV-1 targets $CD4^+$ T-cells, the most abundant white cells in the immune system, decreasing the body's ability to fight infection.

In 1999, Stilianakis and Seydel [8] proposed a system of nonlinear differential equations that divides $CD4^+$ T-cells into four compartments as follows:

1.   Uninfected $CD4^+$ T-cells,

2.   Latently infected cells,

3.   Actively infected cells,

4.   Leukemia cells.

Let $T(t)$, $T_L(t)$, $T_A(t)$ and $T_M(t)$ represent the concentration of healthy CD4$^+$ T-cells at time t, latently infected cells, actively infected cells and leukemia cells, respectively. This model is formulated as follows

$$\frac{dT}{dt} = \Lambda - \mu_T T - kT_A T,$$
$$\frac{dT_L}{dt} = kT_A T - (\mu_L + \alpha_L)T_L,$$
$$\frac{dT_A}{dt} = \alpha_L T_L - (\mu_A + \rho)T_A, \tag{1}$$
$$\frac{dT_M}{dt} = \rho T_A + \beta T_M \left(1 - \frac{T_M}{T_{M_{max}}}\right) - \mu_M T_M.$$

Table **1** summarizes the meaning of parameters and variables. Wang *et al.* [9] have investigated the global dynamics of system (1). Song and Li [7] investigated the dynamics behavior of the following model

$$\frac{dT}{dt} = \Lambda - \mu_T T - k\frac{T_A}{1+\alpha_1 T_A}T,$$
$$\frac{dT_L}{dt} = k\frac{T_A}{1+\alpha_1 T_A}T - (\mu_L + \alpha_L)T_L,$$
$$\frac{dT_A}{dt} = \alpha_L T_L - (\mu_A + \rho)T_A, \tag{2}$$
$$\frac{dT_M}{dt} = \rho T_A + \beta T_M \left(1 - \frac{T_M}{T_{M_{max}}}\right) - \mu_M T_M.$$

Table 1. List of variables and parameters (modified from [2]).

Parameters and Variables	Meaning
Dependent variables	
T	Uninfected CD4$^+$ T-cells population concentration
T_L	Latently infected CD4$^+$ T-cells concentration
T_A	Activity infected CD4$^+$ T-cells concentration
T_M	Leukemic CD4$^+$ T-cells concentration
Parameters and constants	
μ_T	Natural death rate of CD4$^+$ T-cells concentration
μ_L	Blanket death rate of latently infected CD4$^+$ T-cells

(Table 1) cont.....

μ_A	Blanket death rate of activity infected T-cells
μ_M	Death rate of leukemic T-cells
k	Rate CD4$^+$ T-cells become infected with virus
β	Growth rate of leukemic CD4$^+$ T-cells concentration
α_L	Rate latently infected cells become activity infected
ρ	Rate activity infected cells become leukemic
T_{Mmax}	Maximal concentration of leukemic CD4$^+$ T-cells
Λ	Source term for uninfected CD4$^+$ T-cells

Song and Li [7] determined the global dynamics of T-cells infection by using basic reproduction number $\Re_0 = \frac{\alpha_L k \Lambda}{\mu_T(\mu_L + \alpha_L)(\mu_A + \rho)}$. If $\Re_0 \leq 1$, infected T-cells always die out and if $\Re_0 > 1$, the unique infected equilibrium is globally stable in the interior of feasible region [7]. Cai *et al.* [10] considered the general form of models (1) and (2) as

$$\frac{dT}{dt} = \Lambda - \mu_T T - k f(T_A)T,$$
$$\frac{dT_L}{dt} = k f(T_A)T - (\mu_L + \alpha_L)T_L,$$
$$\frac{dT_A}{dt} = \alpha_L T_L - (\mu_A + \rho)T_A,$$
$$\frac{dT_M}{dt} = \rho T_A + \beta T_M \left(1 - \frac{T_M}{T_{Mmax}}\right) - \mu_M T_M,$$

(3)

where $k f(T_A)$ is incidence rate and k is the infection rate. The function $f(T_A)$ should be satisfied in the following properties

$$f(0) = 0, \qquad\qquad \acute{f}(T_A) > 0, \qquad\qquad f''(T_A) \leq 0.$$

Cai *et al.* [10] showed the global stability of the equilibria for the model (3).

For general form, we replace the ordinary derivative by a fractional derivative. Thus, our proposed fractional-order of Human T-cells lymphotropic virus-1 (HTLV-1) infection of CD4$^+$ T-cells (HLTV-1 fractional) has the form

$$D_*^{\alpha}T = \Lambda - \mu_T T - kf(T_A)T,$$
$$D_*^{\alpha}T_L = kf(T_A)T - (\mu_L + \alpha_L)T_L,$$
$$D_*^{\alpha}T_A = \alpha_L T_L - (\mu_A + \rho)T_A,$$
(4)
$$D_*^{\alpha}T_M = \rho T_A + \beta T_M \left(1 - \frac{T_M}{T_{M_{max}}}\right) - \mu_M T_M.$$

where $0 < \alpha \leq 1$. The reason for considering a fractional-order system instead of its integer order counterpart is that fractional-order differential equations are generalizations of integer order differential equations. Using fractional-order differential equations can also help us to reduce the errors arising from the neglected parameters in modeling real life phenomena. In fact, fractional derivative-based approaches establish more advanced and updated models of engineering and biological systems than the ordinary derivative-based approaches do in many applications. One advantage of the fractional-order differential equation is that they provide a powerful instrument for incorporation of memory and hereditary properties of the systems as opposed to the integer order models, where such effects are neglected or difficult to incorporate. Various applications, like in the reaction kinetics of proteins, the anomalous electron transport in amorphous materials, the dielectrical or mechanical relation of polymers, the modeling of glass-forming liquids and others, are successfully performed in numerous papers [11]. However, development still needs to be achieved before the ordinary derivatives could be truly interpreted as a subset of the fractional derivatives [12-14]. We should note that system (4) can be reduced to an integer order system by setting $\alpha = 1$.

Stability analysis of a system in epidemiology and immunology determines the behavior of the system in disease transmission. By stability analysis, one knows when and where the disease spreads by calculating the most wanted quantity known as the basic reproduction number denoted by $\mathfrak{R}_0$. It is the threshold quantity, which shows whether infection spreads or not in the susceptible population. In other words, we can say that the population is free from the disease if $\mathfrak{R}_0 < 1$ and disease spreads in the whole population if $\mathfrak{R}_0 > 1$. Here, we consider the human T-cells lymphotropic virus-1 (HTLV-1) infection of CD4$^+$ T-cells model given by system (1) initially proposed by Wang et al. [9]. They have considered the model in classical form and studied the local as well as global stability results of the equilibria. Furthermore, Song and Li [7] modified the system (1) with system (2) and determined the global dynamics of the modified model as well. Cai et al. [10] extended the systems (1) and (2) in the more general form denoted by (3) and obtained its stability results as well. Arshad et al. [15], proposed a model with the effect of HIV infection on CD4$^+$ T-cell population based on a fractional-order derivative. Their model has studied the dynamical behavior of uninfected, latently infected and actively infected CD4$^+$ T-cells. Also, they have studied the detailed analysis of the extended fractional-order model. Here, we further generalized the model (3) initially proposed by Cai *et al.* [10] using fractional-order derivatives with $0 < \alpha \leq 1$ and obtained the fractional-order system denoted by (4). For the stability analysis of the extended model (4), readers are suggested to go through the studies [7,9,10,15] because the fractional-order differential equations are, at least, as stable as their integer order counterpart [15-19]. So, here the primary objective of the proposed study is to obtain the approximate analytical solution of the extended fractional-order model (4) with the help of well-known method HAM, this is because the more accurate solutions lead towards better cure, control and treatment strategies.

In this chapter, we construct the Homotopy Analysis Method (HAM) solution for nonlinear fractional-order HTLV-1 model (4). It is known that HAM has an advantage over perturbation methods in that it is not dependent on small or large parameters. It is well-known that perturbation methods are based on the existence of small or large parameters, and they cannot be applied to all nonlinear equations. Non-perturbative methods such as δ-expansion and Adomian decomposition method are independent of small parameters [20-21]. However, according to [20-23], both perturbation techniques and non-perturbative methods cannot provide a simple procedure to adjust or control the convergence region and rate of given approximate series. HAM allows us to fine-tune the convergence region and rate of convergence by allowing an auxiliary parameter $\hbar$ to vary [20-28]. The proper choice of the initial condition, the auxiliary linear operator and auxiliary parameter $\hbar$ will ensure the convergence of the HAM solution series [20, 21, 23]. Compared to the Homotopy Perturbation Method (HPM), the HAM solution series will be convergent by considering two factors: the auxiliary linear operator and initial guess [20].

In recent years, HAM and its modifications have been used to solve various nonlinear system and mathematical models in many branches of mathematics and sciences. In [28-29], the HAM was introduced and applied to the modified SIR epidemiological model of computer viruses. In [30-31], Noeiaghdam have applied and compared some approximate analytical method, such as, variational iteration method (VIM), differential transform method (DTM) and homotopy analysis method (HAM) to modified epidemiological model of computer viruses. He also showed all methods are accurate to find the approximate solution to model of computer viruses. Another application of HAM is fractional damped beam equations [32], non-differentible problems on Cantor sets by using local fractional HAM [33], homotopy analysis transfer method (HATM) for fractional model of HIV infection [34], q-HATM for solving smoking epidemic model of fractional-order [35], Viscous Boundary Layer Flow by Homotopy Analysis Decomposition Method [36], fractional homotopy analysis method (FHAM) to fractional HIV infection of $CD4^+$ T-cells [37], nonlinear system by multi-frequency HAM [38], space-time fractional differential equations by HAM [39], solve of integral equations based on the stochastic arithmetic by using HAM [9] and so on.

This chapter is organized as follows:

After the introduction in section 1, section 2 discusses some background on fractional calculus. In section 3, we apply the HAM as expounded by previous researchers in particular [17, 20, 22, 24, 25], to describe the series solution of nonlinear fractional-order of HTLV-1 infection of $CD4^+$ T-cells model (4). We have used the same description in related studies. In this section, we also prove the convergence of HAM. In section 4, numerical results are given to illustrate the capability of HAM. In section 5, we discuss the solution obtained by using HAM. In this section we also improve the solution obtained by applying the least squares method. Finally, in section 6, we give the conclusion of this study.

BASIC DEFINITIONS

The following definitions from [40] are used in this chapter. Details on fractional derivatives can be found in [41-45].

Definition-1: The Reimann-Liouville fractional integral operator of order $\alpha \geq 0$, for $f \in C_\mu, \mu \geq -1$ is defined as follows

$$J_a^\alpha f(x) = \frac{1}{\Gamma(\alpha)} \int_a^x (x-t)^{\alpha-1} f(t)\, dt, \qquad x > 0$$

$$J_a^0 f(x) = f(x).$$

It should be noted that a real function $f(x), x > 0$, belong to the space $C_\mu, \mu \in R$ if there exists a real number $p\ (> \mu)$ such that $f(x) = x^p f_1(x)$, where $f_1(x) \in C[0, \infty)$. Furthermore, the function is said to belong to the space C_μ^n if $f^{(n)} \in C_\mu, n \in N$.

Definition-2: The (left-sided) Caputo fractional derivative of a function $f(x)$ is defined as [41-45, 50, 51].

$$D_*^\alpha f(x) = J_a^{n-\alpha} D_*^n f(x) = \frac{1}{\Gamma(n-\alpha)} \int_a^x (x-t)^{n-\alpha-1} \frac{d^n f(t)}{dt^n} dt,$$

for $n - 1 < \alpha \leq n, n \in N, x > a, f \in C_{-1}^n$.

Properties of the operators J_a^α, D_*^α can be found in [41-45]. It has been mentioned as

$$(J_a^\alpha J_a^\beta f)(x) = (J_a^\beta J_a^\alpha f)(x) = (J_a^{\alpha+\beta} f)(x),$$

$$(J_a^\alpha)(x-a)^\gamma = \frac{\Gamma(\gamma+1)}{\Gamma(\gamma+\alpha+1)}(x-a)^{\gamma+1},$$

$$D_*^\alpha D_*^\beta f(x) = D_*^\beta D_*^\alpha f(x) = D_*^{\alpha+\beta} f(x),$$

$$(J_a^\alpha D_*^\alpha f)(x) = J^n D_*^n f(x) = f(x) - \sum_{k=0}^{\infty} f^{(k)}(a^+) \frac{(x-a)^k}{k!}.$$

for $f \in C_\mu, \mu \geq -1, \alpha, \ \beta \geq 0, a \geq 0, \ n - 1 < \alpha \leq n$ and $\gamma > -1$.

SOLUTION OF HTLV-1 MODEL BY HAM

Firstly, we consider the model (4) in the following form

$$\begin{aligned}
D_*^\alpha T &= \Lambda - \mu_T T - kf(T_A T - \alpha_1 TT_A^2), \\
D_*^\alpha T_L &= kf(T_A T - \alpha_1 TT_A^2) - (\mu_L + \alpha_L)T_L, \\
D_*^\alpha T_A &= \alpha_L T_L - (\mu_A + \rho)T_A, \\
D_*^\alpha T_M &= \rho T_A + \beta T_M \left(1 - \frac{T_M}{T_{M_{max}}}\right) - \mu_M T_M.
\end{aligned} \tag{5}$$

To apply HAM, the HTLV infection of CD4$^+$ T-cells (5) is considered. We denote

$$T(0) = T_0(t) = T_0, \tag{6}$$

$$T_L(0) = T_{L,0}(t) = T_{L,0}, \tag{7}$$

$$T_A(0) = T_{A,0}(t) = T_{A,0}, \tag{8}$$

$$T_M(0) = T_{M,0}(t) = T_{M,0}, \tag{9}$$

The auxiliary linear operators $\mathcal{L}_T$, $\mathcal{L}_{T_L}$, $\mathcal{L}_{T_A}$ and $\mathcal{L}_{T_M}$ are selected as

$$\mathcal{L}_T = D_*^\alpha,$$

$$\mathcal{L}_{T_L} = D_*^\alpha,$$

$$\mathcal{L}_{T_A} = D_*^\alpha,$$

$$\mathcal{L}_{T_M} = D_*^\alpha,$$

satisfying the following properties:

$$\mathcal{L}_T(c_T) = 0,$$

$$\mathcal{L}_{T_L}(c_{T_L}) = 0,$$

$$\mathcal{L}_{T_A}(c_{T_A}) = 0,$$

$$\mathcal{L}_{T_M}(c_{T_M}) = 0,$$

where c_T, c_{T_L}, c_{T_A} and c_{T_M} are integral constants. The homotopy maps are defined as

$$\mathcal{H}_T\left(\widehat{T}(t;q), \widehat{T}_L(t;q), \widehat{T}_A(t;q), \widehat{T}_M(t;q)\right)$$
$$= (1-q)\mathcal{L}_T\left[\widehat{T}(t;q) - T_0(t)\right]$$
$$- q\hbar H_T(t)N_T\left[\widehat{T}(t;q), \widehat{T}_L(t;q), \widehat{T}_A(t;q), \widehat{T}_M(t;q)\right],$$

$$\mathcal{H}_{T_L}\left(\widehat{T}(t;q), \widehat{T}_L(t;q), \widehat{T}_A(t;q), \widehat{T}_M(t;q)\right)$$
$$= (1-q)\mathcal{L}_{T_L}\left[\widehat{T}_L(t;q) - T_{L,0}(t)\right]$$
$$- q\hbar H_{T_L}(t)N_{T_L}\left[\widehat{T}(t;q), \widehat{T}_L(t;q), \widehat{T}_A(t;q), \widehat{T}_M(t;q)\right],$$

$$\mathcal{H}_{T_A}\left(\widehat{T}(t;q), \widehat{T}_L(t;q), \widehat{T}_A(t;q), \widehat{T}_M(t;q)\right)$$
$$= (1-q)\mathcal{L}_{T_A}\left[\widehat{T}_A(t;q) - T_{A,0}(t)\right]$$
$$- q\hbar H_{T_A}(t)N_{T_A}\left[\widehat{T}(t;q), \widehat{T}_L(t;q), \widehat{T}_A(t;q), \widehat{T}_M(t;q)\right],$$

$$\mathcal{H}_{T_M}\left(\widehat{T}(t;q),\widehat{T}_L(t;q),\widehat{T}_A(t;q),\widehat{T}_M(t;q)\right)$$
$$= (1-q)\mathcal{L}_{T_M}\left[\widehat{T}_L(t;q) - \qquad T_{M,0}(t)\right]$$
$$- q\hbar H_{T_M}(t)N_{T_M}\left[\widehat{T}(t;q),\widehat{T}_L(t;q),\widehat{T}_A(t;q),\widehat{T}_M(t;q)\right],$$

where $q \in [0,1]$ is an embedding parameter, $\hbar$ is nonzero auxiliary parameter, H_T, H_{T_L}, H_{T_A} and H_{T_M} are auxiliary functions and N_T, N_{T_L}, N_{T_A} and N_{T_M} are nonlinear operators that are defined as

$$N_T\left(\widehat{T}(t;q),\widehat{T}_L(t;q),\widehat{T}_A(t;q),\widehat{T}_M(t;q)\right) = D_*^\alpha\widehat{T}(t;q) - \Lambda + \mu_T\widehat{T}(t;q) -$$
$$k\left(\widehat{T}_A(t;q)\widehat{T}(t;q) - \alpha_1(\widehat{T}_A(t;q))^2\widehat{T}(t;q)\right),$$

$$N_{T_L}\left(\widehat{T}(t;q),\widehat{T}_L(t;q),\widehat{T}_A(t;q),\widehat{T}_M(t;q)\right) = D_*^\alpha\widehat{T}_L(t;q) -$$
$$k\left(\widehat{T}_A(t;q)\widehat{T}(t;q) - \alpha_1(\widehat{T}_A(t;q))^2\widehat{T}(t;q)\right) + (\mu_L + \alpha_L)\widehat{T}_L(t;q),$$

$$N_{T_A}\left(\widehat{T}(t;q),\widehat{T}_L(t;q),\widehat{T}_A(t;q),\widehat{T}_M(t;q)\right) = D_*^\alpha\widehat{T}_A(t;q) - \alpha_L\widehat{T}_L(t;q) +$$
$$(\mu_A + \rho)\widehat{T}_A(t;q),$$

$$N_{T_M}\left(\widehat{T}(t;q),\widehat{T}_L(t;q),\widehat{T}_A(t;q),\widehat{T}_M(t;q)\right) = D_*^\alpha\widehat{T}_M(t;q) - \rho\widehat{T}_A(t;q) -$$
$$\beta\widehat{T}_M(t;q)\left(1 - \frac{\widehat{T}_M(t;q)}{T_{M_{max}}}\right) - \mu_M\widehat{T}_M(t;q).$$

Clearly, when $q = 0$, we have

$$\mathcal{H}_T\left(\widehat{T}(t;0),\widehat{T}_L(t;0),\widehat{T}_A(t;0),\widehat{T}_M(t;0)\right) = \mathcal{L}_T\left[\widehat{T}(t;0) - T_0(t)\right]$$

$$\mathcal{H}_{T_L}\left(\widehat{T}(t;0),\widehat{T}_L(t;0),\widehat{T}_A(t;0),\widehat{T}_M(t;0)\right) = \mathcal{L}_{T_L}\left[\widehat{T}_L(t;0) - T_{L,0}(t)\right],$$

$$\mathcal{H}_{T_A}\left(\widehat{T}(t;0),\widehat{T}_L(t;0),\widehat{T}_A(t;0),\widehat{T}_M(t;0)\right) = \mathcal{L}_{T_A}\left[\widehat{T}_A(t;0) - T_{A,0}(t)\right],$$

$$\mathcal{H}_{T_M}\left(\widehat{T}(t;0),\widehat{T}_L(t;0),\widehat{T}_A(t;0),\widehat{T}_M(t;0)\right) = \mathcal{L}_{T_M}\left[\widehat{T}_M(t;0) - T_{M,0}(t)\right].$$

If $q = 1$, then

$$\mathcal{H}_T\left(\widehat{T}(t;1),\widehat{T}_L(t;1),\widehat{T}_A(t;1),\widehat{T}_M(t;1)\right) =$$
$$-\hbar H_T(t)N_T\left[\widehat{T}(t;1),\widehat{T}_L(t;1),\widehat{T}_A(t;1),\widehat{T}_M(t;1)\right],$$

$$\mathcal{H}_{T_L}\left(\widehat{T}(t;1),\widehat{T}_L(t;1),\widehat{T}_A(t;1),\widehat{T}_M(t;1)\right) =$$
$$-\hbar H_{T_L}(t)N_{T_L}\left[\widehat{T}(t;1),\widehat{T}_L(t;1),\widehat{T}_A(t;1),\widehat{T}_M(t;1)\right],$$

$$\mathcal{H}_{T_A}\left(\widehat{T}(t;1),\widehat{T}_L(t;1),\widehat{T}_A(t;1),\widehat{T}_M(t;1)\right)$$
$$= -\hbar H_{T_A}(t)N_{T_A}\left[\widehat{T}(t;1),\widehat{T}_L(t;1),\widehat{T}_A(t;1),\widehat{T}_M(t;1)\right],$$

$$\mathcal{H}_{T_M}\left(\widehat{T}(t;1),\widehat{T}_L(t;1),\widehat{T}_A(t;1),\widehat{T}_M(t;1)\right)$$
$$= -\hbar H_{T_M}(t)N_{T_M}\left[\widehat{T}(t;1),\widehat{T}_L(t;1),\widehat{T}_A(t;1),\widehat{T}_M(t;1)\right],$$

Thus, by requiring

$$\mathcal{H}_T\left(\widehat{T}(t;1),\widehat{T}_L(t;1),\widehat{T}_A(t;1),\widehat{T}_M(t;1)\right)$$
$$= \mathcal{H}_{T_L}\left(\widehat{T}(t;1),\widehat{T}_L(t;1),\widehat{T}_A(t;1),\widehat{T}_M(t;1)\right) = 0,$$

$$\mathcal{H}_{T_A}\left(\widehat{T}(t;1),\widehat{T}_L(t;1),\widehat{T}_A(t;1),\widehat{T}_M(t;1)\right)$$
$$= \mathcal{H}_{T_M}\left(\widehat{T}(t;1),\widehat{T}_L(t;1),\widehat{T}_A(t;1),\widehat{T}_M(t;1)\right) = 0.$$

We can obtain

$$(1-q)\mathcal{L}_T\left[\widehat{T}(t;q) - T_0(t)\right] =$$
$$q\hbar H_T(t)N_T\left[\widehat{T}(t;q),\widehat{T}_L(t;q),\widehat{T}_A(t;q),\widehat{T}_M(t;q)\right], \quad (10)$$

$$(1-q)\mathcal{L}_{T_L}\left[\widehat{T}_L(t;q) - T_{L,0}(t)\right] =$$
$$q\hbar H_{T_L}(t)N_{T_L}\left[\widehat{T}(t;q),\widehat{T}_L(t;q),\widehat{T}_A(t;q),\widehat{T}_M(t;q)\right], \quad (11)$$

$$(1-q)\mathcal{L}_{T_A}\left[\widehat{T}_A(t;q) - T_{A,0}(t)\right] =$$
$$q\hbar H_{T_A}(t)N_{T_A}\left[\widehat{T}(t;q),\widehat{T}_L(t;q),\widehat{T}_A(t;q),\widehat{T}_M(t;q)\right], \quad (12)$$

$$(1-q)\mathcal{L}_{T_M}\left[\widehat{T}_L(t;q) - T_{M,0}(t)\right] =$$
$$q\hbar H_{T_M}(t)N_{T_M}\left[\widehat{T}(t;q),\widehat{T}_L(t;q),\widehat{T}_A(t;q),\widehat{T}_M(t;q)\right]. \quad (13)$$

Following [17, 20, 22, 24, 25], if $q = 0$ and $q = 1$, the homotopy equations vary to

$$\widehat{T}(t;0) = T_0(t), \qquad \widehat{T}(t;1) = T(t),$$

$$\widehat{T}_L(t;0) = T_{L,0}(t), \qquad \widehat{T}_L(t;1) = T_L(t),$$

$$\widehat{T}_A(t;0) = T_{A,0}(t), \qquad \widehat{T}_A(t;1) = T_A(t),$$

$$\widehat{T}_M(t;0) = T_{M,0}(t), \qquad \widehat{T}_M(t;1) = T_M(t).$$

As q varies from 0 to 1, the solution of the HTLV-1 model (5) will vary from the initial guesses $T_0(t)$ $T_{L,0}(t)$, $T_{A,0}(t)$ and $T_{M,0}(t)$ to the exact solutions $T(t), T_L(t)$, $T_A(t)$ and $T_M(t)$ of the HTLV-1 model (5). Following [17, 20, 22, 24, 25], expanding $\widehat{T}(t;q), \widehat{T}_L(t;q), \widehat{T}_A(t;q)$ and $\widehat{T}_M(t;q)$ as a Taylor series with respect to q yield

$$\widehat{T}(t;q) = T_0(t) + \sum_{j=1}^{\infty} T_j(t)\, q^j, \tag{14}$$

$$\widehat{T}_L(t;q) = T_{L,0}(t) + \sum_{j=1}^{\infty} T_{L,j}(t)\, q^j, \tag{15}$$

$$\widehat{T}_A(t;q) = T_{A,0}(t) + \sum_{j=1}^{\infty} T_{A,j}(t)\, q^j, \tag{16}$$

$$\widehat{T}_M(t;q) = T_{M,0}(t) + \sum_{j=1}^{\infty} T_{M,j}(t)\, q^j, \tag{17}$$

where

$$T_j(t) = \frac{1}{j!}\frac{\partial^j \widehat{T}(t;q)}{\partial q^j}\bigg|_{q=0}, \qquad T_{L,j}(t) = \frac{1}{j!}\frac{\partial^j \widehat{T}_L(t;q)}{\partial q^j}\bigg|_{q=0},$$

$$T_{A,j}(t) = \frac{1}{j!}\frac{\partial^j \widehat{T}_A(t;q)}{\partial q^j}\bigg|_{q=0}, \qquad T_{M,j}(t) = \frac{1}{j!}\frac{\partial^j \widehat{T}_M(t;q)}{\partial q^j}\bigg|_{q=0}.$$

According to [21], the convergence of the series (14)-(17) strongly depends upon the convergence-control parameter parameter $\hbar$. Note that if $q = 1$ then

$$\widehat{T}(t;1) = T(t) = T_0(t) + \sum_{j=1}^{\infty} T_j(t), \tag{18}$$

$$\widehat{T}_L(t;1) = T_L(t) = T_{L,0}(t) + \sum_{j=1}^{\infty} T_{L,j}(t), \tag{19}$$

$$\widehat{T}_A(t;1) = T_A(t) = T_{A,0}(t) + \sum_{j=1}^{\infty} T_{A,j}(t), \tag{20}$$

$$\widehat{T}_M(t;1) = T_M(t) = T_{M,0}(t) + \sum_{j=1}^{\infty} T_{M,j}(t), \tag{21}$$

According to definitions (14)-(17), the governing equations for the unknowns can be deduced from the zeroth-deformation equations (10)-(13). For further analysis, the vectors

$$\hat{T}_n(t) = \{T_0(t), T_1(t), \dots, T_n(t)\},$$

$$\hat{T}_{L,n}(t) = \{T_{L,0}(t), T_{L,1}(t), \dots, T_{L,n}(t)\},$$

$$\hat{T}_{A,n}(t) = \{T_{A,0}(t), T_{A,1}(t), \dots, T_{A,n}(t)\},$$

$$\hat{T}_{M,n}(t) = \{T_{M,0}(t), T_{M,1}(t), \dots, T_{M,n}(t)\},$$

are defined. Following [20, 21, 23, 24], differentiating (10)-(13) m-times with respect to q, dividing by $m!$ and setting $q = 0$, gives the linear equations

$$D_*^\alpha\left[T_j(t) - \mathcal{X}_j T_{j-1}(t)\right] = \hbar H_T(t) R_{T,j}(\vec{T}_{j-1}, \vec{T}_{L,j-1}, \vec{T}_{A,j-1}, \vec{T}_{M,j-1}), \qquad (22)$$

$$D_*^\alpha\left[T_{L,j}(t) - \mathcal{X}_j T_{L,j-1}(t)\right] = \hbar H_{T_L}(t) R_{T_L,j}(\vec{T}_{j-1}, \vec{T}_{L,j-1}, \vec{T}_{A,j-1}, \vec{T}_{M,j-1}), \quad (23)$$

$$D_*^\alpha\left[T_{A,j}(t) - \mathcal{X}_j T_{A,j-1}(t)\right] = \hbar H_{T_A}(t) R_{T_A,j}(\vec{T}_{j-1}, \vec{T}_{L,j-1}, \vec{T}_{A,j-1}, \vec{T}_{M,j-1}), \quad (24)$$

$$D_*^\alpha\left[T_{M,j}(t) - \mathcal{X}_j T_{M,j-1}(t)\right] = \hbar H_{T_M}(t) R_{T_M,j}(\vec{T}_{j-1}, \vec{T}_{L,j-1}, \vec{T}_{A,j-1}, \vec{T}_{M,j-1}), \quad (25)$$

with initial conditions

$$T_j(0) = 0, \quad T_{L,j}(0) = 0, \quad T_{A,j}(0) = 0, \quad T_{M,j}(0) = 0, \qquad (26)$$

where

$$R_{T,j}(t) = D_*^\alpha T_{j-1}(t) - (1 - \mathcal{X}_j)\Lambda + \mu_T T_{j-1}(t) + k\left(\sum_{i=0}^{j-1} T_i(t) T_{A,j-1-i}(t) - \right.$$
$$\left. \alpha_1 \sum_{i_1=0}^{j-1} T_{A,i_1}(t) \sum_{i_2=0}^{j-1-i_1} T_{i_2}(t) T_{A,j-1-i_1-i_2}(t)\right), \qquad (27)$$

$$R_{T_L,j}(t) = D_*^\alpha T_{L,j-1}(t) - k\left(\sum_{i=0}^{j-1} T_i(t) T_{A,j-1-i}(t) - \right.$$
$$\left. \alpha_1 \sum_{i_1=0}^{j-1} T_{A,i_1}(t) \sum_{i_2=0}^{j-1-i_1} T_{i_2}(t) T_{A,j-1-i_1-i_2}(t)\right) +$$
$$(\mu_L + \alpha_L) T_{L,j-1}(t), \qquad (28)$$

$$R_{T_A,j}(t) = D_*^\alpha T_{A,j-1}(t) - \mu_L T_{L,j-1}(t) + (\mu_A + \rho) T_{A,j-1}(t), \qquad (29)$$

$$R_{T_M,j}(t) = D_*^\alpha T_{M,j-1}(t) - \rho T_{A,j-1}(t) + \frac{\beta}{T_{M_{max}}} \sum_{i=0}^{j-1} T_{M,i}(t) T_{M,j-1-i}(t) +$$
$$\mu_M T_{M,j-1}(t), \tag{30}$$

and

$$\mathcal{X}_j = \begin{cases} 0, & j \le 1 \\ 1, & j > 1. \end{cases}$$

In [18], some cases of $H(t)$ were considered to solve systems of linear equations. In this chapter, by using $H_T = H_{T_L} = H_{T_A} = H_{T_M} = 1$, the solution of the m-order deformation equations (22)-(25) for $m \ge 1$ becomes

$$T_j(t) = \mathcal{X}_j T_{j-1}(t) + \frac{\hbar}{\Gamma(\alpha)} \int_0^t (t-s)^{\alpha-1} R_{T,j}(s) \, ds,$$

$$T_{L,j}(t) = \mathcal{X}_j T_{L,j-1}(t) + \frac{\hbar}{\Gamma(\alpha)} \int_0^t (t-s)^{\alpha-1} R_{T_L,j}(s) \, ds,$$

$$T_{A,j}(t) = \mathcal{X}_j T_{A,j-1}(t) + \frac{\hbar}{\Gamma(\alpha)} \int_0^t (t-s)^{\alpha-1} R_{T_A,j}(s) \, ds,$$

$$T_{M,j}(t) = \mathcal{X}_j T_{M,j-1}(t) + \frac{\hbar}{\Gamma(\alpha)} \int_0^t (t-s)^{\alpha-1} R_{T_M,j}(s) \, ds.$$

If we choose $T_0(t) = T_0$, $T_L(0) = T_{L,0}$, $T_A(0) = T_{A,0}$ and $T_M(0) = T_{M,0}$ as initial guess approximations of $T(t), T_L(t), T_A(t)$ and $T_M(t)$ respectively, then two terms approximations for $T(t), T_L(t), T_A(t)$ and $T_M(t)$ are calculated and presented below

$$T_0 = 1000,$$

$$T_{L,0} = 1.5,$$

$$T_{A,0} = 0,$$

$$T_{M,0} = 250,$$

$$T_1(t; \hbar) = 0,$$

$$T_{L,1}(t;\hbar) = \frac{0.0096\hbar t^{\alpha}}{\alpha\Gamma(\alpha)},$$

$$T_{A,1}(t;\hbar) = -\frac{0.0006\hbar t^{\alpha}}{\alpha\Gamma(\alpha)},$$

$$T_{M,1}(t;\hbar) = \frac{0.058522\hbar t^{\alpha}}{\alpha\Gamma(\alpha)},$$

$$T_{2}(t;\hbar) = -\frac{0.10635 \times 4^{-\alpha}\hbar^{2}t^{2\alpha}}{\alpha\Gamma(\alpha)\Gamma\left(\alpha+\frac{1}{2}\right)},$$

$$T_{L,2}(t;\hbar)$$
$$= \frac{0.0096(1+\hbar)\hbar t^{\alpha}}{\alpha\Gamma(\alpha)} + \hbar^{2}t^{2\alpha}Csc(\pi\alpha)$$
$$\times \frac{\left[\begin{array}{c}\hbar t^{\alpha} \times 1.13097 \times 10^{-7}(1+\alpha)\Gamma(1+2\alpha)\Gamma(2+2\alpha) + (0.094344 + \\ 0.18869\alpha)\Gamma(1+\alpha)\Gamma(2+\alpha)\Gamma(3+\alpha)\end{array}\right]}{(1+\alpha)\left(\frac{1}{2}+\alpha\right)\Gamma(1-\alpha)\Gamma(\alpha)\Gamma^{2}(1+\alpha)\Gamma(1+2\alpha)\Gamma(1+3\alpha)},$$

$$T_{A,2}(t;\hbar) = -\frac{0.0006(1+\hbar)\hbar t^{\alpha}}{\alpha\Gamma(\alpha)} - \frac{6.00224 \times 10^{-5} \times 4^{-\alpha}\hbar^{2}t^{2\alpha}}{\alpha\Gamma(\alpha)\Gamma\left(\alpha+\frac{1}{2}\right)},$$

$$T_{M,2}(t;\hbar) = \frac{0.058522(1+\hbar)\hbar t^{\alpha}}{\alpha\Gamma(\alpha)} + \frac{2.78604 \times 10^{-5} \times 4^{-\alpha}\hbar^{2}t^{2\alpha}}{\alpha\Gamma(\alpha)\Gamma\left(\alpha+\frac{1}{2}\right)}.$$

Finally, we approximate the solution $T(t), T_L(t), T_A(t)$ and $T_M(t)$ of the model (5) by the j-th truncated series

$$\varphi_{T,m}(t;\hbar_1) = T_0 + \sum_{j=1}^{m} T_j\,(t;\hbar_1),$$
$$\varphi_{T_L,m}(t;\hbar_2) = T_{L,0} + \sum_{j=1}^{m} T_{L,j}\,(t;\hbar_2),$$
$$\varphi_{T_A,m}(t;\hbar_3) = T_{A,0} + \sum_{j=1}^{m} T_{A,j}\,(t;\hbar_3),$$
$$\varphi_{T_M,m}(t;\hbar_4) = T_{M,0} + \sum_{j=1}^{m} T_{M,j}\,(t;\hbar_4).$$

$$(31)$$

CONVERGENCE THEOREM

Theorem : As long as the series $T(t) = T_0(t) + \sum_{j=1}^{\infty} T_j(t)$, $T_L(t) = T_{L,0}(t) + \sum_{j=1}^{\infty} T_{L,j}(t)$, $T_A(t) = T_{A,0}(t) + \sum_{j=1}^{\infty} T_{A,j}(t)$ and $T_M(t) = T_{M,0}(t) + \sum_{j=1}^{\infty} T_{M,j}(t)$ converge where $T_j(t)$, $T_{L,j}(t)$, $T_{A,j}(t)$ and $T_{M,j}(t)$ are governed by (22)-(25) under definitions (27)-(30), it must be the solutions of the system (5).

Proof: If the series $\sum_{j=0}^{\infty} T_j(t)$, $\sum_{j=0}^{\infty} T_{L,j}(t)$, $\sum_{j=0}^{\infty} T_{A,j}(t)$ and $\sum_{j=0}^{\infty} T_{M,j}(t)$ are converge, we can write

$S_T = \sum_{j=0}^{\infty} T_j(t)$, $S_{L,T} = \sum_{j=0}^{\infty} T_{L,j}(t)$, $S_{A,T} = \sum_{j=0}^{\infty} T_{A,j}(t)$, $S_{M,T} = \sum_{j=0}^{\infty} T_{M,j}(t)$.

and it holds

$$\lim_{j \to \infty} T_j(t) = \lim_{j \to \infty} T_{L,j}(t) = \lim_{j \to \infty} T_{A,j}(t) = \lim_{j \to \infty} T_{M,j}(t) = 0.$$

From (22)-(25) and using $\mathcal{L}_T = D_*^\alpha$, $\mathcal{L}_{T_L} = D_*^\alpha$, $\mathcal{L}_{T_A} = D_*^\alpha$ and $\mathcal{L}_{T_M} = D_*^\alpha$, we have

$$\hbar \sum_{j=1}^{\infty} R_{T,j}(t) = \sum_{j=1}^{\infty} D_*^\alpha \left[T_j(t) - \mathcal{X}_j T_{j-1}(t) \right],$$

$$= \lim_{n \to \infty} \sum_{j=1}^{n} D_*^\alpha \left[T_j(t) - \mathcal{X}_j T_{j-1}(t) \right],$$

$$= D_*^\alpha \left[\lim_{n \to \infty} \sum_{j=1}^{n} \left(T_j(t) - \mathcal{X}_j T_{j-1}(t) \right) \right],$$

$$= D_*^\alpha \left[\lim_{n \to \infty} T_n(t) \right] = 0.$$

By repeating this procedure, it can be easily shown

$$\hbar \sum_{j=1}^{\infty} R_{T_L,j}(t) = \hbar \sum_{j=1}^{\infty} R_{T_A,j}(t) = \hbar \sum_{j=1}^{\infty} R_{T_M,j}(t) = \mathbf{0}.$$

Since $\hbar \neq \mathbf{0}$, then

$$\sum_{j=1}^{\infty} R_{T,j}(t) = 0, \tag{32}$$

$$\sum_{j=1}^{\infty} R_{T_L,j}(t) = 0, \tag{33}$$

$$\sum_{j=1}^{\infty} R_{A,j}(t) = 0, \tag{34}$$

$$\sum_{j=1}^{\infty} R_{T_M,j}(t) = 0. \tag{35}$$

Substituting (27) into (32) and simplifying it, we obtain

$$\sum_{j=1}^{\infty} R_{T,j}(t) = \sum_{j=1}^{\infty} \Big(D_*^{\alpha} T_{j-1}(t) - (1 - \mathcal{X}_j)\Lambda + \mu_T T_{j-1}(t) +$$
$$k \Big[\sum_{i=0}^{j-1} T_i(t) T_{A,j-1-i}(t) - \alpha_1 \sum_{i_1=0}^{j-1} T_{A,i_1}(t) \sum_{i_2=0}^{j-1-i_1} T_{i_2}(t) T_{A,j-1-i_1-i_2}(t) \Big] \Big),$$

$$= D_*^{\alpha} \sum_{j=0}^{\infty} T_j(t) - \Lambda + \mu_T \sum_{j=0}^{\infty} T_j(t) +$$
$$k \Big[\sum_{j=1}^{\infty} \sum_{i=0}^{j-1} T_i(t) T_{A,j-1-i}(t) - \alpha_1 \sum_{j=0}^{\infty} \sum_{i_1=0}^{j-1} T_{A,i_1}(t) \sum_{i_2=0}^{j-1-i_1} T_{i_2}(t) T_{A,j-1-i_1-i_2}(t) \Big],$$

$$= D_*^{\alpha} S_T - \Lambda + \mu_T S_T +$$
$$k \Big[\sum_{i=1}^{\infty} \sum_{j=i+1}^{\infty} T_i(t) T_{A,j-1-i}(t) - \alpha_1 \sum_{i_1=0}^{\infty} \sum_{j=i_1+1}^{\infty} T_{A,i_1}(t) \sum_{i_2=0}^{j-1-i_1} T_{i_2}(t) T_{A,j-1-i_1-i_2}(t) \Big],$$

$$= D_*^{\alpha} S_T - \Lambda + \mu_T S_T +$$
$$k \Big[\sum_{i=0}^{\infty} T_i(t) \sum_{j=i+1}^{\infty} \sum_{i=0}^{j-1} T_{A,j-1-i}(t) - \alpha_1 \sum_{i_1=0}^{\infty} T_{A,i_1}(t) \sum_{k=0}^{\infty} \sum_{i_2=0}^{k} T_{i_2}(t) T_{A,k-i_2}(t) \Big],$$

$$= D_*^{\alpha} S_T - \Lambda + \mu_T S_T +$$
$$k \Big[\sum_{i=0}^{\infty} T_i(t) \sum_{l=0}^{\infty} \sum_{i=0}^{j-1} T_{A,l}(t) - \alpha_1 \sum_{i_1=0}^{\infty} T_{A,i_1}(t) \sum_{i_2=0}^{\infty} \sum_{k=i_2}^{\infty} T_{i_2}(t) T_{A,k-i_2}(t) \Big],$$

$$= D_*^{\alpha} S_T - \Lambda + \mu_T S_T + k(S_T S_A - \alpha_1 S_A^2 S_T).$$

By repeating the above procedure and substituting (28), (29) and (30) into (33), (34) and (35), respectively, we obtain

$$\sum_{j=1}^{\infty} R_{T_L,j}(t) = \sum_{j=1}^{\infty} \Big(D_*^{\alpha} T_{L,j-1}(t) - k \Big[\sum_{i=0}^{j-1} T_i(t) T_{A,j-1-i}(t) -$$
$$\alpha_1 \sum_{i_1=0}^{j-1} T_{A,i_1}(t) \sum_{i_2=0}^{j-1-i_1} T_{i_2}(t) T_{A,j-1-i_1-i_2}(t) \Big] + (\mu_L +$$
$$\alpha_L) T_{L,j-1}(t) \Big),$$

$$= D_*^{\alpha} S_{L,T} -$$
$$k \Big[\sum_{j=1}^{\infty} \sum_{i=0}^{j-1} T_i(t) T_{A,j-1-i}(t) - \alpha_1 \sum_{j=0}^{\infty} \sum_{i_1=0}^{j-1} T_{A,i_1}(t) \sum_{i_2=0}^{j-1-i_1} T_{i_2}(t) T_{A,j-1-i_1-i_2}(t) +$$
$$\sum_{j=1}^{\infty} (\mu_L + \alpha_L) T_{L,j-1}(t) \Big],$$

$$= D_*^{\alpha} S_{L,T} -$$
$$k \Big[\sum_{i=1}^{\infty} \sum_{j=i+1}^{\infty} T_i(t) T_{A,j-1-i}(t) - \alpha_1 \sum_{i_1=0}^{\infty} \sum_{j=i_1+1}^{\infty} T_{A,i_1}(t) \sum_{i_2=0}^{j-1-i_1} T_{i_2}(t) T_{A,j-1-i_1-i_2}(t) +$$
$$(\mu_L + \alpha_L) S_L \Big],$$

$$= D_*^{\alpha} S_{L,T} -$$
$$k \Big[\sum_{i=0}^{\infty} T_i(t) \sum_{j=i+1}^{\infty} \sum_{i=0}^{j-1} T_{A,j-1-i}(t) - \alpha_1 \sum_{i_1=0}^{\infty} T_{A,i_1}(t) \sum_{k=0}^{\infty} \sum_{i_2=0}^{k} T_{i_2}(t) T_{A,k-i_2}(t) +$$
$$(\mu_L + \alpha_L) S_L \Big],$$

$$= D_*^\alpha S_{L,T} -$$
$$k\Big[\sum_{i=0}^\infty T_i(t) \sum_{l=0}^\infty \sum_{i=0}^{j-1} T_{A,l}(t) - \alpha_1 \sum_{i_1=0}^\infty T_{A,i_1}(t) \sum_{i_2=0}^\infty \sum_{k=i_2}^\infty T_{i_2}(t)\, T_{A,k-i_2}(t) +$$
$$(\mu_L + \alpha_L)S_L\Big],$$

$$= D_*^\alpha S_{L,T} - k(S_T S_A - \alpha_1 S_A^2 S_T) + (\mu_L + \alpha_L)S_L.$$

Similarly,

$$\sum_{j=1}^\infty R_{T_A,j}(t) = \sum_{j=1}^\infty \Big(D_*^\alpha T_{A,j-1}(t) - \alpha_L T_{L,j-1}(t) + (\mu_A + \rho)T_{A,j-1}(t)\Big),$$

$$= D_*^\alpha S_{A,T} - \alpha_L \sum_{j=1}^\infty T_{L,j-1}(t) + (\mu_A + \rho)\sum_{j=1}^\infty T_{A,j-1}(t),$$

$$= D_*^\alpha S_{A,T} - \alpha_L \sum_{j=0}^\infty T_{L,j}(t) + (\mu_A + \rho)\sum_{j=0}^\infty T_{A,j}(t),$$

$$= D_*^\alpha S_{A,T} - \alpha_L S_L + (\mu_A + \rho)S_A.$$

and

$$\sum_{j=1}^\infty R_{M,j}(t) = \sum_{j=1}^\infty \Big(D_*^\alpha T_{M,j-1}(t) - \rho T_{A,j-1}(t) + (\mu_M - \beta)T_{M,j-1}(t) + \frac{\beta}{T_{M max}}\sum_{i=0}^{j-1} T_{M,i}(t)T_{M,j-1-i}(t)\Big),$$

$$= D_*^\alpha \sum_{j=0}^\infty T_{M,j-1}(t) - \rho \sum_{j=1}^\infty T_{A,j-1}(t) + (\mu_M - \beta)\sum_{j=1}^\infty T_{M,j-1}(t) + \frac{\beta}{T_{M max}}\sum_{j=1}^\infty \sum_{i=0}^{j-1} T_{M,i}(t)T_{M,j-1-i}(t),$$

$$= D_*^\alpha S_{M,T} - \rho \sum_{j=0}^\infty T_{A,j}(t) + (\mu_M - \beta)\sum_{j=0}^\infty T_{M,j}(t) + \frac{\beta}{T_{M max}}\sum_{j=1}^\infty \sum_{i=0}^{j-1} T_{M,i}(t)T_{M,j-1-i}(t),$$

$$= D_*^\alpha S_{M,T} - \rho S_A + (\mu_M - \beta)S_M + \frac{\beta}{T_{M max}}\sum_{i=0}^\infty \sum_{j=i+1}^\infty T_{M,i}(t)T_{M,j-1-i}(t),$$

$$= D_*^\alpha S_{M,T} - \rho S_A + (\mu_M - \beta)S_M + \frac{\beta}{T_{M max}}\sum_{i=0}^\infty T_{M,i}(t)\sum_{k=0}^\infty T_{M,k}(t),$$

$$= D_*^\alpha S_{M,T} - \rho S_A + (\mu_M - \beta)S_M + \frac{\beta}{T_{M max}}S_T S_M.$$

From (6)-(9) and (26), it holds that

$$S_T = \sum_{j=0}^\infty T_j(t) = T_0(t) + \sum_{j=1}^\infty T_j(t) = T_0,$$

$$S_{L,T} = \sum_{j=0}^\infty T_{L,j}(t) = T_{L,0}(t) + \sum_{j=1}^\infty T_{L,j}(t) = T_{L,0},$$

$$S_{A,T} = \sum_{j=0}^\infty T_{A,j}(t) = T_{A,0}(t) + \sum_{j=1}^\infty T_{A,j}(t) = T_{A,0},$$

$$S_{M,T} = \sum_{j=0}^\infty T_{M,j}(t) = T_{M,0}(t) + \sum_{j=1}^\infty T_{M,j}(t) = T_{M,0}.$$

Thus, S_T, $S_{L,T}$, $S_{A,T}$ and $S_{M,T}$ satisfy in the system (5) and it must be exact solution for system (5) with initial conditions (6)-(9).

NUMERICAL RESULTS

In this section, to illustrate the capability of the HAM the variables and parameters, described in Table **1**, are considered as follows [10]

$$\Lambda = 6, \quad k = 0.1, \quad \alpha_L = 0.0004, \quad \beta = 0.0003, \quad \mu_L = 0.006, \quad \mu_T = 0.006,$$

$$\mu_A = 0.05, \quad \mu_M = 0.0005, \quad \rho = 0.00004, \quad T_{M_{max}} = 2200, \quad \alpha_1 = 0.001,$$

$$T_0 = 1000, \quad T_{L,0} = 250, \quad T_{A,0} = 1.5, \quad T_{M,0} = 0.$$

To consider the behavior of solution for susceptibles $T(t)$, $T_L(t)$, $T_A(t)$ and $T_M(t)$ for different values of α, $0 < \alpha \leq 1$, we will take advantage of the explicit formula (31) available for α, $0 < \alpha \leq 1$ and consider the following two special cases:

Case 1: By using Mathematica software, we will examine the fractional-order HTLV-1 model (5) along (6)-(9), by setting $\alpha = 0.7$ in Eq. (5). The partial sums (31) are determined and in particular 6th-order approximations for susceptibles $T(t)$, $T_L(t)$, $T_A(t)$ and $T_M(t)$ respectively, were calculated and presented below.

$$\varphi_{T,6}(t;\hbar) = \sum_{i=0}^{5} T_i(t) = 1000 - 0.483026\hbar^2\, t^{1.4} - 0.966052\,\hbar^3\, t^{1.4} -$$
$$0.724539\,\hbar^4\, t^{1.4} - 0.19321\,\hbar^5\, t^{1.4} - 0.016884\,\hbar^3\, t^{2.1} -$$
$$0.0249164\,\hbar^4\, t^{2.1} - 0.00973723\,\hbar^5\, t^{2.1} - 0.00274638\,\hbar^4\, t^{2.8} -$$
$$0.0021904\,\hbar^5\, t^{2.8} - 0.0000239355\,\hbar^5\, t^{3.5} - 1.11755 \times 10^{-9}\,\hbar^6\, t^{3.5} -$$
$$7.50719 \times 10^{-12}\,\hbar^6\, t^{4.2}, \tag{36}$$

$$\varphi_{T_L,6}(t;\hbar) = \sum_{i=0}^{5} T_{L,i}(t) = 250 + 0.0528263\,\hbar\, t^{0.7} + 0.105653\,\hbar^2\, t^{0.7} +$$
$$0.105653\,\hbar^3\, t^{0.7} + 0.0528263\,\hbar^4\, t^{0.7} + 0.0105653\,\hbar^5\, t^{0.7} +$$
$$0.483521\,\hbar^2\, t^{1.4} + 0.967041\,\hbar^3\, t^{1.4} + 0.725281\,\hbar^4\, t^{1.4} +$$
$$0.193408\,\hbar^5\, t^{1.4} + 0.017159\,\hbar^3\, t^{2.1} + 0.0257385\,\hbar^4\, t^{2.1} +$$
$$0.0102955\,\hbar^5\, t^{2.1} + 4.92922 \times 10^{-8}\,\hbar^6\, t^{2.1} + 0.00275772\,\hbar^4\, t^{2.8} +$$
$$0.00220618\,\hbar^5\, t^{2.8} + 4.43071 \times 10^{-10}\,\hbar^6\, t^{2.8} + 0.00002536\,\hbar^5\, t^{3.5} +$$
$$1.1187 \times 10^{-9}\,\hbar^6\, t^{3.5} + 8.35929 \times 10^{-12}\,\hbar^6\, t^{4.2}, \tag{37}$$

$$\varphi_{T_A,6}(t;\hbar) = \sum_{i=0}^{5} T_{A,i}(t) = 1.5 - 0.00330164\,\hbar\, t^{0.7} - 0.00660328\,\hbar^2\, t^{0.7} -$$
$$0.00660328\,\hbar^3\, t^{0.7} - 0.00330164\,\hbar^4\, t^{0.7} - 0.000660328\,\hbar^5\, t^{0.7} -$$
$$0.00027262\,\hbar^2\, t^{1.4} - 0.00054524\,\hbar^3\, t^{1.4} - 0.00040893\,\hbar^4\, t^{1.4} -$$

$0.000109048\ \hbar^5\ t^{1.4} - 0.000117032\ \hbar^3\ t^{2.1} - 0.000175548\ \hbar^4\ t^{2.1} -$
$0.000070219\ \hbar^5\ t^{2.1} - 2.9775 \times 10^{-6}\ \hbar^4\ t^{2.8} - 2.38203 \times$
$10^{-6}\ \hbar^5\ t^{2.8} - 2.76919 \times 10^{-11}\ \hbar^6\ t^{2.8} - 1.0106 \times 10^{-7}\ \hbar^5\ t^{3.5} -$
$6.30746 \times 10^{-13}\ \hbar^6\ t^{3.5} - 5.70125 \times 10^{-14}\ \hbar^6\ t^{4.2},$ (38)

$$\varphi_{TM,6}(t;\hbar) = \sum_{i=0}^{5} T_{M,i}(t) = 0 + 0.322035\ \hbar\ t^{0.7} + 0.64407\ \hbar^2\ t^{0.7} +$$
$0.64407\ \hbar^3\ t^{0.7} + 0.322035\ \hbar^4\ t^{0.7} + 0.064407\ \hbar^5\ t^{0.7} +$
$0.000126543\ \hbar^2\ t^{1.4} + 0.000253085\ \hbar^3\ t^{1.4} +$
$0.000189814\ \hbar^4\ t^{1.4} + 0.000050617\ \hbar^5\ t^{1.4} + 2.85431 \times$
$10^{-8}\ \hbar^3\ t^{2.1} + 4.28147 \times 10^{-8}\ \hbar^4\ t^{2.1} + 1.71259 \times$
$10^{-8}\ \hbar^5\ t^{2.1} + 1.0981 \times 10^{-9}\ \hbar^4\ t^{2.8} + 8.78481 \times$
$10^{-10}\ \hbar^5\ t^{2.8} + 9.63715 \times 10^{-12}\ \hbar^5\ t^{3.5} + 4.47021 \times$
$10^{-16}\ \hbar^6\ t^{3.5} + 3.01697 \times 10^{-18}\ \hbar^6\ t^{4.2}.$ (39)

Case 2: In this case, we will examine the fractional-order HTLV-1 model (5) along (6)-(9) by setting $\alpha = 0.8$ in Eq.(5) and similar results are obtained for $\alpha = 0.9$ and $\alpha = 1$. The partial sums (31) are determined and in particular 6th-order approximations for susceptibles $T(t)$, $T_L(t)$, $T_A(t)$ and $T_M(t)$ respectively, were calculated and presented below.

$$\varphi_{T,6}(t;\hbar) = \sum_{i=0}^{5} T_i(t) = 1000 - 0.419691\ \hbar^2\ t^{1.6} - 0.839381\ \hbar^3\ t^{1.6} -$$
$0.629536\ \hbar^4\ t^{1.6} - 0.167876\ \hbar^5\ t^{1.6} - 0.0124462\ \hbar^3\ t^{2.4} -$
$0.0183674\ \hbar^4\ t^{2.4} - 0.00717788\ \hbar^5\ t^{2.4} - 0.00166137\ \hbar^4\ t^{3.2} -$
$0.00132504\ \hbar^5\ t^{3.2} - 0.0000115901\ \hbar^5\ t^4 - 5.9329 \times 10^{-10}\ \hbar^6\ t^4 -$
$3.12889 \times 10^{-12}\ \hbar^6\ t^{4.8},$ (40)

$$\varphi_{TL,6}(t;\hbar) = \sum_{i=0}^{5} T_{L,i}(t) = 250 + 0.0515362\ \hbar\ t^{0.8} + 0.103072\ \hbar^2\ t^{0.8} +$$
$0.103072\ \hbar^3\ t^{0.8} + 0.0515362\ \hbar^4\ t^{0.8} + 0.0103072\ \hbar^5\ t^{0.8} +$
$0.42012\ \hbar^2\ t^{1.6} + 0.840241\ \hbar^3\ t^{1.6} + 0.630181\ \hbar^4\ t^{1.6} +$
$0.168048\ \hbar^5\ t^{1.6} + 0.0126489\ \hbar^3\ t^{2.4} + 0.0189734\ \hbar^4\ t^{2.4} +$
$0.00758941\ \hbar^5\ t^{2.4} + 3.9802 \times 10^{-8}\ \hbar^3\ t^{2.4} + 0.00166824\ \hbar^4\ t^{3.2} +$
$0.00133459\ \hbar^5\ t^{3.2} + 2.93712 \times 10^{-10}\ \hbar^6\ t^{3.2} +$
$0.0000122802\ \hbar^5\ t^4 + 5.93898 \times 10^{-10}\ \hbar^6\ t^4 + 3.48403 \times$
$10^{-12}\ \hbar^6\ t^{4.8},$ (41)

$$\varphi_{TA,6}(t;\hbar) = \sum_{i=0}^{5} T_{A,i}(t) = 250 - 0.00322101\ \hbar\ t^{0.8} - 0.00644203\ \hbar^2\ t^{0.8} -$$
$0.00644203\ \hbar^3\ t^{0.8} - 0.00322101\ \hbar^4\ t^{0.8} -$
$0.000644203\ \hbar^5\ t^{0.8} - 0.000236873\ \hbar^2\ t^{1.6} -$
$0.000473747\ \hbar^3\ t^{1.6} - 0.00035531\ \hbar^4\ t^{1.6} -$
$0.0000947494\ \hbar^5\ t^{1.6} - 0.0000862709\ \hbar^3\ t^{2.4} -$

$$0.000129406\ \hbar^4\ t^{2.4} - 0.0000517625\ \hbar^5\ t^{2.4} - 1.80192 \times$$
$$10^{-6}\ \hbar^4\ t^{3.2} - 1.44155 \times 10^{-6}\ \hbar^5\ t^{3.2} -\ 1.8357 \times$$
$$10^{-11}\ \hbar^6\ t^{3.2} - 4.8962 \times 10^{-8}\ \hbar^5\ t^4 - 3.34853 \times 10^{-13}\ \hbar^6\ t^4 -$$
$$2.37617 \times 10^{-14}\ \hbar^6\ t^{4.8}, \tag{42}$$

$$\varphi_{T_M,6}(t;\hbar) = \sum_{i=0}^{5} T_{M,i}(t) =\ 0\ + 0.314171\ \hbar\ t^{0.8} + 0.628342\ \hbar^2\ t^{0.8} +$$
$$0.628342\ \hbar^3\ t^{0.8} + 0.314171\ \hbar^4\ t^{0.8} + 0.0628342\ \hbar^5\ t^{0.8} +$$
$$0.00010995\ \hbar^2\ t^{1.6} + 0.0002199\ \hbar^3\ t^{1.6} +$$
$$0.000164925\ \hbar^4\ t^{1.6} +\ 0.00004398\ \hbar^5\ t^{1.6} + 2.12656 \times$$
$$10^{-8}\ \hbar^3\ t^{2.4} + 3.18984 \times\ 10^{-8}\ \hbar^4\ t^{2.4} + 1.27594 \times$$
$$10^{-8}\ \hbar^5\ t^{2.4} + 6.64605 \times 10^{-8}\ \hbar^4\ t^{3.2} +\ 5.31684 \times$$
$$10^{-10}\ \hbar^5\ t^{3.2} + 4.67073 \times 10^{-12}\ \hbar^5\ t^4 + 2.37316 \times$$
$$10^{-16}\ \hbar^6\ t^4 +\ 1.25742 \times 10^{-18}\ \hbar^6\ t^{4.8}. \tag{43}$$

AN OPTIMAL HOMOTOPY ANALYSIS APPROACH OF SOLUTIONS

According to [17, 46, 47], the homotopy terms depend on both the physical variable t and the convergence control parameter $\hbar$. The artificial parameter $\hbar$ can be freely chosen to adjust and control the interval of convergence, and even more, to increase the convergence at a reasonable rate, fortunately at the quickest rate. This concept plays a key role in the HAM and is generally used to gain sufficiently accurate approximations with the smallest number of homotopy terms in the homotopy series (18)-(21). In fact, the use of such an auxiliary parameter clearly distinguishes the HAM from other perturbation-like analytical techniques. According to convergence theorem in Section 3.1, the homotopy series solution contains the auxiliary parameter $\hbar$ which provides a simple way to adjust and control the convergence of the series (18)-(21)

Interval of Convergence and Optimal Value from an Appropriate Ratio

Let us consider $j + 1$ homotopy terms $T_0(t), T_1(t), ..., T_j(t),$ $T_{L,0}(t), T_{L,1}(t), ..., T_{L,j}(t),$ $T_{A,0}(t), T_{A,1}(t), ..., T_{A,j}(t)$ and $T_{M,0}(t), T_{M,1}(t), ..., T_{M,j}(t)$ of an homotopy series [17, 46, 47]

$$T(t) = T_0(t) + \sum_{j=1}^{\infty} T_j(t), \tag{44}$$

$$T_L(t) = T_{L,0}(t) + \sum_{j=1}^{\infty} T_{L,j}(t), \tag{45}$$

$$T_A(t) = T_{A,0}(t) + \sum_{j=1}^{\infty} T_{A,j}(t), \tag{46}$$

$$T_M(t) = T_{M,0}(t) + \sum_{j=1}^{\infty} T_{M,j}(t). \tag{47}$$

Therefore, for a preassigned value of parameter $\hbar$, the finite number of terms in homotopy series does not affect its convergence. It is sufficient to keep track of magnitudes of the ratio defined by

$$\left|\frac{T_j(t)}{T_{j-1}(t)}\right|, \qquad \left|\frac{T_{L,j}(t)}{T_{L,j-1}(t)}\right|, \qquad \left|\frac{T_{A,j}(t)}{T_{A,j-1}(t)}\right|, \qquad \left|\frac{T_{M,j}(t)}{T_{M,j-1}(t)}\right|, \tag{48}$$

and observing whether it remains less than unity for increasing values of j. The optimal value for the convergence control parameter $\hbar$ can be determined by taking (48) and requiring this ratio to be as close to zero as possible. For such a value, the rate of convergence of the homotopy series (44)-(47) will be the fastest (and as a consequence, the remainder of the series will rapidly decay). If the ratio is less than unity for a prescribed $\hbar$, then the convergence of HAM is guaranteed [17, 46, 47]. In other words, this is a sufficient condition for the convergence of the homotopy analysis method. This implies that in the cases where the limit for the ratio in (48) cannot be reached or tends to unity, the method may still converge or fail to do so. It is appropriate to search for an optimum value of $\hbar$, i.e., a value of $\hbar$ that gives rise to a ratio (48) as small as possible. Taking a time interval Ω, the ratios

$$\beta_T = \frac{\int_\Omega \left[T_j(t)\right]^2 dt}{\int_\Omega \left[T_{j-1}(t)\right]^2 dt},$$

$$\beta_{T_L} = \frac{\int_\Omega \left[T_{L,j}(t)\right]^2 dt}{\int_\Omega \left[T_{L,j-1}(t)\right]^2 dt},$$

$$\beta_{T_A} = \frac{\int_\Omega \left[T_{A,j}(t)\right]^2 dt}{\int_\Omega \left[T_{A,j-1}(t)\right]^2 dt},$$

$$\beta_{T_M} = \frac{\int_\Omega \left[T_{M,j}(t)\right]^2 dt}{\int_\Omega \left[T_{M,j-1}(t)\right]^2 dt},$$

represent a more convenient way of evaluating the convergence control parameter $\hbar$. In fact, given an order of approximation, the curves of ratio β_T, β_{T_L}, β_{T_A} and β_{T_M} versus $\hbar$ indicate not only the effective region for the convergence control parameter $\hbar$, but also the optimal value of $\hbar$ that corresponds to the minimum of β_T, β_{T_L}, β_{T_A} and β_{T_M}. Now, plotting β_T, β_{T_L}, β_{T_A} and β_{T_M} versus $\hbar$, as well as by solving [17, 46, 47].

$$\frac{\int_\Omega [T_j(t)]^2 dt}{\int_\Omega [T_{j-1}(t)]^2 dt} < 1, \qquad \frac{d\beta_T}{dt} = 0,$$

$$\frac{\int_\Omega [T_{L,j}(t)]^2 dt}{\int_\Omega [T_{L,j-1}(t)]^2 dt} < 1, \qquad \frac{d\beta_{T_L}}{dt} = 0,$$

$$\frac{\int_\Omega [T_{A,j}(t)]^2 dt}{\int_\Omega [T_{A,j-1}(t)]^2 dt} < 1, \qquad \frac{d\beta_{T_A}}{dt} = 0,$$

$$\frac{\int_\Omega [T_{M,j}(t)]^2 dt}{\int_\Omega [T_{M,j-1}(t)]^2 dt} < 1, \qquad \frac{d\beta_{T_M}}{dt} = 0.$$

The interval of convergence and the optimum value for parameter $\hbar$ can be simultaneously achieved. As an illustration at th order of approximations $M = 5$, the curves of ratio β versus $\hbar$, corresponding to $T(t)$, $T_L(t)$, $T_A(t)$ and $T_M(t)$ (β_T vs $\hbar_T$, β_{T_L} vs $\hbar_L$, β_{T_A} vs $\hbar_A$ and β_{T_M} vs $\hbar_M$ respectively), are displayed in Figs. (1-4) for $\alpha = 0.7, 0.8, 0.9, 1$. For better presentation, in Table 2, we exhibit these intervals of convergence of $\hbar_T$, $\hbar_L$, $\hbar_A$, and $\hbar_M$ for $\alpha = 0.7, 0.8, 0.9, 1$. It is to be noted that these valid regions ensure the convergence of the obtained series. This represents a central advantage in the study of the convergence of HAM.

Table 1. Interval of convergence of $\hbar_T, \hbar_L, \hbar_A,$ and $\hbar_M$ correspond to Figs. 1-4 for $\alpha = 0.7, 0.8, 0.9, 1$.

$\alpha/\hbar$	$\hbar_T$	$\hbar_L$	$\hbar_A$	$\hbar_M$
0.7	$(-1.71147, -0.24680)$	$(-1.71186, -0.29351)$	$(-0.98411, 0)$	$(-0.99440, 0)$
0.8	$(-1.72041, -0.24723)$	$(-1.72269, -0.29947)$	$(-1.87299, 0)$	$(-0.99567, 0)$
0.9	$(-1.72742, -0.24764)$	$(-1.73172, -0.30654)$	$(-1.89792, 0)$	$(-1.99972, 0)$
1	$(-1.72880, -0.24789)$	$(-1.73929, -0.31488)$	$(-1.91855, 0)$	$(-0.99485, 0)$

SQUARED RESIDUAL ERROR AND DIFFERENT ORDERS OF APPROXIMATION

A procedure to check the convergence of a homotopy-series solution is to substitute this series into the original governing equations and initial conditions, and then to evaluate the corresponding squared residual errors-the more quickly the residual error decays to zero, the faster the homotopy-series converges [17, 46, 47]. In this context, an error analysis is performed in the following lines. We substitute (36)-(39) or (40)-(43) into into HTLV-1 model (5) and obtain the residual functions as follows

$$ER_T(t; \hbar_T) = \frac{1}{\Gamma(1-\alpha)} \int_0^t \frac{\varphi_T'(s;\hbar_T)}{(t-s)^\alpha} \, ds - \Lambda + \mu_T \varphi_T(t; \hbar_T) + k\left[\varphi_T(t; \hbar_T)\varphi_{T_A}(t; \hbar_T) - \alpha_1 \varphi_T(t; \hbar_T)\varphi_{T_A}^2(t; \hbar_T)\right],$$

$$ER_L(t; \hbar_L) = \frac{1}{\Gamma(1-\alpha)} \int_0^t \frac{\varphi_{T_L}'(s;\hbar_L)}{(t-s)^\alpha} \, ds - k\left[\varphi_T(t; \hbar_L)\varphi_{T_A}(t; \hbar_L) - \alpha_1 \varphi_T(t; \hbar_L)\varphi_{T_A}^2(t; \hbar_L)\right],$$

$$ER_A(t; \hbar_A) = \frac{1}{\Gamma(1-\alpha)} \int_0^t \frac{\varphi_{T_A}'(s;\hbar_A)}{(t-s)^\alpha} \, ds - \alpha_L \varphi_{T_L}(t; \hbar_A) + (\mu_A + \rho)\varphi_{T_A}(t; \hbar_A),$$

$$ER_M(t; \hbar_M) = \frac{1}{\Gamma(1-\alpha)} \int_0^t \frac{\varphi_M'(s;\hbar_M)}{(t-s)^\alpha} \, ds - \rho \varphi_{T_A}(t; \hbar_M) - \beta \varphi_{T_M}(t; \hbar_M)\left(1 - \frac{\varphi_{T_M}(t;\hbar_M)}{T_{M_{max}}}\right) + \mu_M \varphi_{T_M}(t; \hbar_M).$$

Yabushita *et al.* [48], suggested an optimization method for convergence control parameters. Their work is based on the squared residual error. Inspired by their approach, and following the studies carried out in [17, 20, 46, 47, 49], we define the square residual error for the jth-order approximation to be:

$$RT(\hbar_T) = \int_0^1 \left(ER_1(t; \hbar_T)\right)^2 dt, \tag{49}$$

$$RT_L(\hbar_L) = \int_0^1 \left(ER_2(t; \hbar_L)\right)^2 dt, \tag{50}$$

$$RT_A(\hbar_A) = \int_0^1 \left(ER_3(t; \hbar_A)\right)^2 dt, \tag{51}$$

$$RT_M(\hbar_M) = \int_0^1 \left(ER_4(t; \hbar_M)\right)^2 dt. \tag{52}$$

Values of $\hbar_T$, $\hbar_L$, $\hbar_A$, and $\hbar_M$ for which the $RT(\hbar_T)$, $RT_L(\hbar_L)$, $RT_A(\hbar_A)$ and $RT_M(\hbar_M)$ are minimum can be obtained. Thus, we have

$$\frac{dRT(\hbar_T)}{d\hbar_T} = 0, \quad \frac{dRT_L(\hbar_L)}{d\hbar_L} = 0, \quad \frac{dRT_A(\hbar_A)}{d\hbar_A} = 0, \quad \frac{dRT_M(\hbar_M)}{d\hbar_M} = 0.$$

In Table 3, the optimal values of $\hbar_T^*$, $\hbar_L^*$, $\hbar_A^*$, and $\hbar_M^*$, the optimal values of β_T^*, $\beta_{T_L}^*$, $\beta_{T_A}^*$ and $\beta_{T_M}^*$, and also minimum values of square residual errors $RT(\hbar_T)$, $RT_L(\hbar_L)$, $RT_A(\hbar_A)$ and $RT_M(\hbar_M)$ are given for $\alpha = 0.7, 0.8, 0.9$ and 1.

Table 3. Optimum values of $\hbar^*, \beta^*$ and squared residual error of $RT(\hbar_T)$, $RT_L(\hbar_L)$, $RT_A(\hbar_A)$ and $RT_M(\hbar_M)$ for $\alpha = 0.7, 0.8, 0.9, 1$.

α	$\alpha = 0.7$	$\alpha = 0.8$	$\alpha = 0.9$	$\alpha = 1$
$\hbar_T^*$	-0.98754	-0.98985	-0.98988	-0.99810
β_T^*	0.000099133	0.000076619	0.00017559	0.00018257
$RT(\hbar_T^*)$	6.35405×10^{-13}	6.22654×10^{-14}	7.72065×10^{-14}	6.64329×10^{-14}
$\hbar_L^*$	-0.99053	-0.99484	-0.99699	-0.99712
$\beta_{T_L}^*$	0.000043749	0.00019953	0.00023211	0.000097520
$RT(\hbar_L^*)$	7.78521×10^{-12}	1.88658×10^{-12}	4.10119×10^{-13}	6.03224×10^{-13}
$\hbar_A^*$	-0.94075	-0.92721	-0.89893	-0.88921
$\beta_{T_A}^*$	0.016255	0.018157	0.026277	0.025778
$RT(\hbar_A^*)$	4.26417×10^{-16}	5.68864×10^{-16}	2.00311×10^{-15}	2.21428×10^{-15}
$\hbar_M^*$	-0.89571	-0.87707	-1.10576	-0.89913
$\beta_{T_M}^*$	0.010849	0.015084	0.011221	0.010157

(Table 3) cont.....

$RT(\hbar_M^*)$	5.13745×10^{-13}	2.66940×10^{-12}	6.08894×10^{-13}	3.69101×10^{-13}

In Tables **4-7**, the values of absolute residual errors $ER_T(t;\hbar_T^*)$, $ER_L(t;\hbar_L^*)$, $ER_A(t;\hbar_A^*)$ and $ER_M(t;\hbar_M^*)$ have been calculated for various $t \in (0,1)$, and $\alpha = 0.7, 0.8, 0.9, 1$, respectively. From the tables, it can be seen that the HAM provides us an accurate approximate solution for the nonlinear system of fractional-order HTLV-1 model (5).

Table 4. The absolute residual errors $ER_T(t;\hbar_T^*)$, $ER_L(t;\hbar_L^*)$, $ER_A(t;\hbar_A^*)$ and $ER_M(t;\hbar_M^*)$ for various $t \in (0,1)$ and $\alpha = 0.7$.

t/ER	ER_T	ER_L	ER_A	ER_M
0.1	1.40568×10^{-7}	1.64291×10^{-8}	1.79329×10^{-11}	7.88474×10^{-13}
0.3	2.55308×10^{-7}	1.78756×10^{-7}	2.08339×10^{-10}	7.03368×10^{-13}
0.5	4.31284×10^{-8}	7.63818×10^{-7}	5.33124×10^{-10}	4.38955×10^{-13}
0.7	2.08654×10^{-7}	2.24500×10^{-6}	6.04791×10^{-10}	2.37116×10^{-13}
0.9	4.92499×10^{-9}	5.24975×10^{-6}	2.22593×10^{-10}	1.61573×10^{-12}

Table 5. The absolute residual errors $ER_T(t;\hbar_T^*)$, $ER_L(t;\hbar_L^*)$, $ER_A(t;\hbar_A^*)$ and $ER_M(t;\hbar_M^*)$ for various $t \in (0,1)$ and $\alpha = 0.8$.

t/ER	ER_T	ER_L	ER_A	ER_M
0.1	9.726698×10^{-8}	5.26442×10^{-9}	9.23163×10^{-12}	3.45584×10^{-13}
0.3	3.32035×10^{-7}	6.11915×10^{-8}	6.89925×10^{-11}	3.13791×10^{-13}
0.5	3.38280×10^{-7}	2.85546×10^{-7}	2.23422×10^{-10}	2.14843×10^{-13}

(Table 5) cont.....

0.7	8.30641×10^{-8}	9.41453×10^{-7}	3.07631×10^{-10}	7.85111×10^{-14}
0.9	2.49539×10^{-7}	2.45619×10^{-6}	6.77032×10^{-11}	7.63205×10^{-13}

Table 6. The absolute residual errors $ER_T(t; \hbar_T^*)$, $ER_L(t; \hbar_L^*)$, $ER_A(t; \hbar_A^*)$ and $ER_M(t; \hbar_M^*)$ for various $t \in (0, 1)$ and $\alpha = 0.9$.

t/ER	ER_T	ER_L	ER_A	ER_M
0.1	5.30462×10^{-8}	1.98656×10^{-9}	9.79494×10^{-13}	1.44235×10^{-13}
0.3	2.57851×10^{-7}	2.32115×10^{-8}	2.12529×10^{-11}	1.32053×10^{-13}
0.5	3.57289×10^{-7}	1.08841×10^{-7}	8.67243×10^{-11}	9.65749×10^{-14}
0.7	1.96415×10^{-7}	3.85266×10^{-7}	1.43762×10^{-10}	2.24320×10^{-14}
0.9	2.47724×10^{-7}	1.10146×10^{-6}	1.14618×10^{-11}	3.39213×10^{-13}

Table 7. The absolute residual errors $ER_T(t; \hbar_T^*)$, $ER_L(t; \hbar_L^*)$, $ER_A(t; \hbar_A^*)$ and $ER_M(t; \hbar_M^*)$ for various $t \in (0, 1)$ and $\alpha = 0.1$.

t/ER	ER_T	ER_L	ER_A	ER_M
0.1	3.53409×10^{-13}	1.58084×10^{-12}	7.89000×10^{-17}	1.24702×10^{-12}
0.3	2.56657×10^{-11}	6.42202×10^{-11}	5.62552×10^{-14}	1.27237×10^{-12}
0.5	1.90173×10^{-10}	5.41522×10^{-10}	1.20571×10^{-12}	1.30146×10^{-12}
0.7	3.13427×10^{-10}	2.13897×10^{-9}	9.07842×10^{-12}	1.34092×10^{-12}
0.9	1.48048×10^{-9}	5.06836×10^{-9}	9.10088×10^{-11}	1.40390×10^{-12}

The square residual errors $RT(\hbar_T)$, $RT_L(\hbar_L)$, $RT_A(\hbar_A)$ and $RT_M(\hbar_M)$ (defined in (49)-(52)) have been plotted in Fig. (**5**) for various $\hbar$. Whereas Figs. (**6-9**) give plots of β-HAM solutions for $T(t)$, $T_L(t)$, $T_A(t)$, $T_M(t)$, at time t with respect to $\hbar$

for $t = 0.3, 0.6$ and 1 respectively. By considering these figures, it can be concluded that the solution obtained by using HAM gives an analytical solution with high order of accuracy involving only few iterations.

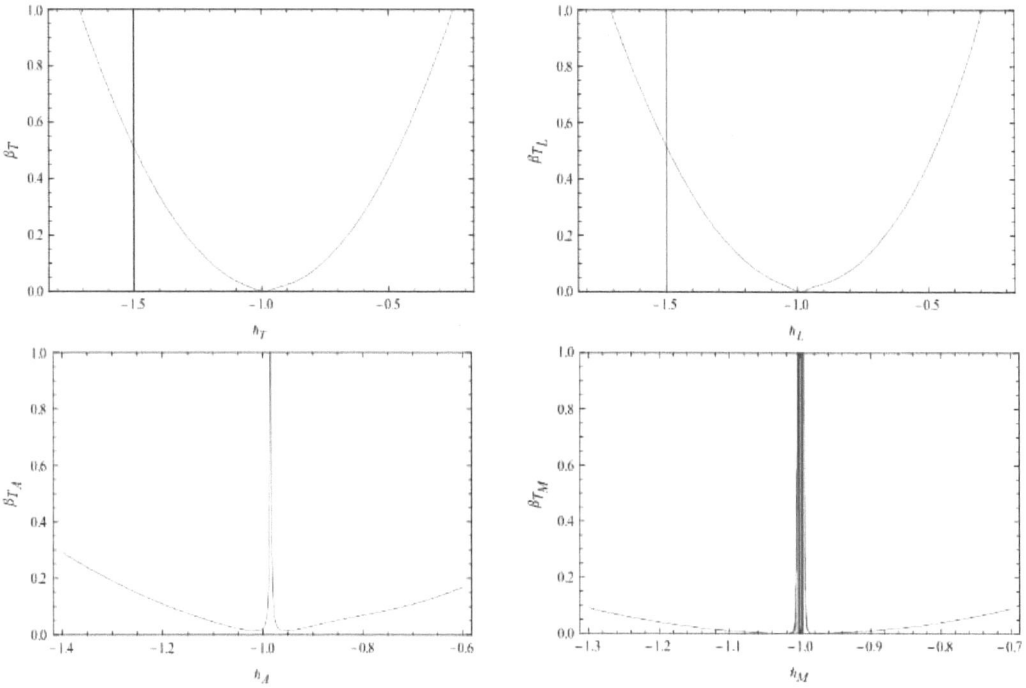

Fig. (1). (From left to right) The curves of ratios β_T, β_{T_L}, β_{T_A}, β_{T_M} versus $\hbar_T$, $\hbar_L$, $\hbar_A$, and $\hbar_M$ respectively, corresponding to 6th-order approximation of solutions $T(t)$, $T_L(t)$, $T_A(t)$ and $T_M(t)$ for $\alpha = 0.7$. The optimum value of $\hbar$ and $\hbar^*$ gives rise to the minimum values of β-curves.

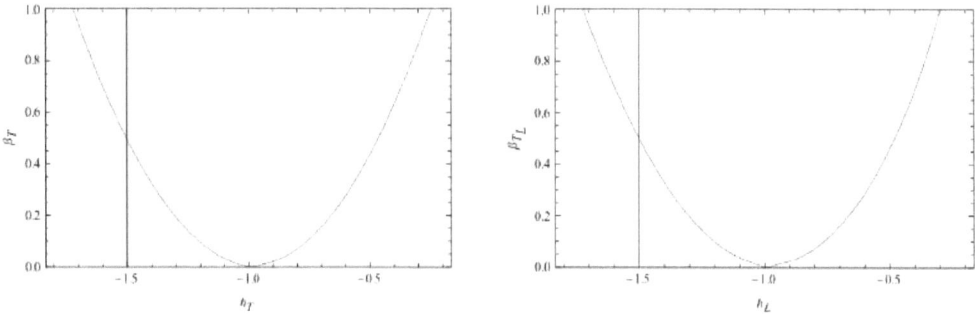

(Fig. 2) contd.....

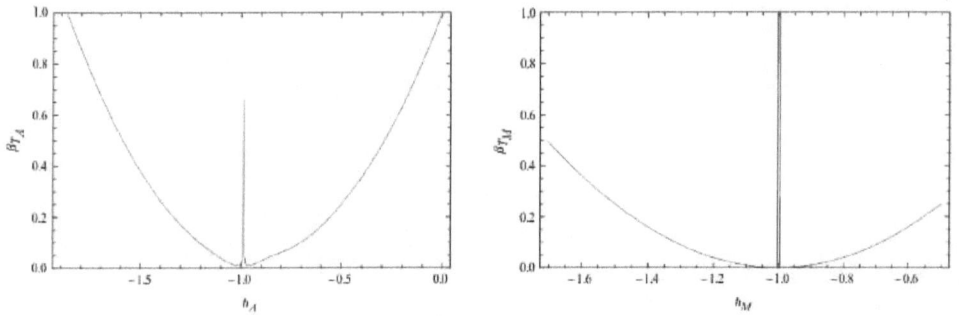

Fig. (2). (From left to right) The curves of ratios β_T, β_{T_L}, β_{T_A}, β_{T_M} versus $\hbar_T$, $\hbar_L$, $\hbar_A$, and $\hbar_M$ respectively, corresponding to 6th-order approximation of solutions $T(t)$, $T_L(t)$, $T_A(t)$ and $T_M(t)$ for $\alpha = 0.8$. The optimum value of $\hbar$ and $\hbar^*$gives rise to the minimum values of β-curves.

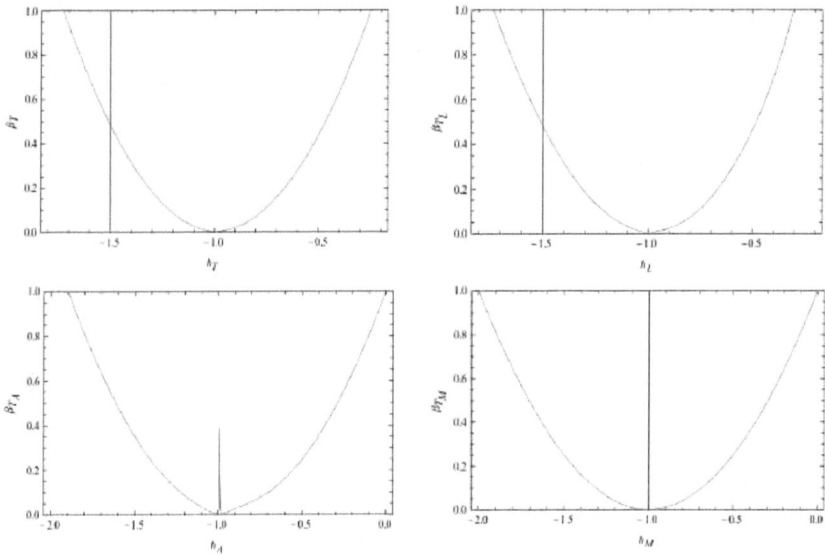

Fig. (3). (From left to right) The curves of ratios β_T, β_{T_L}, β_{T_A}, β_{T_M} versus $\hbar_T$, $\hbar_L$, $\hbar_A$, and $\hbar_M$ respectively, corresponding to 6th-order approximation of solutions $T(t)$, $T_L(t)$, $T_A(t)$ and $T_M(t)$ for $\alpha = 0.9$. The optimum value of $\hbar$ and $\hbar^*$gives rise to the minimum values of β-curves.

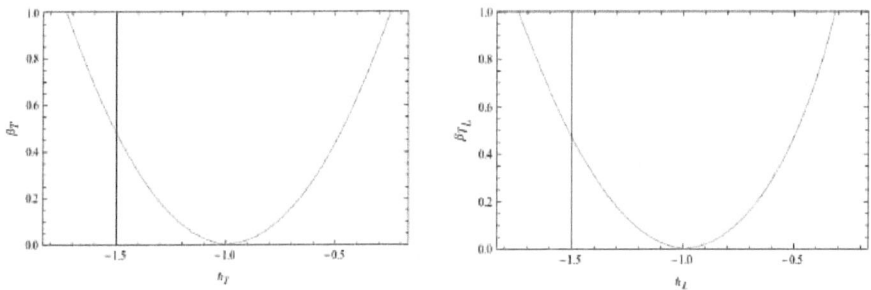

(Fig. 4) contd.....

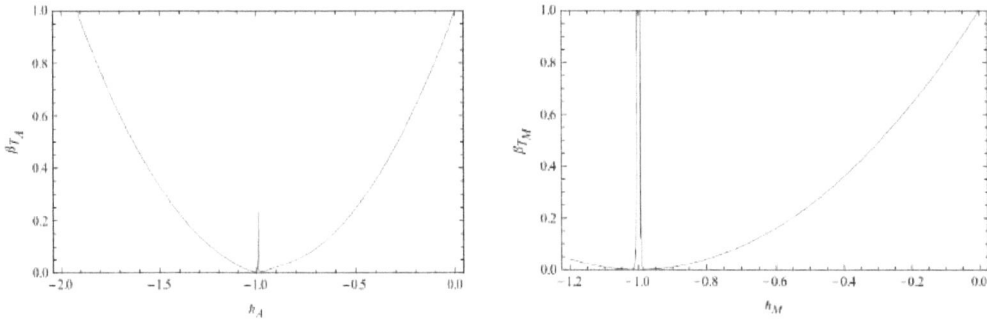

Fig. (4). (From left to right) The curves of ratios β_T, β_{T_L}, β_{T_A}, β_{T_M} versus $\hbar_T$, $\hbar_L$, $\hbar_A$, and $\hbar_M$ respectively, corresponding to 6th-order approximation of solutions $T(t)$, $T_L(t)$, $T_A(t)$ and $T_M(t)$ for $\alpha = 1$. The optimum value of $\hbar$ and $\hbar^*$ gives rise to the minimum values of β-curves.

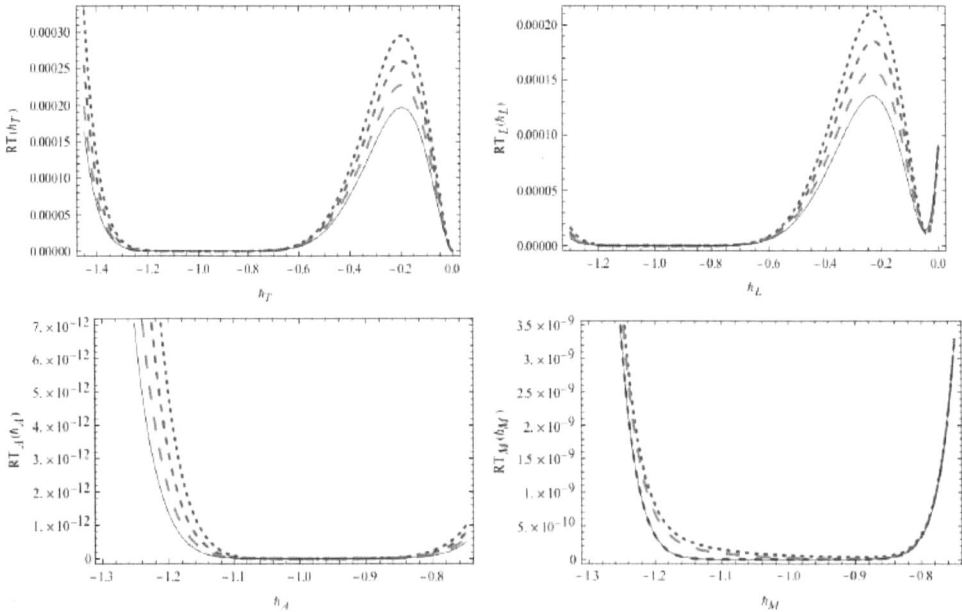

Fig. (5). Squared residual error functions RT, RT_L, RT_A and RT_M versus $\hbar_T$, $\hbar_L$, $\hbar_A$ and $\hbar_M$ respectively for $\alpha = 0.7, 0.8, 0.9$ and 1. These functions correspond to approximate solution of $T(t)$, $T_L(t)$, $T_A(t)$ and $T_M(t)$ by using β-HAM. In these figures, dotted small line corresponds to $\alpha = 0.7$, dotted medium line corresponds to $\alpha = 0.8$, dotted large line corresponds to $\alpha = 0.9$ and solid black line corresponds to $\alpha = 1$. Each optimum values $\hbar_T^*$, $\hbar_L^*$, $\hbar_A^*$ and $\hbar_M^*$, rise minimum values of the $RT(\hbar_T^*)$, $RT_L(\hbar_L^*)$, $RT_A(\hbar_A^*)$ and $RT_M(\hbar_M^*)$.

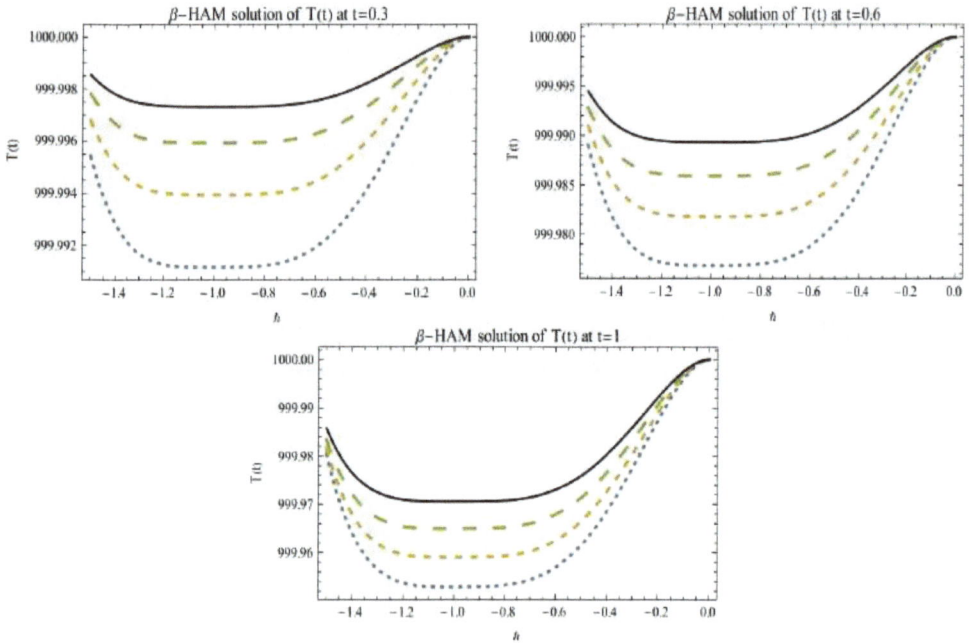

Fig. (6). (From left to right) Plot of β-HAM solution for $T(t)$ that shows concenteration of healthy CD4$^+$ T-cells at time t with respect to $\hbar$ for $t = 0.3, 0.6$ and 1. In these figures, dotted small line corresponds to $\alpha = 0.7$, dotted medium line corresponds to $\alpha = 0.8$, dotted large line corresponds to $\alpha = 0.9$ and solid black line corresponds to $\alpha = 1$.

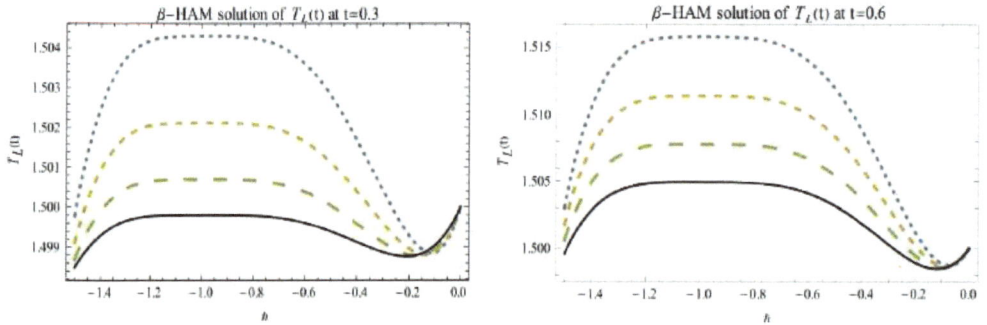

(Fig. 7) contd.....

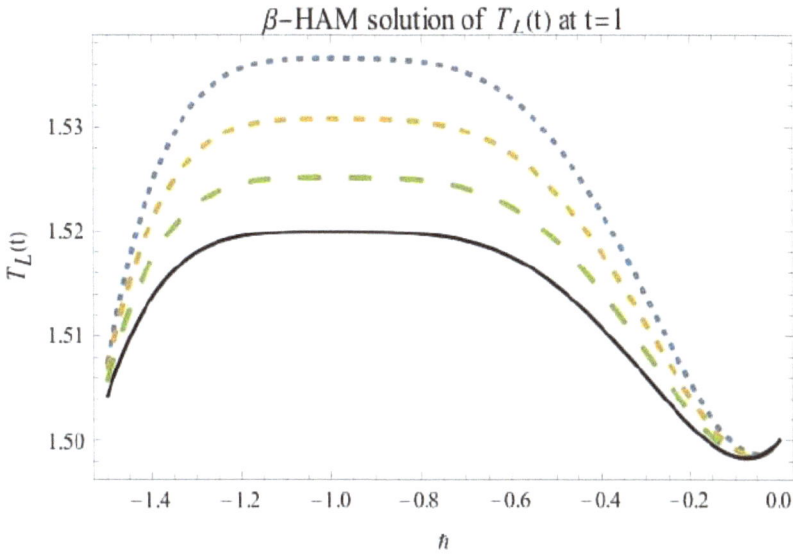

Fig. (7). (From left to right) Plot of β-HAM solution for $T_L(t)$ that shows lately infected CD4$^+$ T-cells at time t with respect to $\hbar$ for $t = 0.3, 0.6$ and 1. In these figures, dotted small line corresponds to $\alpha = 0.7$, dotted medium line corresponds to $\alpha = 0.8$, dotted large line corresponds to $\alpha = 0.9$ and solid black line corresponds to $\alpha = 1$.

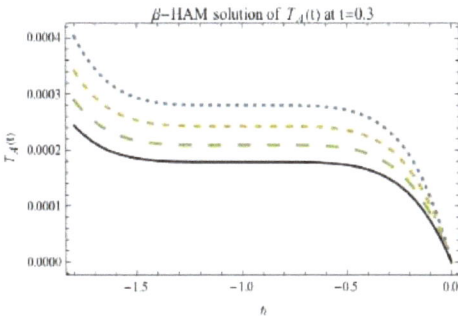

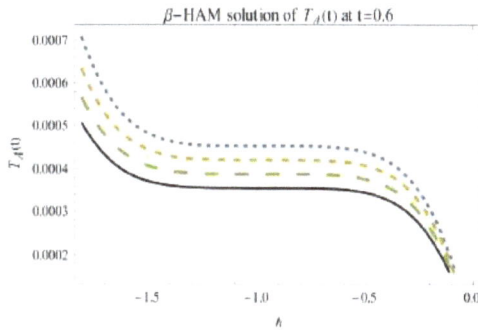

(Fig. 7) contd.....

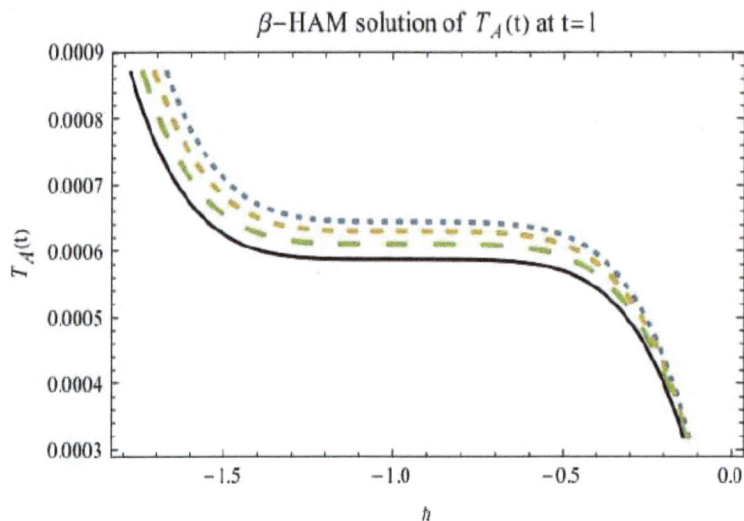

Fig. (8). (From left to right) Plot of β-HAM solution for $T_A(t)$ that shows actively infected CD4$^+$ T-cells at time t with respect to $\hbar$ for $t = 0.3, 0.6$ and 1. In these figures, dotted small line corresponds to $\alpha = 0.7$, dotted medium line corresponds to $\alpha = 0.8$, dotted large line corresponds to $\alpha = 0.9$ and solid black line corresponds to $\alpha = 1$.

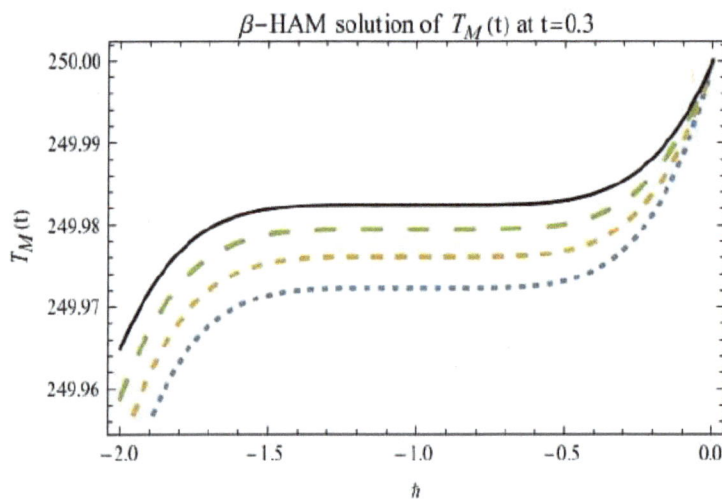

(Fig. 9) contd.....

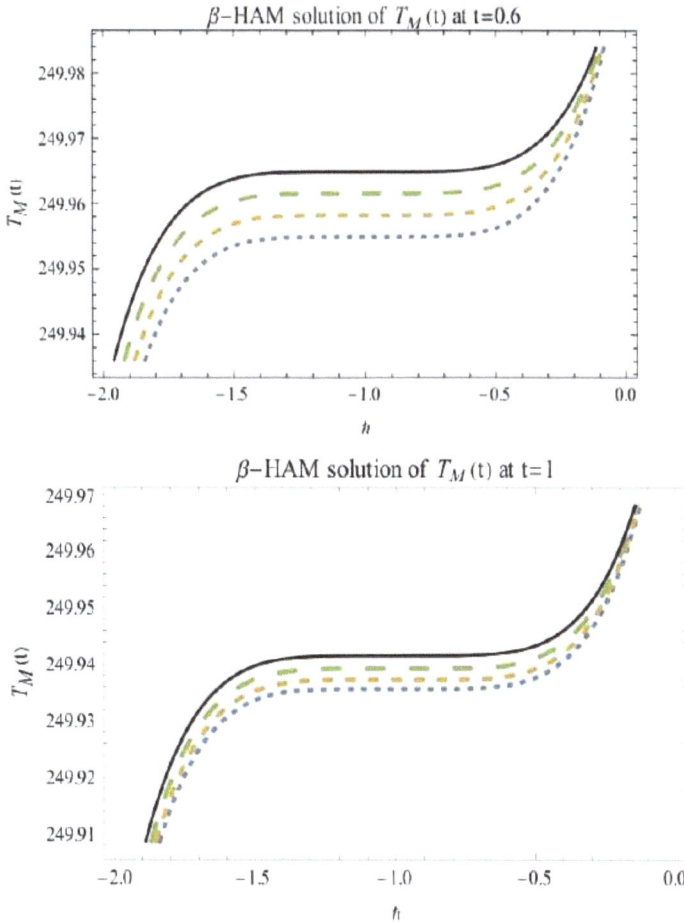

Fig. (9). (From left to right) Plot of β-HAM solution for $T_M(t)$ that shows leukemia T-cells at time t with respect to $\hbar$ for $t = 0.3, 0.6$ and 1. In these figures, dotted small line corresponds to $\alpha = 0.7$, dotted medium line corresponds to $\alpha = 0.8$, dotted large line corresponds to $\alpha = 0.9$ and solid black line corresponds to $\alpha = 1$.

CONCLUSION

This chapter introduced the use of homotopy analysis method (HAM) for solving nonlinear fractional-order of Human T-cells Lymphotropic Virus-1 (HTLV-1) infection of CD4$^+$ T-cells. We first modified the system of ordinary differential equations proposed by Stiliankis and Seydel [8]. We then developed HAM to describe the solution of modified system obtained in the form of infinite series. It is important to note that this method has some auxiliary parameters and functions. One of these parameters is the convergence control parameter $\hbar$ which can be applied to adjust and control the convergence region of obtained series solutions. Thus, by plotting several β-curves and finding the regions of convergence, we showed the advantages and abilities of the HAM. Another advantage of HAM over the perturbation methods is that it does not depend on small or large parameters. The main advantages of using HAM as a solution technique is its reliability, easy to handle, that utilizes a simple process to adjust and control the convergence region of the obtained infinite series solution. It uses an auxiliary parameter and allows to obtain a one-parametric family of explicit series solutions. Although, the HAM on Caputo nonlinear fractional-order HTLV-1 infection model provides sufficient information to comprehend the epidemic transmission process and helps to determine the crucial factors for its spread, but for more detailed analysis, one needs to have new tools to disclose unnoticed behavior of such nonlinear epidemiological systems and thus the operators known as Caputo-Fabrizio, Atangana-Baleanu, Atangana-Gomez, Atangana beta-derivative, truncated M-derivative, fractal-fractional and others can be used in future research work. The convergence theorem of HAM to the HTLV-1 was also proved in the present chapter to demonstrate the efficiency of the method. The residual and absolute errors were applied to show the efficiency and accuracy of the method. The results obtained show that the HAM is an accurate and effective technique for obtaining the approximate solution of the proposed nonlinear fractional-order of (HTLV-1) infection of CD4$^+$ T-cells.

CONSENT FOR PUBLICATION

Not applicable.

CONFLICT OF INTEREST

The authors declare that they have no conflict of interest.

ACKNOWLEDGEMENTS

This study was supported by the Research Fund for International Scientists (RFIS), National Natural Science Foundation of China (Grant No. 12150410306) and the China Postdoctoral Science Foundation (Grant No. 2019M663653). The funding body did not play any role in the design of the study and in writing the chapter.

REFERENCES

[1]   R. F. Edlich, J. A. Arnette, and F. M. Williams, "Global epidemic of human T-cell lymphotropic virus type-I (HTLV-1)", J. Emerg. Med., vol. 18, no. 1, pp. 109-119, 2000.

[2]   P. Katri, S. Ruan, "Dynamics of human T-cell lymphotropic virus I (HTLV-1) infection of CD4$^+$ T-cells", C. R. Biologies, vol. 327, no. 11, pp. 1009-1016, 2004.

[3]   P. A. Naik, J. Zu, and K. M. Owolabi, "Modeling the mechanics of viral kinetics under immune control during primary infection of HIV-1 with treatment in fractional-order", Physica A, vol. 545, pp. 123816, 2020.

[4]   P. A. Naik, M. Yavuz, and J. Zu, "The role of prostitution on HIV transmission with memory: A modeling approach", Alexandria Eng. J., vol. 59, no. 4, pp. 2513-2531, 2020.

[5]   P. A. Naik, J. Zu, and K. M. Owolabi, "Global dynamics of a fractional-order model for the transmission of HIV epidemic with optimal control", Chaos Soliton. Fract., vol. 138, pp. 109826, 2020.

[6]   P. A. Naik, K. M. Owolabi, M. Yavuz, and J. Zu, "Chaotic dynamics of a fractional-order HIV-1 model involving AIDS-related cancer cells", Chaos Soliton. Fract., vol. 140, pp. 110272, 2020.

[7]   X. Song and Y. Li, "Global stability and periodic solution of a model for HTLV-1 infection and ATL progression", Appl. Math. Comput., vol. 180, pp. 401-410, 2006.

[8]   N. I. Stilianakis and J. Seydel, "Modeling the T-cell dynamics and pathogenesis of HTLV-1 infection", Bull. Math. Biol., vol. 61, pp. 935-947, 1999.

[9]   L. Wang, M. Y. Li, and D. Kirschner, "Mathematical analysis of the global dynamics of a model for HTLV-1 infection and ATL progression", Math. Biosci., vol. 179, pp. 207-217, 2002.

[10]   L. Cai, X. Li, and M. Ghosh, "Global dynamics of a mathematical model for HTLV-1 infection of CD4$^+$ T-cells", Appl. Math. Model., vol. 35, pp. 3587-359, 2011.

[11]   Z. U. A. Zafar, K. Rehan, and M. Mushtaq, "HIV/AIDS epidemic fractional-order model", J. Diff. Eq. Appl., vol. 23, no. 1, pp. 1-19, 2017.

[12]   A. Atangana and D. Baleanu, "New fractional derivatives with nonlocal and nonsingular kernel: theory and applications to heat transfer model", Ther. Sci., vol. 20, no. 2, pp. 763-769, 2016.

[13]   A. Atangana and I. Koca, "Chaos in a simple nonlinear system with Atangana-Baleanu derivatives with fractional-order", Chaos Soliton. Fract., vol. 89, pp. 447-454, 2016.

[14]   A. Atangana, "Fractal-fractional differentiation and integration: connecting fractal calculus and fractional calculus to predict complex system", Chaos Soliton. Fract., vol. 102, pp. 396-406, 2017.

[15]   S. Arshad, D. Baleanu, W. Bu, and Y. Tang, "Effect of HIV infection on CD4$^+$ T-cell population based on a fractional-order model", Adv. Diff. Eq., vol. 92, pp. 1-14, 2017.

[16]   A. H. Ganie, "Some new approach of spaces of non-integral order", J. Nonlinear Sci. Appl., vol. 14, pp. 89-96, 2021.

[17]   P. A. Naik, J. Zu, and M. Ghoreishi, "Estimating the approximate analytical solution of HIV viral dynamic model by using homotopy analysis method", Chaos Soliton. Fract., vol. 131, pp. 109500, 2020.

[18]   M. H. Khani, J. Rashidinia, and S. Z. Borujeni, "Application of different h(t) in homotopy analysis methods for solving systems of linear equations", Adv. Linear Alg. Matris Theor., vol. 5, pp. 129-137, 2015.

[19]   D. Matignon, "Stability results for fractional differential equations with applications to control processing, computational engineering in systems and application", Multiconference, IMACS, IEEE-SMC, IEEE Xplore, Lille, France, vol. 2, pp. 963-968, 1996.

[20]   M. Ghoreishi, A. I. B. M. Ismail, A. K. Alomari, "Application of homotopy analysis method for solving a model for HIV infection of CD4$^+$ T-cells", Math. Comput. Model., vol. 54, pp. 3007-3015, 2011.

[21]   S. J. Liao, "Beyond Perturbation: Introduction to the Homotopy Analysis Method", CRC Press, Boca Raton,Chapman and Hall, 2003.

[22]   M. Ghoreishi, A. I. B. M. Ismail, and A. Rashid, "On the convergence of the homotopy analysis method for inner-resonance of tangent nonlinear cushioning packaging system with critical components", Abst. Appl. Anal., vol. 2013, pp. 424510, 2013.

[23]   S. J. Liao, "Comparison between the homotopy analysis method and homotopy perturbation method", Appl. Math. Comput., vol. 169, pp. 1186-1194, 2005.

[24]   M. Ghoreishi, A. I. B. M. Ismail, A. K. Alomari, and A. S. Bataineh, "The comparison between homotopy analysis method and optimal homotopy asymptotic method for nonlinear age-structured population models", Commun. Nonlinear Sci. Numer. Simul., vol. 17, no. 3, pp. 1163-1177, 2012.

[25]   M. Ghoreishi, A. I. B. M. Ismail, and A. K. Alomari, "Comparison between homotopy analysis method and optimal homotopy asymptotic method for nth-order integro-differential equation", Math. Methods Appl. Sci., vol. 34, pp. 1833-1842, 2011.

[26]   M. Ghoreishi, A. I. B. M. Ismail, and A. Rashid, "Solution of a strongly coupled reaction-diffusion system by the homotopy analysis method", Bull. Belg. Math. Soc. Simon Stevin, vol. 18, pp. 471-481, 2011.

[27]   P. A. Naik, J. Zu, and M. Ghoreishi, "Stability analysis and approximate solution of SIR epidemic model with Crowley-Martin type functional response and holling type-II treatment rate by using homotopy analysis method", J. Appl. Anal. Comput., vol. 10, no. 4, pp. 1482-1515, 2020.

[28]   S. Noeiaghdam, M. Suleman, and H. Budak, "Solving a modified nonlinear epidemiological model of computer viruses by homotopy analysis method", Math. Sci., vol. 12, pp. 211-222, 2018.

[29]   P. A. Naik, K. Owolabi, J. Zu, and M. Naik, "Modeling the transmission dynamics of COVID-19 pandemic in Caputo type fractional derivative", J. Multiscale Model., vol. 12, no. 3, pp. 2150006, 2021.

[30]   S. Noeiaghdam, "A novel technique to solve the modified epidemiological model of computer viruses", SeMA J., vol. 76, pp. 97-108, 2019.

[31]   S. Noeiaghdam, M. A. F. Araghi, and S. Abbasbandy, "Valid implementation of Sinc-collocation method to solve the fuzzy Fredholm integral equation" J. Comput. Appl. Math., vol. 370, pp. 112632, 2019.

[32]   R. M. Jena, S. Chakraverty, and S. K. Jena, "Dynamic response analysis of fractionally damped beams subjected to external loads using homotopy analysis method", J. Appl. Comput. Mech., vol. 5, no. 2, pp. 355-366, 2019.

[33]   S. Maitama and W. Zhao, "Local fractional homotopy analysis method for solving non-differentiable problems on Cantor sets", Adv. Diff. Eq., vol. 2019, pp. 127, 2019.

[34]   H. Buluta, D. Kumar, J. Singh, R. Swroop, and H. M. Baskonus, "Analytic study for a fractional model of HIV infection of CD4$^+$ T lymphocyte cells", Math. Nat. Sci., vol. 2, pp. 33-43, 2018.

[35]   P. Veeresha, D. G. Prakasha, and H. M. Baskonus, "Solving smoking epidemic model of fractional-order using a modifed homotopy analysis transform method", Math. Sci., vol. 13, pp. 115-128, 2019.

[36]   S. Alao, R. A. Oderinu, F. O. Akinpelu, and E. I. Akinola, "Homotopy analysis decomposition method for the solution of viscous boundary layer flow due to a moving Sheet", J. Adv. Math. Comput. Sci., vol. 32, no. 5, pp. 1-7, 2019.

[37]   S. Aljhani, M. S. M. Noorani, and A. K. Alomari, "Numerical solution of fractional-order hiv model using homotopy method", Discrete Dyn. Nat. Soc., vol. 2020, pp. 2149037, 2020.

[38]   Y. H. Qiang, Y. H. Qian, and X. Y. Guo, "Periodic solutions of delay nonlinear system by multi-frequency homotopy analysis method", J. Low Freq. Noise V. A., vol 38, no. 3, pp. 1-16, 2019.

[39]   A. Demir, M. A. Bayrak, and E. Ozbilge, "A new approach for the approximate analytical solution of space-time fractional differential equations by the homotopy analysis method", Adv. Math. Phy., vol. 2019, pp. 5602565, 2019.

[40]   J. Xie and M. Yi, "Numerical research of nonlinear system of fractional Volterra–Fredholm integral–differential equations via Block-Pulse functions and error analysis", J. Comput. Appl. Math., vol. 345, pp. 159-167, 2019.

[41]   K. S. Miller and B. Ross, "An Introduction to the Fractional Calculus and Fractional Differential Equations", John Wiley and Sons, New York, 1993.

[42]   P. A. Naik, M. Yavuz, S. Qureshi, J. Zu, and S. Townley, "Modeling and analysis of COVID-19 epidemics with treatment in fractional derivatives using real data from Pakistan", Eur. Phys. J. Plus, vol. 135, no. 10, pp. 795, 2020.

[43]   I. Podlubny, "Fractional Differential Equations", Academic Press, SanDiego, 1999.

[44]   M. Shateri and D. D. Ganji, "Solitary wave solutions for a time-fraction generalized hirota-satsuma coupled kdv equation by a new analytical technique", Inter. J. Diff. Eq., vol. 2010, pp. 954674, 2010.

[45]   S. Yanga, A. Xiao, and H. Sua, "Convergence of the variational iteration method for solving multi-order fractional differential equations", Comput. Math. Appl., vol. 60, pp. 2871-2879, 2010.

[46]   J. Duarte, C. Januario, N. Martins, C. Ramos, C. Rodrigues, and J. Sardanyes, "Optimal homotopy analysis of a chaotic HIV-1 model incorporating AIDS-related cancer cells", Numer. Algor., vol. 77, pp. 261-288, 2018.

[47]   S. J. Liao, "Advances in the Homotopy Analysis Method", World Scientific, 2014.

[48]   K. Yabushita, M. Yamashita, and K. Tsuboi, "An analytical solution of projectile motion with the quadratic resistance law using the homotopy analysis method", J. Phy. A: Math. Theor., vol. 40, no. 29, pp. 8403-8416, 2007.

[49]   S. J. Liao, "An optiomal homotopy-analysis approach for strongly nonlinear differential equation", Commun. Nonlinear Sci. Numer. Simulat., vol. 15, pp. 2003-2016, 2010.

[50]   P. Veeresha, "A numerical approach to the coupled atmospheric ocean model using a fractional operator", Math. Model. Numer. Simul. Appl., vol. 1, no. 1, pp. 1-10, 2021.

[51]   Z. Hammouch, M. Yavuz, and N. Özdemir, "Numerical solutions and synchronization of a variable-order fractional chaotic system", Math. Model. Numer. Simul. Appl., vol. 1, no. 1, pp. 11-23, 2021.

CHAPTER 7

Behavior Analysis and Asymptotic Stability of the Traveling Wave Solution of the Kaup-Kupershmidt Equation for Conformable Derivative

Hülya Durur[1], Asıf Yokuş[2]* and Mehmet Yavuz[3]

[1]*Department of Computer Engineering, Faculty of Engineering, Ardahan University, Ardahan 75000, Turkey*

[2]*Department of Mathematics, Faculty of Science, Firat University, Elazig 23100, Turkey*

[3]*Department of Mathematics and Computer Sciences, Necmettin Erbakan University, 42090 Konya, Turkey*

Abstract: This article suggests solving the traveling wave solutions of the time-fractional Kaup-Kupershmidt (KK) equation *via* $1/G'$-expansion and sub-equation methods. Non-local fractional derivatives have some advantages over local fractional derivatives. The most important of these advantages are the chain rule and the Leibniz rule. The conformable derivative, which has a local fractional derivative feature, is taken into account in this study. Different types of traveling wave solutions of the time-fractional KK equation have been produced by using the important benefits of the time-dependent conformable derivative operator. These wave types are dark, singular, rational, trigonometric and hyperbolic type solitons. 2D, 3D and contour graphics are presented by giving arbitrary values to the constants in the solutions produced by analytical methods. These presented graphs represent the shape of the standing wave at any given moment. Besides, the advantages and disadvantages of the two analytical methods are discussed and presented in the result and discussion section. In addition, wave behavior analysis for different velocity values of the dark soliton produced by the analytical method is analyzed by simulation. The conditional convergence and asymptotic stability of the dark soliton discussed are analyzed. Computer software is also used in operations such as drawing graphs, complex operations, and solving algebraic equation systems.

Keywords: $1/G'$-expansion method, Asymptotic stability, Conformable derivative, Sub-equation method, Time-fractional Kaup-Kupershmidt equation.

*Corresponding author Asıf Yokuş: Department of Mathematics, Faculty of Science, Firat University, Elazig 23100, Turkey; E-mail: asfyokus@yahoo.com

Mehmet Yavuz & Necati Özdemir (Eds.)

INTRODUCTION

Recently, a number of works on nonlinear partial differential equations (NLPDEs) have been raised, as they may be employed in many disciplines, including engineering, physical, chemical and biological sciences. Most of these works have focused on attaining analytical solutions for fractional PDEs. However, fractional derivative definitions such as Caputo and Riemann-Liouville, are not always capable of reaching analytic solutions, for they do not meet some basic principles of known integer order derivatives. Some fractional derivatives are impossible to solve with these definitions.

Though the fractional derivative originated in the 17th century, interest in this subject has increased in recent years, and many studies have been made on this subject. This is because physical systems are often referred to as fractional derivatives. In the literature, several studies related to fractional derivatives, has continued to increase. Various fractional derivative definitions have been made from the 1730s to this time. Recently, Khalil et al. presented a simple, understandable and intriguing definition of the fractional derivative called the congruent fractional derivative [1].

Definition: For $t > 0$ and $\alpha \in (0,1]$, an α-th order "conformable derivative" of a function is defined by (Khalil et al. 2014) as

$$T_\alpha(f)(t) = \lim_{\varepsilon \to \infty} \frac{f\left(t + \varepsilon t^{1-\alpha}\right) - f(t)}{\varepsilon},$$

for $f : [0,\infty) \to R$.

Theorem: Let $\alpha \in (0,1]$, $t > 0$ and g, f be α-differentiable functions. Then

a) $T_\alpha(dg + cf) = dT_\alpha(g) + cT_\alpha(f)$, for all $d, c \in R$.

b) $T_\alpha(t^p) = pt^{p-\alpha}$ for all $p \in R$.

c) $T_\alpha(\lambda) = 0$ for all constant functions $f(t) = \lambda$.

d) $T_\alpha(gf) = gT_\alpha(f) + f\,T_\alpha(g)$.

e) $T_\alpha\left(\dfrac{f}{g}\right) = \dfrac{gT_\alpha(g) - fT_\alpha(f)}{g^2}$.

f) If f is a differentiable function, then $T_\alpha(f)(t) = t^{1-\alpha} \dfrac{df}{dt}$.

Kaup has proposed first the significant diffuse classical KK equation [2] and was later presented by Kupershmidt [3]. This paper is about the investigation of the KK equation. The KK equation is used to study the operation of behavioral capillary gravitational waves and nonlinear scattered waves.

The fifth order NLPDE is as follows:

$$D_t^\alpha u(x,t) + ruu_{xxx} + bpu_x u_{xx} + cu^2 u_x + u_{xxxxx} = 0, \tag{1}$$

where c, b and r are real constants, $0 < \alpha \leq 1$ which is the parameter representing the order of fractional time derivative. We write Eq. (1) for $c=45$, $b=-15$ and $r=-15$, in the form below [4]

$$D_t^\alpha u(x,t) - 15uu_{xxx} - 15pu_x u_{xx} + 45u^2 u_x + u_{xxxxx} = 0. \tag{2}$$

Recently, great research based on the work of the classic KK equation have been done. The classical KK equation is integrable at p = 5/2 [5] and is known to have bilinear representations [4].

As a result of the effects of surface tension on phase velocity, a capillary wave is formed that travels along the phase boundary of a liquid. Besides, a longer wavelength occurring on the surface of the fluid will cause the formation of gravity-capillary waves that are affected by both the surface tension and the effect of gravity and the fluid property. As is known, the modeling of physical events is done with differential equations. Obviously, solutions of differential equations play an important role in illuminating physical phenomena. In this study, we consider the time-fractional KK equation, which is used in the modeling of capillary waves and gravity capillary waves, which have an important place physically. If the constants in the solutions we have presented gain physical meaning, it will be much more valuable.

There are many studies in the literature regarding the time-fractional KK equation. For example; the time-fractional KK equation has been solved via 2-D Legendre multiwavelet method [6], Lie point symmetries of the time-fractional KK equation are found and its invariant solutions are determined with the help of infinitesimal generators [7], with the help of extended G'/G-expansion and improved G'/G-

expansion methods, some exact solutions of the time-fractional KK equation have been made [8], the similarity reduction and Lie point symmetries of the time-fractional KK equation were obtained, and then conservation laws were established by Ibragimov's method [9].

Many effective schemes have been proposed in the literature to produce analytical solutions of nonlinear wave models. Some of these are Hirota bilinear method [10,11], $(1/G')$-expansion method [12-15], F-expansion method [16], Clarkson–Kruskal direct method [17], $(m+1/G')$-expansion method [18], the sinh-Gordon function method [19], Haar wavelet method [20,21] and so on [22-40].

In this study, $1/G'$-expansion and sub-equation methods, which are important instruments in mathematics to produce analytical solutions, are used. Besides, it will be discussed that both analytical methods are effective, reliable and useful in generating analytical solutions.

DESCRIPTION OF THE SUB EQUATION METHOD

In this chapter, we define the description of this method [41]. We consider the nonlinear fractional PDE

$$T\left(D_t^\alpha u, u, u_x, u_{xx}, \ldots\right) = 0. \tag{3}$$

$u(x,t)$ is any function and D_t^α depends on time conformable derivative $u(x,t)$. This method to be announced in steps:

Step 1: First, *via* the wave transmutation, we attain the following equalities.

$$u = u(\xi) = u(x,t), \qquad \xi = kx + v\frac{t^\alpha}{\alpha},$$

where, k, v are scalars to be examined after and $\alpha \in (0,1]$. Eq. (3) may be written *via* hain rule as the following ODE [42]:

$$G\left(u, u', u'', \ldots\right) = 0, \tag{4}$$

wherein the prime shows derivative known by ξ.

Step 2: Suppose that Eq. (4) has a solution as follows

$$u(\xi) = \sum_{i=0}^{n} a_i \phi^i(\xi), \quad a_n \neq 0, \tag{5}$$

where $a_i \, (0 \leq i \leq n)$ are scalars to be determined. n represents a positive integer to find by using the balancing procedure in Eq. (4) and $\phi(\xi)$ gratifies the ODE below

$$\phi'(\xi) = \mu + \left(\phi(\xi) \right)^2, \tag{6}$$

here μ is a scalar. In the formulas below, some special solutions are given for Eq. (6).

$$\phi(\xi) = \begin{cases} -\sqrt{-\mu} \tanh\left(\sqrt{-\mu}\xi\right), & \mu < 0 \\ -\sqrt{-\mu} \coth\left(\sqrt{-\mu}\xi\right), & \mu < 0 \\ \sqrt{\mu} \tan\left(\sqrt{\mu}\xi\right), & \mu > 0 \\ -\sqrt{\mu} \cot\left(\sqrt{\mu}\xi\right), & \mu > 0 \\ -\dfrac{1}{\xi + W}, & \mu = 0 \ (W \text{ is a constant}). \end{cases} \tag{7}$$

Step 3: Eqs. (5) and (6) are replaced into Eq. (4) and the coefficients $\phi^i(\xi)$ are equal to zero. This procedure $a_i \, (i = 0,1,\ldots,n)$ gives a nonlinear algebraic system.

Step 4: The constants in Eq. (5) are calculated with the help of ready-made package programs of the nonlinear algebraic system obtained in the third step. These calculated constants are written in the appropriate places. Then, the wave transform is operated in reverse to arrive at the traveling wave solution of Eq. (3).

DESCRIPTION OF THE $1/G'$-EXPANSION METHOD

Consider the form of NLPDEs,

$$W\left(u, \frac{\partial^\alpha u}{\partial t^\alpha}, \frac{\partial u}{\partial x}, \frac{\partial^2 u}{\partial x^2}, \ldots\right) = 0. \tag{8}$$

Let $u = u(x,t) = u(\xi)$, $\xi = kx + v\dfrac{t^\alpha}{\alpha}$, $v \neq 0$, here v is the velocity of the wave and constant. We may transform it the following nODE for $u(\xi)$:

$$\theta(u, u', u'', \ldots) = 0. \tag{9}$$

The solution of Eq. (9) is supposed that with the form

$$u(\xi) = a_0 + \sum_{i=1}^{n} a_i \left(\frac{1}{G'}\right)^i, \tag{10}$$

where a_i, $i = \{1, \ldots, n\}$ are scalars, $G = G(\xi)$ ensures following second-order IODE

$$G'' + \lambda G' + \tau = 0, \tag{11}$$

here τ and λ are constants to be determined,

$$\frac{1}{G'(\xi)} = \frac{1}{-\dfrac{\tau}{\lambda} + A\cosh[\xi\lambda] - A\sinh[\xi\lambda]}, \tag{12}$$

where A is the integral constant. The wanted derivatives of Eq. (10) are calculated and written into Eq. (9), obtaining a polynomial with $(1/G')$. Equating the coefficients of this polynomial to zero, an algebraic system of equations is created. The Equation is solved via the package program and the default Eq. (9) is put in its place in the solution function. Eventually, the solutions of Eq. (8) are found.

APPLICATION OF SUB EQUATION METHOD

We consider Eq. (2). By using $u(x,t) = u(\xi) = u$, $\xi = kx + v\dfrac{t^\alpha}{\alpha}$, Eq. (2) is transformed into an ODE

$$vu' + 45ku^2u' - 15k^3pu'u'' - 15k^3uu^{(3)} + k^5u^{(5)} = 0. \tag{13}$$

Once these Eq. (13) are integrated concerning ξ, we can obtain the following equation

$$vu + 15ku^3 + \frac{15}{2}k^3u'^2 - \frac{15}{2}k^3pu'^2 - 15k^3uu'' + k^5u^{(4)} = 0. \tag{14}$$

In Eq. (14), integral constant is taken as zero, we get the balancing term $n = 2$ and by considering in Eq. (5),

$$u(\xi) = a_0 + a_1\phi(\xi) + a_2\left(\phi(\xi)\right)^2, \tag{15}$$

if Eq. (15) is written in Eq. (14) and if necessary adjustments are made, the following systems of equations can be written:

$$
\begin{aligned}
\left(\phi(\xi)\right)^0: \quad & va_0 + 15ka_0^3 + \frac{15}{2}k^3\mu^2a_1^2 - \frac{15}{2}k^3p\mu^2a_1^2 + 16k^5\mu^3a_2 - 30k^3\mu^2a_0a_2 = 0, \\
\left(\phi(\xi)\right)^1: \quad & va_1 + 16k^5\mu^2a_1 - 30k^3\mu a_0a_1 + 45ka_0^2a_1 - 30k^3p\mu^2a_1a_2 = 0, \\
\left(\phi(\xi)\right)^2: \quad & -15k^3\mu a_1^2 - 15k^3p\mu a_1^2 + 45ka_0a_1^2 + va_2 + 136k^5\mu^2a_2 - 120k^3\mu a_0a_2 \\
& + 45ka_0^2a_2 - 30k^3p\mu^2a_2^2 = 0, \\
\left(\phi(\xi)\right)^3: \quad & 40k^5\mu a_1 - 30k^3a_0a_1 + 15ka_1^3 - 90k^3\mu a_1a_2 - 60k^3p\mu a_1a_2 + 90ka_0a_1a_2 = 0, \\
\left(\phi(\xi)\right)^4: \quad & -\frac{45}{2}k^3a_1^2 - \frac{15}{2}k^3pa_1^2 + 240k^5\mu a_2 - 90k^3a_0a_2 + 45ka_1^2a_2 - 60k^3\mu a_2^2 \\
& - 60k^3p\mu a_2^2 + 45ka_0a_2^2 = 0, \\
\left(\phi(\xi)\right)^5: \quad & 24k^5a_1 - 90k^3a_1a_2 - 30k^3pa_1a_2 + 45ka_1a_2^2 = 0, \\
\left(\phi(\xi)\right)^6: \quad & 120k^5a_2 - 60k^3a_2^2 - 30k^3pa_2^2 + 15ka_2^3 = 0.
\end{aligned} \tag{16}
$$

a_0, a_1, a_2 and v, k, p, μ, constants are obtained from Eq. (16) the system utilizing a software program.

Case 1. If

$$a_0 = \frac{1}{15}\left(15k^2\mu + \sqrt{105}k^2\mu\right), \quad a_1 = 0, \quad a_2 = 2k^2, \quad v = 2\left(-11k^5\mu^2 + \sqrt{105}k^5\mu^2\right), \quad p = 1, \tag{17}$$

replacing values in Eq. (17) into Eq. (15), we get dark soliton for Eq. (2)

$$u_1(x,t) = \frac{1}{15}\left(15k^2\mu + \sqrt{105}k^2\mu\right) - 2k^2\mu \tanh\left[\sqrt{-\mu}\left(kx + \frac{2t^\alpha\left(-11k^5\mu^2 + \sqrt{105}k^5\mu^2\right)}{\alpha}\right)\right]^2. \tag{18}$$

The solution presented with Eq. (18) is known as dark soliton in the literature.

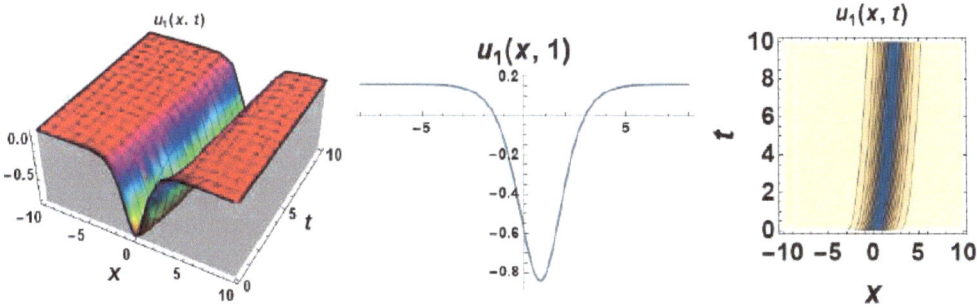

Fig. (1). 3-D, 2-D and contour graphs of Eq. (18) for $\alpha = 0.5, k = 1, \mu = -0.5$.

The graph presented in Fig. (**1**) is a dark soliton representing a standing wave at any given time.

Case 2: If

$$a_0 = \frac{1}{15}\left(15k^2\mu + \sqrt{105}k^2\mu\right), \quad a_1 = 0, \quad a_2 = 2k^2, \quad v = 2\left(-11k^5\mu^2 + \sqrt{105}k^5\mu^2\right), \quad p = 1, \tag{19}$$

replacing values in Eq. (19) into Eq. (15), we get singular soliton for Eq. (2)

$$u_2(x,t) = \frac{1}{15}\left(15k^2\mu + \sqrt{105}k^2\mu\right) - 2k^2\mu \coth\left[\sqrt{-\mu}\left(kx + \frac{2t^\alpha\left(-11k^5\mu^2 + \sqrt{105}k^5\mu^2\right)}{\alpha}\right)\right]^2. \tag{20}$$

The solution presented with Eq. (20) is known as singular soliton in the literature.

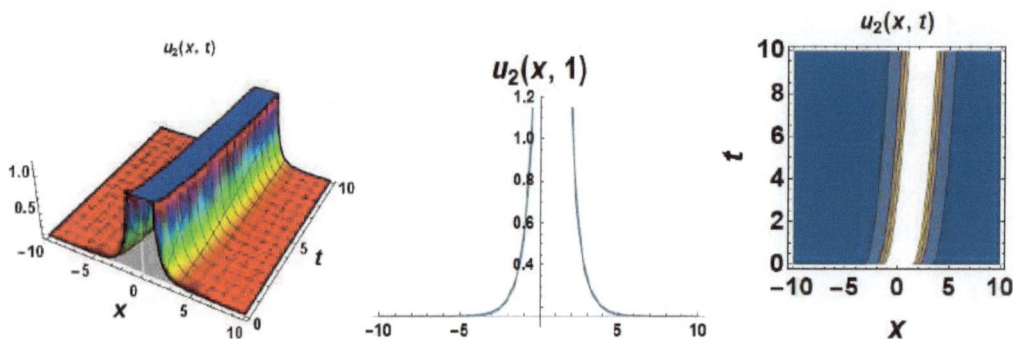

Fig. (2). 3-D, 2-D and contour graphs of Eq. (20) for $\alpha = 0.5$, $k = 1$, $\mu = -0.5$.

The graph presented in Fig. (2) is a singular soliton representing a standing wave at any given time

Case 3: If

$$a_0 = \frac{1}{15}\left(15k^2\mu + \sqrt{105}k^2\mu\right), \quad a_1 = 0, \quad a_2 = 2k^2, \quad v = 2\left(-11k^5\mu^2 + \sqrt{105}k^5\mu^2\right), \quad p = 1, \tag{21}$$

replacing values in Eq. (21) into Eq. (15), we get trigonometric oscillating traveling wave solution for Eq. (2)

$$u_3(x,t) = \frac{1}{15}\left(15k^2\mu + \sqrt{105}k^2\mu\right) + 2k^2\mu \tan\left[\sqrt{\mu}\left(kx + \frac{2t^\alpha\left(-11k^5\mu^2 + \sqrt{105}k^5\mu^2\right)}{\alpha}\right)\right]^2. \tag{22}$$

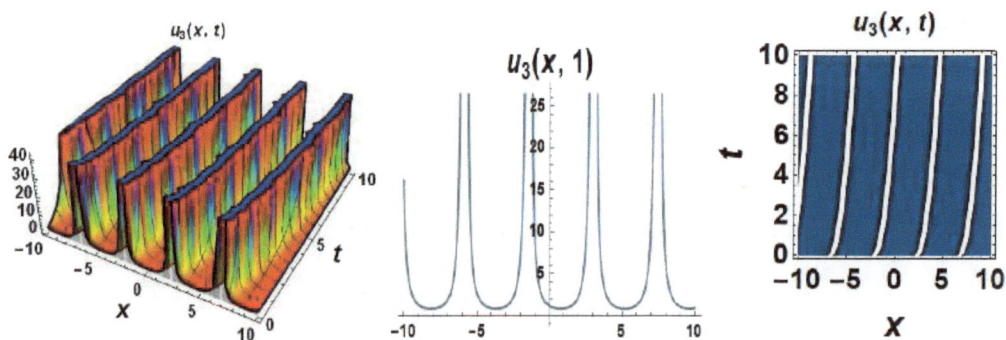

Fig. (3). 3-D, 2-D and contour graphs of Eq. (22) for $\alpha = 0.5$, $k = 1$, $\mu = 0.5$.

Case 4: If

$$a_0 = \frac{1}{15}\left(15k^2\mu + \sqrt{105}k^2\mu\right), \quad a_1 = 0, \quad a_2 = 2k^2, \quad v = 2\left(-11k^5\mu^2 + \sqrt{105}k^5\mu^2\right), \quad p = 1, \qquad \textbf{(23)}$$

replacing values in Eq. (23) into Eq. (15), we get trigonometric oscillating traveling wave solution for Eq. (2)

$$u_4(x,t) = \frac{1}{15}\left(15k^2\mu + \sqrt{105}k^2\mu\right) + 2k^2\mu\cot\left[\sqrt{\mu}\left(kx + \frac{2t^\alpha\left(-11k^5\mu^2 + \sqrt{105}k^5\mu^2\right)}{\alpha}\right)\right]^2. \qquad \textbf{(24)}$$

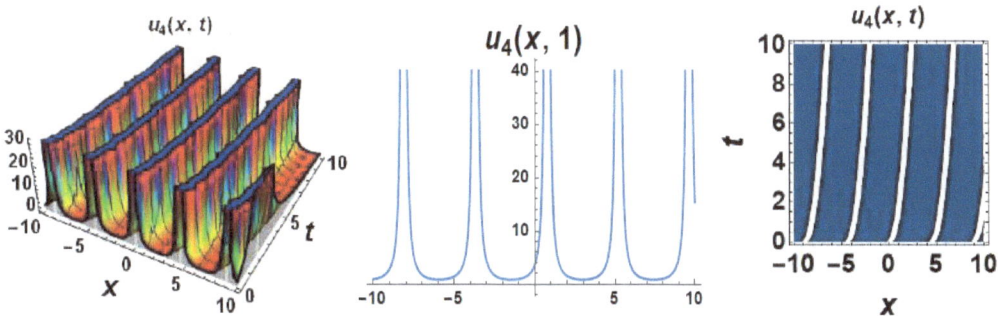

Fig. (4). 3-D, 2-D and contour graphs of Eq. (24) for $\alpha = 0.5, k = 1, \mu = 0.5$.

The graphics presented in Figs. **(3)** and **(4)** are the trigonometric oscillating traveling wave solution representing the stationary wave at any time.

Case 5: If

$$\mu = 0, \quad a_0 = \frac{1}{15}\left(15k^2\mu + \sqrt{105}k^2\mu\right), \quad a_1 = 0, \quad a_2 = 2k^2, \quad v = 2\left(-11k^5\mu^2 + \sqrt{105}k^5\mu^2\right), \quad p = 1, \qquad \textbf{(25)}$$

replacing values in Eq. (25) into Eq. (15), we get rational solution for Eq. (2)

$$u_5(x,t) = \frac{2k^2}{\left(W + kx\right)^2}. \qquad \textbf{(26)}$$

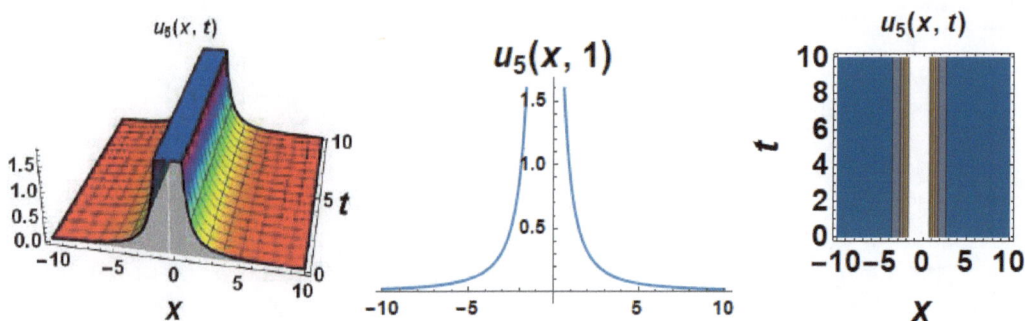

Fig. (5). 3-D, 2-D and contour graphs of Eq. (26) for $W = 0.5, k = 1, \mu = 0$.

The graphic presented in Fig. (**5**) is the rational solution representing the stationary wave at any time, which in this graphic contains singular point.

Case 6: If

$$a_0 = 2k^2\mu, \quad a_1 = 0, \quad a_2 = 2k^2, \quad v = -16k^5\mu^2, \quad p = 1, \tag{27}$$

replacing values in Eq. (27) into Eq. (15), we get dark soliton for Eq. (2)

$$u_6(x,t) = 2k^2\mu - 2k^2\mu \tanh\left[\sqrt{-\mu}\left(kx - \frac{16k^5 t^\alpha \mu^2}{\alpha}\right)\right]^2. \tag{28}$$

The solution presented with Eq. (28) is known as dark soliton in the literature.

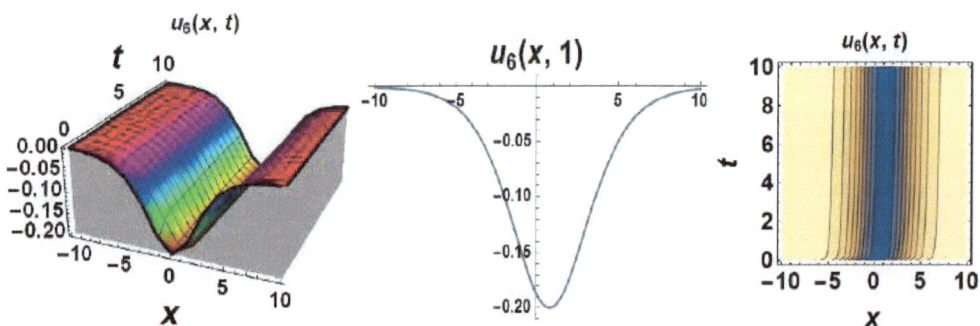

Fig. (6). 3-D, 2-D and contour graphs of Eq. (28) for $\alpha = 0.2, k = 1, \mu = -0.1$.

The graph presented in Fig. (**6**) is a dark soliton representing a standing wave at any given time.

Case 7: If

$$a_0 = 2k^2\mu, \quad a_1 = 0, \quad a_2 = 2k^2, \quad v = -16k^5\mu^2, \quad p = 1, \tag{29}$$

replacing values in Eq. (29) into Eq. (15), we get singular soliton for Eq. (2)

$$u_7(x,t) = 2k^2\mu - 2k^2\mu \coth\left[\sqrt{-\mu}\left(kx - \frac{16k^5 t^\alpha \mu^2}{\alpha}\right)\right]^2. \tag{30}$$

The solution presented with Eq. (30) is known as singular soliton in the literature.

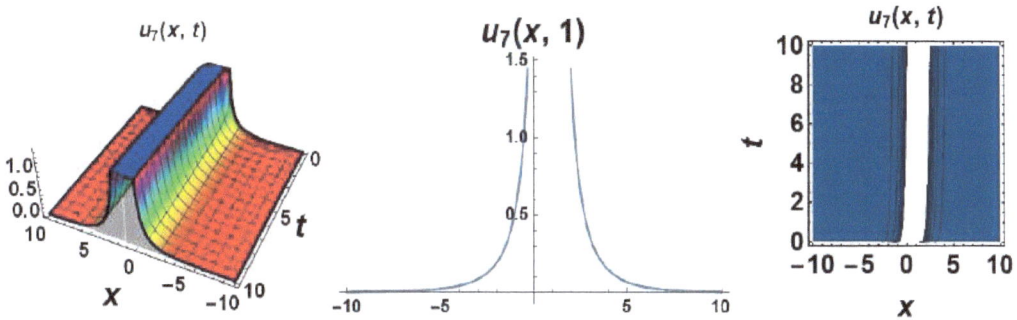

Fig. (7). 3-D, 2-D and contour graphs of Eq. (30) for $\alpha = 0.2$, $k = 1$, $\mu = -0.1$.

The graph presented in Fig. (**7**) is a singular soliton representing a standing wave at any given time.

Case 8: If

$$a_0 = 2k^2\mu, \quad a_1 = 0, \quad a_2 = 2k^2, \quad v = -16k^5\mu^2, \quad p = 1, \tag{31}$$

replacing values in Eq. (31) into Eq. (15), we get trigonometric oscillating traveling wave solution for Eq. (2)

$$u_8(x,t) = 2k^2\mu + 2k^2\mu \tan\left[\sqrt{\mu}\left(kx - \frac{16k^5 t^\alpha \mu^2}{\alpha}\right)\right]^2. \tag{32}$$

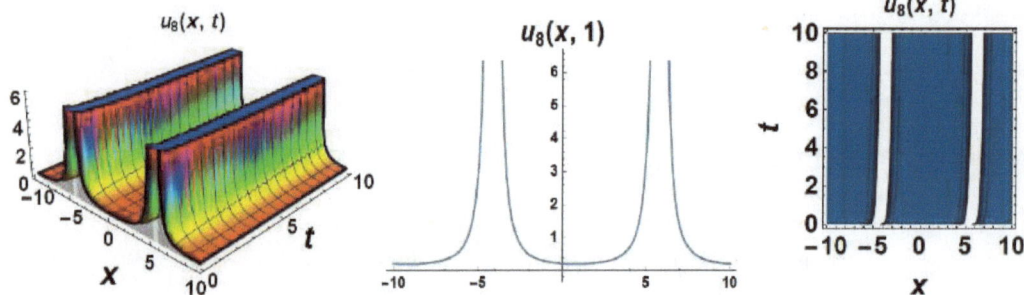

Fig. (8). 3-D, 2-D and contour graphs of Eq. (32) for $\alpha = 0.2$, $k = 1$, $\mu = 0.1$.

Case 9: If

$$a_0 = 2k^2\mu, \quad a_1 = 0, \quad a_2 = 2k^2, \quad v = -16k^5\mu^2, \quad p = 1, \tag{33}$$

replacing values in Eq. (33) into Eq. (15), we get trigonometric oscillating traveling wave solution for Eq. (2)

$$u_9(x,t) = 2k^2\mu + 2k^2\mu\cot\left[\sqrt{\mu}\left(kx - \frac{16k^5t^\alpha\mu^2}{\alpha}\right)\right]^2. \tag{34}$$

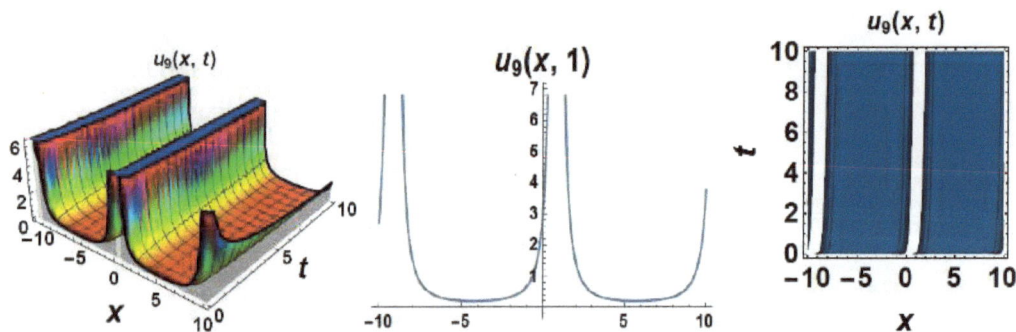

Fig. (9). 3-D, 2-D and contour graphs of Eq. (34) for $\alpha = 0.2$, $k = 1$, $\mu = 0.1$.

The graphics presented in Figs. **(8)** and **(9)** are the trigonometric oscillating traveling wave solution representing the stationary wave at any time.

APPLICATION OF THE $\left(\dfrac{1}{G'}\right)$-EXPANSION METHOD

Let us take Eq. (14) here, $n = 2$ is obtained according to the homogeneous balance principle. In Eq. (10), the following situation is obtained:

$$u(\xi) = a_0 + a_1\left(\frac{1}{G'}\right) + a_2\left(\frac{1}{G'}\right)^2, \quad a_1 \neq 0 \quad or \quad a_2 \neq 0. \tag{35}$$

Substituting Eq. (35) in Eq. (14) and the coefficients of Eq. (2) are set to zero, the following systems of equations can be written:

$$\left(\frac{1}{G'[\xi]}\right)^0 : \quad va_0 + 15ka_0^3 = 0,$$

$$\left(\frac{1}{G'[\xi]}\right)^1 : \quad va_1 + k^5\lambda^4 a_1 - 15k^3\lambda^2 a_0 a_1 + 45ka_0^2 a_1 = 0,$$

$$\left(\frac{1}{G'[\xi]}\right)^2 : \quad 15k^5\lambda^3\tau a_1 - 45k^3\lambda\tau a_0 a_1 - \frac{15}{2}k^3\lambda^2 a_1^2 - \frac{15}{2}k^3 p\lambda^2 a_1^2$$
$$+ 45ka_0 a_1^2 + va_2 + 16k^5\lambda^4 a_2 - 60k^3\lambda^2 a_0 a_2 + 45ka_0^2 a_2 = 0,$$

$$\left(\frac{1}{G'[\xi]}\right)^3 : \quad 50k^5\lambda^2\tau^2 a_1 - 30k^3\tau^2 a_0 a_1 - 30k^3\lambda\tau a_1^2 - 15k^3 p\lambda\tau a_1^2$$
$$+ 15ka_1^3 + 130k^5\lambda^3\tau a_2 - 150k^3\lambda\tau a_0 a_2 - 45k^3\lambda^2 a_1 a_2$$
$$- 30k^3 p\lambda^2 a_1 a_2 + 90ka_0 a_1 a_2 = 0,$$

$$\left(\frac{1}{G'[\xi]}\right)^4 : \quad 60k^5\lambda\tau^3 a_1 - \frac{45}{2}k^3\tau^2 a_1^2 - \frac{15}{2}k^3 p\tau^2 a_1^2 + 330k^5\lambda^2\tau^2 a_2$$
$$- 90k^3\tau^2 a_0 a_2 - 135k^3\lambda\tau a_1 a_2 - 60k^3 p\lambda\tau a_1 a_2$$
$$+ 45ka_1^2 a_2 - 30k^3\lambda^2 a_2^2 - 30k^3 p\lambda^2 a_2^2 + 45ka_0 a_2^2 = 0,$$

$$\left(\frac{1}{G'[\xi]}\right)^5 : \quad 24k^5\tau^4 a_1 + 336k^5\lambda\tau^3 a_2 - 90k^3\tau^2 a_1 a_2 - 30k^3 p\tau^2 a_1 a_2$$
$$- 90k^3\lambda\tau a_2^2 - 60k^3 p\lambda\tau a_2^2 + 45ka_1 a_2^2 = 0,$$

$$\left(\frac{1}{G'[\xi]}\right)^6 : \quad 120k^5\tau^4 a_2 - 60k^3\tau^2 a_2^2 - 30k^3 p\tau^2 a_2^2 + 15ka_2^3 = 0.$$

Case1. If

$$a_0 = \frac{1}{60}\left(15k^2\lambda^2 - \sqrt{105}k^2\lambda^2\right), \quad a_1 = 2k^2\lambda\tau, \quad a_2 = 2k^2\tau^2, \quad v = \frac{1}{8}\left(-11k^5\lambda^4 + \sqrt{105}k^5\lambda^4\right), \quad p = 1, \tag{36}$$

modifying values Eq. (36) in Eq. (35) and we have the hyperbolic type traveling wave solution for Eq. (2):

$$u_1(x,t) = \frac{1}{60}\left(15k^2\lambda^2 - \sqrt{105}k^2\lambda^2\right) + \tag{37}$$

$$\frac{2k^2\tau^2}{\left(-\dfrac{\tau}{\lambda} + A\cosh\left[\lambda\left(kx + \dfrac{t^\alpha\left(-11k^5\lambda^4 + \sqrt{105}k^5\lambda^4\right)}{8\alpha}\right)\right] - A\sinh\left[\lambda\left(kx + \dfrac{t^\alpha\left(-11k^5\lambda^4 + \sqrt{105}k^5\lambda^4\right)}{8\alpha}\right)\right]\right)^2} +$$

$$\frac{2k^2\lambda\tau}{\left(-\dfrac{\tau}{\lambda} + A\cosh\left[\lambda\left(kx + \dfrac{t^\alpha\left(-11k^5\lambda^4 + \sqrt{105}k^5\lambda^4\right)}{8\alpha}\right)\right] - A\sinh\left[\lambda\left(kx + \dfrac{t^\alpha\left(-11k^5\lambda^4 + \sqrt{105}k^5\lambda^4\right)}{8\alpha}\right)\right]\right)}.$$

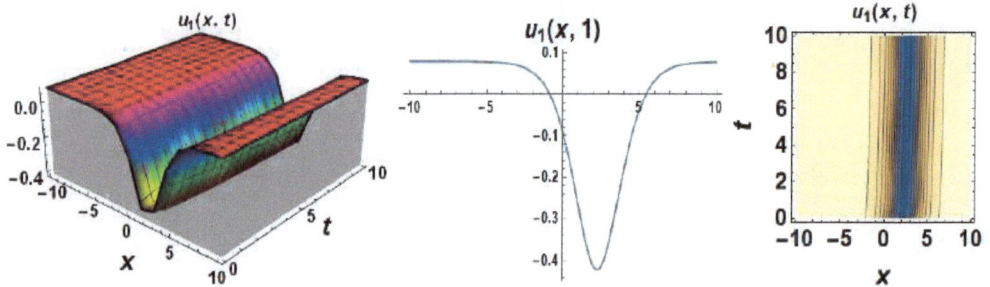

Fig. (10). 3-D, 2-D and contour graphs of Eq. (37) for $k = 1, \alpha = 0.5, \tau = -0.5, \lambda = 1, A = 4$.

The graphic presented in Fig. (**10**) is hyperbolic type traveling wave representing the stationary wave at any time.

Case 2: If

$$a_0 = \frac{1}{60}\left(15k^2\lambda^2 + \sqrt{105}k^2\lambda^2\right), \quad a_1 = \frac{1}{5}\left(15k^2\lambda\tau + \sqrt{105}k^2\lambda\tau\right), \quad a_2 = \frac{1}{5}\left(15k^2\tau^2 + \sqrt{105}k^2\tau^2\right),$$

$$v = \frac{1}{8}\left(-11k^5\lambda^4 - \sqrt{105}k^5\lambda^4\right), \quad p = \frac{1}{15}\left(30 - \sqrt{105}\right),$$

$$\tag{38}$$

replacing values in Eq. (38) into Eq. (35), we get hyperbolic traveling wave solution for Eq. (2)

$$u_2(x,t) = \frac{1}{60}\left(15k^2\lambda^2 + \sqrt{105}k^2\lambda^2\right) + \tag{39}$$

$$\frac{15k^2\tau^2 + \sqrt{105}k^2\tau^2}{\left(5\left(-\frac{\tau}{\lambda} + A\cosh\left[\lambda\left(kx + \frac{t^\alpha\left(-11k^5\lambda^4 - \sqrt{105}k^5\lambda^4\right)}{8\alpha}\right)\right] - A\sinh\left[\lambda\left(kx + \frac{t^\alpha\left(-11k^5\lambda^4 - \sqrt{105}k^5\lambda^4\right)}{8\alpha}\right)\right]\right)\right)^2} +$$

$$\frac{15k^2\lambda\tau + \sqrt{105}k^2\lambda\tau}{\left(5\left(-\frac{\tau}{\lambda} + A\cosh\left[\lambda\left(kx + \frac{t^\alpha\left(-11k^5\lambda^4 - \sqrt{105}k^5\lambda^4\right)}{8\alpha}\right)\right] - A\sinh\left[\lambda\left(kx + \frac{t^\alpha\left(-11k^5\lambda^4 - \sqrt{105}k^5\lambda^4\right)}{8\alpha}\right)\right]\right)\right)}.$$

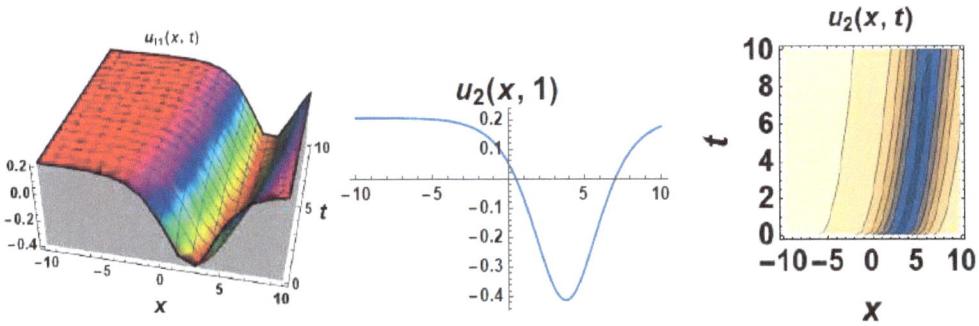

Fig. (11). 3-D, 2-D and contour graphs of Eq. (39) for $k = 1$, $\alpha = 0.5$, $\tau = -0.5$, $\lambda = 0.7$, $A = 4$.

The graphic presented in Fig. (**11**) is hyperbolic type traveling wave representing the stationary wave at any time.

RESULTS AND DISCUSSIONS

In this study, we successfully applied the $1/G'$-expansion and sub-equation methods to the time-fractional KK equation, which is a mathematical model of capillary wave and gravitational capillary wave events, which have an important place in physics. Different types of traveling wave solutions were obtained by both analytical methods. These traveling wave solutions are presented in the form of dark, singular, rational, trigonometric and hyperbolic type solitons. We found the solutions of the time-fractional KK equation for $p = 1$. These solutions found are different from the solutions presented in the study [5].

If the parameters in these solutions gain physical meaning, the importance of the study will become even more prominent. Shapes representing the standing wave at any given moment were presented. Besides, since the time-fractional KK equation we have examined includes fractional derivatives, a different parameter is added to the interpretation of the physical event. While investigating the solutions of the equation of time-fractional KK, some analytical methods were applied, but no analytical solution was produced. This is because different analytical methods produce solutions in different formats.

Sine-Gordon expansion (SGEM) and sinh-Gordon function methods (ShGFM) of the time-fractional KK equation were applied but the solution could not be reached. In the future, solutions of the time-fractional KK equation can be produced with a different perspective on SGEM and ShGFM. The classical ShGFM produces the solution function as follows. Therefore, we can say that the solution of the time-fractional KK equation cannot be obtained in the following format.

$$u(\xi) = \sum_{i=1}^{n}\left(\tan\left[\sqrt{p}\left(d+\xi\right)\right]\right)^{i-1} * \left(B_i \sec\left[\sqrt{p}\left(d+\xi\right)\right] + A_i \tan\left[\sqrt{p}\left(d+\xi\right)\right]\right) + A_0.$$

Similarly, the solution of the classical SGEM is produced as follows.

$$u(\xi) = \sum_{i=1}^{n}\left(\tanh\left[\sqrt{p}\left(d+\xi\right)\right]\right)^{i-1} * \left(B_i Sech\left[\sqrt{p}\left(d+\xi\right)\right] + A_i \tanh\left[\sqrt{p}\left(d+\xi\right)\right]\right) + A_0.$$

We can say that there is no solution to the time-fractional KK equation in this solution format.

Comparison of Methods

a) Common Aspects

1. Used in solutions of NLPDEs.

2. The ordinary differential equation has been obtained by performing classical wave transformation.

3. By using the balancing principle, the balancing term is obtained.

4. Algebraic equation system has been created.

b) Different Aspects

1. Base equations are different. While Eq. (6) is used as the base equation in the sub equation method, Eq. (11) is used in the $1/G'$-expansion method.

2. The serial solutions accepted are different. The serial solution accepted in the sub equation method is Eq. (5) while the serial solution accepted in the $1/G'$-expansion method is Eq. (10).

3. The solutions produced by both methods are different. The solutions obtained with the sub equation method are dark, singular, rational and trigonometric solitons, while in the $1/G'$-expansion method, they are hyperbolic type traveling wave solutions.

Advantages and Disadvantages

1. Quantitatively, the sub equation method produces five different types of solutions, while Eq. (7), $1/G'$-expansion method produces one type of solution Eq. (12).

2. Qualitatively, the sub equation method is more advantageous than the $1/G'$-expansion method. Because the conversion redundancy and processing complexity is more in the $1/G'$-expansion method.

3. NLPDEs traveling wave solutions can be produced by both methods.

Sub equation and $1/G'$-expansion methods, which are important instruments in obtaining traveling wave solutions of NLPDEs in solitons theory, are reliable, easy, applicable, effective and useful methods. NLPDEs, which have solutions in Eqs. (7), (12) format in the future, are recommended methods to obtain the traveling wave solution.

It is known that in the traditional wave transformation, k represents the wave number and v represents the velocity of the wave. When Eq. (16) is taken into consideration, although the v wave velocity seems to depend on the μ and k parameters, the α caused by the conformable fractional derivative operator also affects the wave velocity. Considering Eq. (7), μ is a physical parameter that affects both the amplitude of the wave, the number of wave and the wave velocity. In order to present a different perspective to our discussion, let us analyze the effect of k representing the wave number on both the behavior of the wave and the velocity with the simulation below.

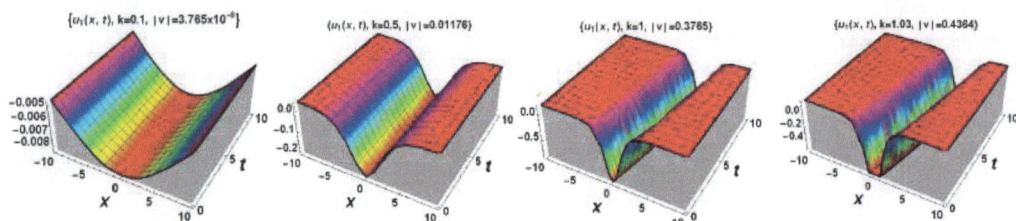

Fig. (12). Simulation graphs of Eq. (16) for $\alpha = 0.8$, $\mu = -0.5$.

Physically, whether the wave velocity is negative or positive affects the direction of the wave. For this reason, the absolute value of wave velocity and amplitude has been taken into account in the discussion. As seen in Fig. (**12**), we can observe that as the velocity of the dark soliton, which is presented with Eq. (16), its wavelength decreases and its amplitude increases. Also, considering the traditional wave transformation, wave velocity is directly related to t time. While waiting for the wave to advance as the velocity increases in the simulation, it can be observed that the wave does not progress in the simulation. This is because the velocity is very small. Although the velocity is small, it can be said that the breaking event of the traveling wave occurs for $|v| = 0.4364$ when the velocity is increased with small changes. In order to observe the progress of the wave more clearly in line with the mathematical analysis made, it may be possible to observe the progress of the wave in the later time period of t. We can observe this situation more clearly in the simulation below.

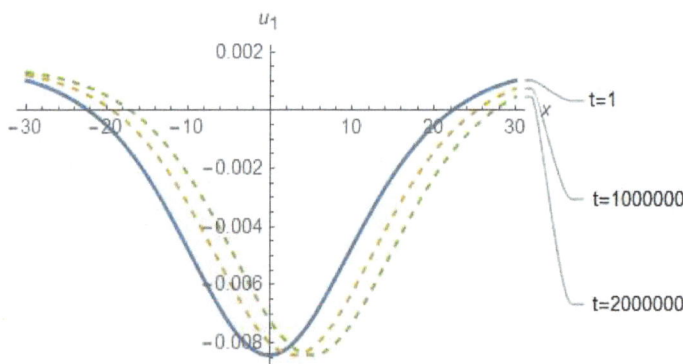

Fig. (13). Simulation graphs of Eq. (16) for $\alpha = 0.8$, $\mu = -0.5$, $k = 0.1$.

In Fig. (**13**), the progress of the dark soliton can be observed more clearly in the time period $t = 1, t = 1000000$ and $t = 2000000$. Also, as seen in Fig. (**12**), the velocity of the wave is quite small. It is concluded that t must be chosen high in order to observe the progress of the wave. Similar arguments can be made for other traveling wave solutions.

Let us present the following limits in order to examine the spatial and time asymptotic stability of the dark soliton presented by Eq. (16), which we have discussed.

$$\lim_{x \to \mp\infty} u_1(x,t) = \frac{1}{15}\left(-15 + \sqrt{105}\right)k^2\mu, \quad \frac{t^\alpha}{\alpha} \in R \quad \text{and} \quad k\sqrt{-\mu} > 0, \qquad (\mathbf{6.1})$$

$$\lim_{t \to \mp\infty} u_1(x,t) = \frac{1}{15}\left(-15 + \sqrt{105}\right)k^2\mu, \quad x \in R, \quad \alpha > 0 \quad \text{and} \quad k\sqrt{-\mu} > 0.$$

Considering Eq. (6.1), $u_1(x,t)$ is conditionally convergent for $t \to \mp\infty$, $x \to \mp\infty$. In addition, the following limit can be written for $k = 0.1$ and $\mu = -0.5$, which is suitable for discussion

$$\lim_{x \to \mp\infty} u_1(x,t) = \lim_{t \to \mp\infty} u_1(x,t) = 0.001584.$$

In this case, the dark soliton presented in Eq. (16) is asymptotically stable. In other words, the traveling wave, which undertakes the transfer of energy from one point to another, becomes stable near the equilibrium point after a certain period of time.

CONCLUSION

In this article, dark, singular, rational, trigonometric solitons and hyperbolic type traveling wave solutions have been produced by successfully applying the sub-equation method and $1/G'$-expansion method for the time-fractional KK equation. Special solutions are obtained when the constants in these solutions are matched in a real number. These special solutions were presented with 2D, 3D and contour graphics. Besides, by using the chain rule advantage of the conformable derivative with local properties, different interpretations were made by adding one more parameter to the capillary wave and gravity capillary wave events. The advantages and disadvantages of both analytical methods have been discussed. Besides, common and different aspects of both analytical methods were examined. In addition, the simulation of the wave behavior for different values of the velocity is presented, provided that the values of the dark soliton, except for the velocity parameter, which are presented with Eq. (16), are constant. The conditional convergence and asymptotic stability of Eq. (16) traveling wave solution have been examined.

Many complex operations such as drawing graphics and solving algebraic equation systems have been overcome by computer software programs. It was concluded that both analytical methods could be recommended for NLPDEs in the future.

CONSENT FOR PUBLICATION

Not applicable.

CONFLICT OF INTEREST

The authors declare no conflict of interest, financial or otherwise.

ACKNOWLEDGEMENT

Declared none.

REFERENCES

[1]   Khalil, R., Al Horani, M., Yousef, A., & Sababheh, M. (2014). A new definition of fractional derivative. Journal of Computational and Applied Mathematics, 264, 65-70.

[2]   Kaup, D. J. (1980). On the inverse scattering problem for cubic eigenvalue problems of the class $\psi xxx + 6Q\psi x + 6R\psi = \lambda\psi$. Studies in applied mathematics, 62(3), 189-216.

[3]   Kupershmidt, B. A. (1984). A super Korteweg-de Vries equation: an integrable system. Physics letters A, 102(5-6), 213-215.

[4]   Prakasha, D. G., Malagi, N. S., Veeresha, P., & Prasannakumara, B. C. (2020). An efficient computational technique for time-fractional Kaup-Kupershmidt equation. Numerical Methods for Partial Differential Equations.

[5]   Fan, E. (2003). Uniformly constructing a series of explicit exact solutions to nonlinear equations in mathematical physics. Chaos, Solitons & Fractals, 16(5), 819-839.

[6]   Gupta, A. K., & Ray, S. S. (2016). The comparison of two reliable methods for accurate solution of time-fractional Kaup–Kupershmidt equation arising in capillary gravity waves. Mathematical Methods in the Applied Sciences, 39(3), 583-592.

[7]   Jafari, H., Kadkhoda, N., Azadi, M., & Yaghobi, M. (2017). Group classification of the time-fractional Kaup-Kupershmidt equation. Scientia Iranica, 24(1), 302-307.

[8]   Sahoo, S., Ray, S. S., & Abdou, M. A. (2020). New exact solutions for time-fractional Kaup-Kupershmidt equation using improved (G'/G)-expansion and extended (G'/G)-expansion methods. Alexandria Engineering Journal, 59(5), 3105-3110.

[9]   Zhao, Z., & Han, B. (2018). Symmetry analysis and conservation laws of the time fractional Kaup-Kupershmidt equation from capillary gravity waves. Mathematical Modelling of Natural Phenomena, 13(2), 24.

[10]   Manafian, J., Ilhan, O. A., Avazpour, L., & Alizadeh, A. A. (2020). N-lump and interaction solutions of localized waves to the (2+1)-dimensional asymmetrical Nizhnik–Novikov–Veselov equation arise from a model for an incompressible fluid. Mathematical Methods in the Applied Sciences.

[11]   Ismael, H. F., Bulut, H., Park, C., & Osman, M. S. (2020). M-Lump, N-soliton solutions, and the collision phenomena for the (2+1)-dimensional Date-Jimbo-Kashiwara-Miwa equation. Results in Physics, 103329.

[12]   Yokuş, A., & Durur, H. (2019). Complex hyperbolic traveling wave solutions of Kuramoto-Sivashinsky equation using (1/G') expansion method for nonlinear dynamic theory. Balıkesir Üniversitesi Fen Bilimleri Enstitüsü Dergisi, 21(2), 590-599.

[13]   Durur, H., Yokuş, A., & Kaya D., Hyperbolic Type Traveling Wave Solutions of Regularized Long Wave Equation. Bilecik Şeyh Edebali Üniversitesi Fen Bilimleri Dergisi, 7(2).

[14]   Yokus, A., Durur, H., & Ahmad, H. (2020). Hyperbolic type solutions for the couple Boiti-Leon-Pempinelli system. Facta Universitatis, Series: Mathematics and Informatics, 35(2), 523-531.

[15]   Durur, H., & Yokuş, A. Vakhnenko-Parkes Denkleminin Hiperbolik Tipte Yürüyen Dalga Çözümü. Erzincan Üniversitesi Fen Bilimleri Enstitüsü Dergisi, 13(2), 550-556.

[16]   Gao, W., Silambarasan, R., Baskonus, H. M., Anand, R. V., & Rezazadeh, H. (2020). Periodic waves of the non-dissipative double dispersive micro strain wave in the micro structured solids. Physica A: Statistical Mechanics and its Applications, 545, 123772.

[17]   Su-Ping, Q., & Li-Xin, T. (2007). Modification of the Clarkson–Kruskal Direct Method for a Coupled System. Chinese Physics Letters, 24(10), 2720.

[18]   Durur, H., Ilhan, E., & Bulut, H. (2020). Novel Complex Wave Solutions of the (2+1)-Dimensional Hyperbolic Nonlinear Schrödinger Equation. Fractal and Fractional, 4(3), 41.

[19]   Yokuş, A., Durur, H., Abro, K. A., & Kaya, D. (2020). Role of Gilson–Pickering equation for the different types of soliton solutions: a nonlinear analysis. The European Physical Journal Plus, 135(8), 1-19.

[20]   Pervaiz, N., & Aziz, I. (2020). Haar wavelet approximation for the solution of cubic nonlinear Schrodinger equations. Physica A: Statistical Mechanics and its Applications, 545, 123738.

[21]   Kirs, M., Karjust, K., Aziz, I., Õunapuu, E., & Tungel, E. (2018). Free vibration analysis of a functionally graded material beam: evaluation of the Haar wavelet method. Proceedings of the Estonian Academy of Sciences, 67(1).

[22]   Duran, S. (2021). Breaking theory of solitary waves for the Riemann wave equation in fluid dynamics. International Journal of Modern Physics B, 2150130. https://doi.org/10.1142/S0217979221501307

[23]   Yokus, A., Kuzu, B., & Demiroğlu, U. (2019). Investigation of solitary wave solutions for the (3+1)-dimensional Zakharov–Kuznetsov equation. International Journal of Modern Physics B, 33(29), 1950350.

[24]   Durur, H. (2020). Different types analytic solutions of the (1+1)-dimensional resonant nonlinear Schrödinger's equation using (G'/G)-expansion method. Modern Physics Letters B, 34(03), 2050036.

[25]   Duran, S., & Karabulut, B. (2021). Nematicons in liquid crystals with Kerr Law by sub-equation method. Alexandria Engineering Journal. https://doi.org/10.1016/j.aej.2021.06.077

[26] Ahmad, H., Khan, T. A., Durur, H., Ismail, G. M., & Yokus, A. (2020). Analytic approximate solutions of diffusion equations arising in oil pollution. Journal of Ocean Engineering and Science.

[27] Duran, S. (2021). Dynamic interaction of behaviors of time-fractional shallow water wave equation system. Modern Physics Letters B, 2150353. https://doi.org/10.1142/S021798492150353X

[28] Abro, K. A., Atangana, A., & Gomez-Aguilar, J. F. (2021). Role of bi-order Atangana–Aguilar fractional differentiation on Drude model: an analytic study for distinct sources. Optical and Quantum Electronics, 53(4), 1-14.

[29] Yokus, A., Durur, H., Ahmad, H., Thounthong, P., & Zhang, Y. F. (2020). Construction of exact traveling wave solutions of the Bogoyavlenskii equation by (G'/G, 1/G)-expansion and (1/G')-expansion techniques. Results in Physics, 19, 103409.

[30] Yavuz, M., Ozdemir, N., & Baskonus, H. M. (2018). Solutions of partial differential equations using the fractional operator involving Mittag-Leffler kernel. The European Physical Journal Plus, 133(6), 1-11.

[31] Duran, S. (2020). Exact Solutions for Time-Fractional Ramani and Jimbo—Miwa Equations by Direct Algebraic Method. Advanced Science, Engineering and Medicine, 12(7), 982-988.

[32] Yavuz, M. (2018). Novel solution methods for initial boundary value problems of fractional order with conformable differentiation. An International Journal of Optimization and Control: Theories & Applications (IJOCTA), 8(1), 1-7.

[33] Pinar, Z., Rezazadeh, H., & Eslami, M. (2020). Generalized logistic equation method for Kerr law and dual power law Schrödinger equations. Optical and Quantum Electronics, 52(504),1-16.

[34] Veeresha, P. (2021). A Numerical Approach to the Coupled Atmospheric Ocean Model Using a Fractional Operator. Mathematical Modelling and Numerical Simulation with Applications (MMNSA), 1(1), 1-10.

[35] Hammouch, Z., Yavuz, M., & Özdemir, N. (2021). Numerical Solutions and Synchronization of a Variable-Order Fractional Chaotic System. Mathematical Modelling and Numerical Simulation with Applications (MMNSA), 1(1), 11-23.

[36] Yokuş, A. (2021). Construction of Different Types of Traveling Wave Solutions of the Relativistic Wave Equation Associated with the Schrödinger Equation. Mathematical Modelling and Numerical Simulation with Applications (MMNSA), 1(1), 24-31.

[37] Liu, J. G., Eslami, M., Rezazadeh, H., & Mirzazadeh, M. (2020). The dynamical behavior of mixed type lump solutions on the (3+1)-dimensional generalized Kadomtsev-Petviashvili-Boussinesq equation. International Journal of Nonlinear Sciences and Numerical Simulation, 21(7-8), 661-665.

[38] Aljohani, A. F., Alqurashi, B. M., & Kara, A. H. (2021). Solitons, travelling waves, invariance, conservation laws and 'approximate' conservation of the extended Jimbo-Miwa equation. Chaos, Solitons & Fractals, 144, 110636.

[39] Abro, K. A., & Atangana, A. (2021). A computational technique for thermal analysis in coaxial cylinder of one-dimensional flow of fractional Oldroyd-B nanofluid. International Journal of Ambient Energy, (just-accepted), 1-17.

[40] Durur, H., Kurt, A., & Tasbozan, O. (2020). New Travelling Wave Solutions for KdV6 Equation Using Sub Equation Method. Applied Mathematics and Nonlinear Sciences, 5(1), 455-460.

[41] Abdeljawad, T. (2015). On conformable fractional calculus. Journal of computational and Applied Mathematics, 279, 57-66.

[42]   Yavuz, M., & Yokus, A. (2020). Analytical and numerical approaches to nerve impulse model of fractional-order. Numerical Methods for Partial Differential Equations, 36(6), 1348-1368.

CHAPTER 8

Mathematical Analysis of a Rumor Spreading Model within the Frame of Fractional Derivative

Chandrali Baishya[1], Sindhu J. Achar[1] and **P. Veeresha[2,*]**

[1]*Department of Studies and Research in Mathematics, Tumkur University, Tumkur 572103, India*

[2]*Center for Mathematical Needs, Department of Mathematics, CHRIST (Deemed to be University), Bengaluru 560029, India*

Abstract: Rumor spreading is a trivial social practice, which has a long history of affecting society both in a positive and negative way, and modelling of transmission of rumors has been an attractive area for social and, of late, for physical scientists. In this chapter, we have modified the rumor-spreading model by incorporating fractional derivatives in the Caputo sense. To analyze the spread of rumors in social as well as virtual networks, we have considered four populations, namely, ignorant, spreader, recaller, and stifler. The existence and uniqueness, and boundedness of the solutions of the present model have been exhibited theoretically. Numerically, we have experimented with the effect of fractional derivatives and the density of one population on the other population by demonstrating the impact of rumor spread with the change of various parameters.

Keywords: Adams-Bashforth-Moulton method, Caputo fractional derivative, Mathematical model, Rumor spreading.

INTRODUCTION

While there are lots of happenings that entertain us for the relaxation of our lives in this world, it is unfortunately a part of human nature to constantly prefer a chunk more. As a result, we tend to consider events that can be genuinely now no longer true. Sometimes we do it collectively for a few activities, that refer to an object, event, or subject matter of public attention and that is how rumors start. Rumors have existed as a large model of social verbal exchange and a trivial social phenomenon throughout human evolutionary history.

*Corresponding author P. Veeresha: Center for Mathematical Needs, Department of Mathematics, CHRIST (Deemed to be University), Bengaluru 560029, India; E-mail: pundikala.veeresha@christuniversity.in

Mehmet Yavuz & Necati Özdemir (Eds.)

People spread rumors for a variety of reasons, including raising awareness, slandering others, creating momentum, diverting attention, and inciting panic. Due to the speedy advancement in various online platforms such as Facebook, Twitter, WhatsApp, *etc.*, the spread of rumors moves from verbal to digital, and as a result, their transmission becomes faster than ever before. Rumors may thrill us, scare us or entertain us temporarily or for a long period. Many hearsay stories about some great personalities who failed in mathematics give us hope that bouncing back may happen with us also. All-time fascinating stories about alien life, confirm that we are not alone in this universe. Rumors about celebrities, politicians, and scientific, financial, and social happenings change our way of looking at events/personality. Presently, a huge racket of rumor industry is running surrounding the happenings of Covid-19 news. These misinformed and misguiding stories harm nations, and they affect the measures implemented to control the situations. Many bloggers are running their business based on rumors only. Websites like www.snopes.com, www.thecut.com, www.cisa.gov, www.factcheck.org *etc.* are some websites, which check the factuality of rumors. With the advent of the internet, rumors are now shared via instant messengers, emails, or publishing blogs. Till the late nineties, studies of the impact of rumors were the areas for only social scientists. Kimmel, in his book "Rumors and Rumor Control," has discussed the understanding and controlling of rumors [1]. But, of late, mathematicians and computer scientists have shown remarkable interest in modelling the transmission of rumors in their research works. Despite an interesting similarity between rumor spreading and the epidemic model [2, 3], rumor spreading dynamics have received less attention than epidemic spreading. Daley and Kendall were the first to investigate the issue of rumor transmission and established the DK model [4-6], where they looked into the subject of rumors by splitting the people into three groups: ignorants, spreaders, and stiflers, which corresponded respectively to those who were unaware of the rumor, those who had heard the rumor and aggressively disseminated it, and those who had heard the rumor but had stopped spreading it. The Maki-Thompson (MK) model appeared next, as one of the DK model's modifications [7]. The rumor spreading was performed via direct contacts of the spreaders with others in the MK model, which has been extensively utilized for quantitative studies of rumor spreading [8-10]. In the study [11], Zhao *et al.* considered the case of online blogging and analyzed the rumor spread with consideration of the forgetting mechanism. The author has presented a model to analyze the impact of rumors on market [12]. Based on the fulfillment of two specific criteria simultaneously for a rumor to surely invade, Galam has used a majority rule reaction-diffusion dynamics to model rumor [13]. More information on rumor spreading model can be found in [14-22] and in the references therein.

Fractional calculus, being one of the most happening areas of research in recent times, has become an exciting trend for scientists, mathematicians, and engineers to explore their areas of study in a fractional sense. This is due to the potentiality of the fractional derivatives to portray the natural and man-made complex phenomena more realistically. The invention of fractional calculus takes us way back to the year 1695, when L'Hopital asked Leibniz about the possibility that n could be something other than an integer in $(d^n f)/(dt^n)$. The search for an operator that continuously transforms f into its nth derivative or anti-derivative opens the door to a vast area of studies called fractional calculus. Many fractional derivatives such as Caputo, Gru'nwald Letnikov, Riemann-Liouville, Jumarie, Caputo-Fabrizio, Atangana-Baleanu are inverted thereafter and their theories are explored in a wide range [23-26]. The fascinating results obtained by analyzing the physical model by incorporating various fractional derivatives in the field of science and technology can be observed in some work presented in [27-33]. The author has analyzed Noyes–Field model for the nonlinear Belousov–Zhabotinsky reaction by incorporating fractional derivative [34]. In the studies [35, 36], authors have presented new existence results of fractional integro-differential equations in Atangana–Baleanu sense. Some very interesting works on non-singular fractional derivatives can be found in [37-39]. Mirzazadeh *et al.* [40], examined a sixth-order dispersive (3+1)-dimensional nonlinear time-fractional Schrödinger equation with cubic-quintic-septic nonlinearities. Nisar *et al.* [41], considered nonlinear Hilfer neutral fractional derivatives with a non-dense domain and analyzed controllability results. Some interesting results are derived by authors [42-54] using numerical and modified schemes. Iyiola *et al.* [55], in their work, have analyzed a generalized Chagas vectors re-infestation model of fractional order type and presented some interesting findings. Moreover, [56-62] are also interesting papers presenting illustrative applications of fractional order modeling.

Even though the rumor spreading model is an analogy to the epidemic model, a widely studied model in the literature of mathematical epidemiology, till date, no research work can be observed where these models are treated in a fractional derivative sense. Fractional-order derivative, being the generalization of the integer-order derivative, is capable of demonstrating better results in modeling real phenomena and due to this reason, in this chapter, we plan to study the burning topic like rumor spread in social as well as virtual networks in the frameworks of fractional derivatives. To model the rumor spread, the populace is classified into four classes in this chapter: ignorants, spreaders, recallers, and stiflers and analyzed the model both theoretically and numerically. Fractional derivative is considered in the Caputo sense. The impact of a fractional derivative is observed, while experimenting with the evolution of densities of various groups under the influence of parameters such as rate of spreader converting to recaller, recaller converting to spreader, believing the rumor and not believing the rumor. We believe that the involvement of the Caputo fractional derivative in the rumor spread model will initiate a new way of looking into the existing models in this area. The chapter is organized as follows:

In Section 2, we have started some essential theorems from the literature, which are used to prove the theorems derived in the chapter. The fractional rumor spreading model, which is mathematically analyzed in this chapter, is presented in Section 3. The existence and uniqueness, and boundedness of solutions are examined theoretically in Section 4 and Section 5, respectively. In Section 6, numerical simulations are performed with the assistance of the Adams-Bashforth-Moulton method and Mathematica software and investigated the effects of various parameters under the observation of fractional derivatives.

SOME ESSENTIAL THEOREMS

Definition 1 [24] *Suppose $g(t)$ is k times continuously differentiable function and $g^{(k)}(t)$ is integrable in $[t_0, T]$. Then the Caputo fractional derivative of order α for a function $g(t)$ is defined as*

$$ {}_{t_0}^{C}D_t^{\alpha} g(t) = \frac{1}{\Gamma(k-\alpha)} \int_{t_0}^{t} \frac{g^{(k)}(\tau)}{(t-\tau)^{\alpha+1-k}} d\tau, $$

where $\Gamma(\cdot)$ refers to Gamma function, $t > a$ and k is a positive integer such that $k - 1 < \alpha < k$.

Definition 2 [24] *The Riemann–Liouville fractional order integral operator is defined by*

$$ J_x^{\alpha} f(x) = \frac{1}{\Gamma(\alpha)} \int_0^x \frac{f(t)}{(x-t)^{\alpha-1}} dt, \alpha > 0, $$

$$ J^0 f(x) = f(x). $$

For Riemann–Liouville fractional order integral operator, we have

$$ J_x^{\alpha} x^n = \frac{\Gamma(n+1)}{\Gamma(n+1+\alpha)} x^{n+\alpha}. $$

Lemma 1 [56] *Consider the system*

$$_{t_0}^{C}D_t^{\alpha}x(t) = g(t,x), t > t_0,$$

with the initial condition $x(t_0)$, *where* $0 < \mu \leq 1$ *and* $g: [t_0, \infty] \times \Omega \to \mathbb{R}^n, \Omega \in \mathbb{R}^n$. *If* $g(t,x)$ *satisfies the locally Lipchitz condition with respect to* x, *then there exists a unique solution of Eq. (1) on* $[t_0, \infty) \times \Omega$.

Lemma 2 [57] *Let* $g(t)$ *be a continuous function on* $[t_0, +\infty)$ *satisfying*

$$_{t_0}^{C}D_t^{\alpha}g(t) \leq -\lambda g(t) + \xi, \qquad g(t_0) = f_{t_0},$$

where $0 < \alpha \leq 1, (\lambda, \xi) \in \mathbb{R}^2$ *and* $\lambda \neq 0$ *and* $t_0 \geq 0$ *is the initial time. Then*

$$g(t) \leq (g(t_0) - \frac{\xi}{\lambda})E_{\alpha}[-\lambda(t-t_0)^{\alpha}] + \frac{\xi}{\lambda}.$$

MODEL FORMULATION

When a rumor starts, there will always be two groups: one who believes in the rumor and the other who does not. When someone hears a rumor but refuses to propagate it, he is referred to as a stifler. However, as more individuals approached him about the rumor, his refusal to believe it may be disturbed a little. Finally, by the tendency of the group of individuals, he will be persuaded of the rumor and begins spreading it like a spreader. That means, a proportion of stiflers in a latent state does indicate the termination of rumor, and will awaken if given a chance. Again, many times, a spreader will be passive about a rumor for some time. But, as he meets a dynamic spreader, he will recall the rumor and become an active spreader. Motivated by the work of Wang *et al.* [58], we have considered the rumor spreading model, where the interaction among four groups namely ignorants (R_i), spreaders (R_s), recallers (R_r) and stiflers (R_{sfl}), is projected by a Caputo fractional evolution equation. The flow of influence of one group on the other is depicted in Fig. (**1**). When a spreader comes into touch with an ignorant, the ignorant have the option of believing the rumor and becoming a spreader at a rate of β, or not believing it and becoming a stifler at a rate of ρ, where $\beta + \rho \leq 1$. When a spreader meets another spreader, he becomes a recaller at a rate of μ. When a spreader comes into contact with a stifler, it becomes a stifler at a rate of γ. When a recaller makes contact with a spreader, at a rate of δ, the recaller becomes a spreader.

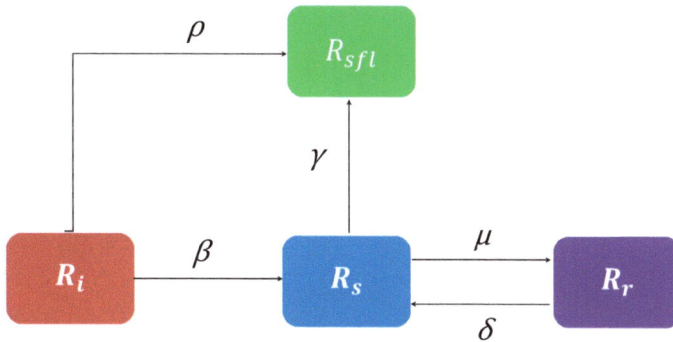

Fig. (1). Flow of moving one population to other [58].

The interaction among the four groups in the act of rumor spreading is demonstrated by the fractional differential equation (2) below:

$$_{t_0}^{C}D_t^{\alpha}R_i(t) = -(\beta + \rho)\kappa R_i(t)R_s(t),$$

$$\begin{aligned}_{t_0}^{C}D_t^{\alpha}R_s(t) = &\; \beta\kappa R_i(t)R_s(t) - \mu\kappa R_s(t)R_s(t) - \gamma\kappa R_s(t)R_{sfl}(t) \\ &+ \delta\kappa R_r(t)R_s(t),\end{aligned}$$

$$_{t_0}^{C}D_t^{\alpha}R_r(t) = \mu\kappa R_s(t)R_s(t) - \delta\kappa R_r(t)R_s(t),$$

$$_{t_0}^{C}D_t^{\alpha}R_{sfl}(t) = \rho\kappa R_i(t)R_s(t) + \gamma\kappa R_s(t)R_{sfl}(t).$$

(1)

EXISTENCE AND UNIQUENESS

The existence and uniqueness of solutions of the fractional rumor spreading system (2) is established in this Section.

Theorem 4.1   *In the region* $\mho \times [t_0, T]$, *where* $\mho = \{(R_i,\ R_s,\ R_r,\ R_{sfl}) \in \mathbb{R}^4 : max\{|R_i|,\ |R_s|,\ |R_r|,\ |R_{sfl}|\} \leq \mathbb{M}\}$ *and* $T < +\infty$, *for system (2), the solution exists and is unique.*

Proof. Let us consider $W = (R_i,\ R_s,\ R_r,\ R_{sfl})$ and $\overline{W} = (\overline{R}_i,\ \overline{R}_s,\ \overline{R}_r,\ \overline{R}_{sfl})$ and a mapping

$$A(W) = (A_1(W), A_2(W), A_3(W), A_4(W)),$$

where

$$A_1(W) = -(\beta + \rho)\kappa R_i R_s,$$

$$A_2(W) = \beta\kappa R_i R_s - \mu\kappa R_s R_s - \gamma\kappa R_s R_{sfl} + \delta\kappa R_r R_s,$$

$$A_3(W) = \mu\kappa R_s R_s - \delta\kappa R_r R_s,$$

$$A_4(W) = \rho\kappa R_i R_s + \gamma\kappa R_s R_{sfl}.$$

To show that there exist some ζ such that

$$||A(W) - A(\overline{W})|| \leq \zeta ||W - \overline{W}||.$$

Consider

$$||A(W) - A(\overline{W})|| = |A_1(W) - A_1(\overline{W})| + |A_2(W) - A_2(\overline{W})|$$

$$+ |A_3(W) - A_3(\overline{W})| + |A_4(W) - A_4(\overline{W})|$$

$$= |-(\beta + \rho)\kappa R_i R_s + (\beta + \rho)\kappa \overline{R}_i \overline{R}_s| + |\beta\kappa R_i R_s - \mu\kappa R_s R_s$$

$$- \gamma\kappa R_s R_{sfl} + \delta\kappa R_r R_s - \beta\kappa \overline{R}_i \overline{R}_s + \mu\kappa \overline{R}_s \overline{R}_s + \gamma\kappa \overline{R}_s \overline{R}_{sfl}$$

$$- \delta\kappa \overline{R}_r \overline{R}_s| + |\mu\kappa R_s R_s - \delta\kappa R_r R_s - \mu\kappa \overline{R}_s \overline{R}_s + \delta\kappa \overline{R}_r \overline{R}_s|$$

$$+ |\rho\kappa R_i R_s + \gamma\kappa R_s R_{sfl} - \rho\kappa \overline{R}_i \overline{R}_s + \gamma\kappa \overline{R}_s \overline{R}_{sfl}|$$

$$= |-(\beta + \rho)\kappa(R_i R_s - \overline{R}_i \overline{R}_s)| + |\beta\kappa(R_i R_s - \overline{R}_i \overline{R}_s)$$

$$- \mu\kappa(R_s R_s - \overline{R}_s \overline{R}_s) - \gamma\kappa(R_s R_{sfl} - \overline{R}_s \overline{R}_{sfl}) + \delta\kappa(R_r R_s$$

$$- \overline{R}_r \overline{R}_s)| + |\mu\kappa(R_s R_s - \overline{R}_s \overline{R}_s) - \delta\kappa(R_r R_s - \overline{R}_r \overline{R}_s)|$$

$$+ |\rho\kappa(R_i R_s - \overline{R}_i \overline{R}_s) + \gamma\kappa(R_s R_{sfl} - \overline{R}_s \overline{R}_{sfl})|$$

$$\leq 2(\beta + \rho)\kappa |R_i R_s - \overline{R}_i \overline{R}_s| + 2\mu\kappa |R_s R_s - \overline{R}_s \overline{R}_s|$$

$$+ 2\delta\kappa|R_r R_s - \bar{R}_r \bar{R}_s| + 2\gamma\kappa|R_s R_{sfl} - \bar{R}_s \bar{R}_{sfl}|$$

$$\leq (2(\beta + \rho)\kappa\mathbb{M})|R_i - \bar{R}_i| + (2(\beta + \rho)\kappa\mathbb{M} + 4\mu\kappa\mathbb{M} + 2\delta\kappa\mathbb{M}$$

$$+ 2\gamma\kappa\mathbb{M})|R_s - \bar{R}_s| + 2\delta\kappa\mathbb{M}|R_r - \bar{R}_r| + 2\gamma\kappa\mathbb{M}|R_{sfl} - \bar{R}_{sfl}|.$$

This implies

$$||A(W) - A(\bar{W})|| \leq (2(\beta + \rho)\kappa\mathbb{M})|R_i - \bar{R}_i| + (2\kappa\mathbb{M}(\beta + \rho + \delta + \gamma + 2\mu))$$

$$\times |R_s - \bar{R}_s| + 2\delta\kappa\mathbb{M}|R_r - \bar{R}_r| + 2\gamma\kappa\mathbb{M}|R_{sfl} - \bar{R}_{sfl}|$$

$$= \mho_1|R_i - \bar{R}_i| + \mho_2|R_s - \bar{R}_s| + \mho_3|R_r - \bar{R}_r| + \mho_4|R_{sfl} - \bar{R}_{sfl}|,$$

where

$$\mho_1 = 2(\beta + \rho)\kappa\mathbb{M},$$

$$\mho_2 = 2\kappa\mathbb{M}(\beta + \rho + \delta + \gamma + 2\mu),$$

$$\mho_3 = 2\delta\kappa\mathbb{M},$$

$$\mho_4 = 2\gamma\kappa\mathbb{M}.$$

Let $\zeta = \max\{\mho_1, \ \mho_2, \ \mho_3, \ \mho_4\}$. Then we have,

$$||A(W) - A(\bar{W})|| \leq \zeta ||W - \bar{W}||.$$

Hence, by Lemma 1 the solution exists and is unique.

BOUNDEDNESS

In this section, we establish that the solutions of system (2) are bounded.

Theorem 5.1 *The solutions of system (2) are uniformly bounded.*

Proof. Let us define a function

$$R(t) = R_i(t) + R_s(t) + R_r(t) + R_{sfl}(t).$$

By using the Lemma 2, we get

$$\begin{aligned}
{}_{t_0}^{C}D_t^{\alpha}R(t) + \beta R(t) &= {}_{t_0}^{C}D_t^{\alpha}\big[R_i(t) + R_s(t) + R_r(t) + R_{sfl}(t)\big] \\
&\quad + \beta[R_i(t) + R_s(t) + R_r(t) + R_{sfl}(t)] \\
&= -(\beta + \rho)\kappa R_i(t)R_s(t) + \beta\kappa R_i(t)R_s(t) \\
&\quad - \mu\kappa R_s(t)R_s(t) - \gamma\kappa R_s(t)R_{sfl}(t) + \delta\kappa R_r(t)R_s(t) \\
&\quad + \mu\kappa R_s(t)R_s(t) - \delta\kappa R_r(t)R_s(t) + \rho\kappa R_i(t)R_s(t) \\
&\quad + \gamma\kappa R_s(t)R_{sfl}(t) + (\beta)(R_i(t) + R_s(t) + R_r(t) \\
&\quad + R_{sfl}(t))
\end{aligned}$$

$$\leq 4\beta\mathbb{M}.$$

The solution exists and is unique in

$$\mathfrak{V} = \{(R_i, R_s, R_r, R_{sfl}) : \; max\{|R_i|, |R_s|, |R_r|, |R_{sfl}|\} \leq \mathbb{M}\}.$$

The above inequality yields

$$ {}_{t_0}^{C}D_t^{\alpha}R(t) + \beta R(t) \leq 4\beta\mathbb{M}.$$

By the Lemma 2, we get

$$ {}_{t_0}^{C}D_t^{\alpha}R(t) \leq \left(R(t_0) - \frac{1}{\beta}(4\beta\mathbb{M})\right)E_\alpha[-\eta(t - t_0)^\alpha] + \frac{1}{\beta}(4\beta\mathbb{M}) \to 4\beta\mathbb{M}, t$$

$$\to \infty.$$

Therefore, all the solution of system (2) that initiates in $\mathfrak{V}$ remained bounded in

$$\Theta = \{(R_i, R_s, R_r, R_{sfl}) \in \mathfrak{V}_+ | R(t) \leq 4\beta\mathbb{M} + \epsilon, \; \epsilon > 0\}.$$

NUMERICAL SIMULATION

In this section, we have investigated model (2) numerically. To solve system (2), we use the Adams-Bashforth-Moulton method [59] and Mathematica software. The method is derived for a fractional differential equation of the form.

$$_{t_0}^{C}D_t^{\alpha}u(t) = g(t,u(t)), 0 \le t \le T,$$

with initial condition

$$u^{(k)}(0) = u_0^{(k)}, k = 0,1,2,\ldots, \lceil \alpha \rceil - 1,$$

which is equivalent to the Volterra integral equation

$$u(t) = \sum_{k=0}^{\lceil \alpha \rceil - 1} u_0^{(k)} \frac{t^k}{k!} + \frac{1}{\Gamma(\alpha)} \int_0^t (t-\tau)^{\alpha-1} g(\tau, u(\tau))d\tau. \qquad (2)$$

To integrate right-hand side of equation (3), Diethelm *et al.* [59] used the Adams-Bashforth-Moulton scheme. Setting $h = \frac{T}{N}, t_n = nh, n = 0,1,2,\cdots, N \in \mathbb{Z}_+$, for model (2), the corrector equations by the Adams-Moulton method are:

$$R_{i_{n+1}} = R_{i_0} + \frac{h^\alpha}{\Gamma(\alpha+2)}\left(-(\beta+\rho)\kappa R_{i_{n+1}}^P R_{s_{n+1}}^P\right)$$

$$+\frac{h^\alpha}{\Gamma(\alpha+2)}\sum_{k=0}^{n} a_{k,n+1}(-(\beta+\rho)\kappa R_{i_k}R_{s_k}),$$

$$R_{s_{n+1}} = R_{s_0} + \frac{h^\alpha}{\Gamma(\alpha+2)}\left(\beta\kappa R_{i_{n+1}}^P R_{s_{n+1}}^P - \mu\kappa R_{s_{n+1}}^P R_{s_{n+1}}^P - \gamma\kappa R_{s_{n+1}}^P R_{sfl_{n+1}}^P\right.$$

$$\left.+\delta\kappa R_{r_{n+1}}^P R_{s_{n+1}}^P\right) + \frac{h^\alpha}{\Gamma(\alpha+2)}\sum_{k=0}^{n} a_{k,n+1}(\beta\kappa R_{i_k}R_{s_k} - \mu\kappa R_{s_k}R_{s_k}$$

$$-\gamma\kappa R_{s_k}R_{sfl_k} + \delta\kappa R_{r_k}R_{s_k}),$$

$$R_{r_{n+1}} = R_{r_0} + \frac{h^\alpha}{\Gamma(\alpha+2)}\left(\mu\kappa R_{s_{n+1}}^P R_{r_{n+1}}^P - \delta\kappa R_{r_{n+1}}^P\right)R_{s_{n+1}}^P\right)$$

$$+\frac{h^\alpha}{\Gamma(\alpha+2)}\sum_{k=0}^{n} a_{k,n+1}\left(\mu\kappa R_{s_k}R_{r_k} - \delta\kappa R_{r_k}\right),$$

$$R_{sfl_{n+1}} = R_{sfl_0} + \frac{h^\alpha}{\Gamma(\alpha+2)}\left(\rho\kappa R_{i_{n+1}}^P R_{s_{n+1}}^P + \gamma\kappa R_{s_{n+1}}^P R_{sfl_{n+1}}^P\right)$$

$$+ \frac{h^\alpha}{\Gamma(\alpha+2)}\sum_{k=0}^{n} a_{k,n+1}\left(\rho\kappa R_{i_k}R_{s_k} + \gamma\kappa R_{s_k}\right),$$

and the predictor equations are

$$R_{i_{n+1}}^P = R_{i_0} + \frac{h^\alpha}{\Gamma(\alpha+1)}\Sigma_{k=0}^n b_{k,n+1}\left(-(\beta+\rho)\kappa R_{i_k}R_{s_k}\right),$$

$$R_{s_{n+1}}^P = R_{s_0} + \frac{h^\alpha}{\Gamma(\alpha+1)}\Sigma_{k=0}^n b_{k,n+1}\left((\beta\kappa R_{i_k}R_{s_k} - \mu\kappa R_{s_k}R_{s_k}\right.$$

$$\left. - \gamma\kappa R_{s_k}R_{sfl_k} + \delta\kappa R_{r_k}R_{s_k})\right),$$

$$R_{r_{n+1}}^P = R_{r_0} + \frac{h^\alpha}{\Gamma(\alpha+1)}\Sigma_{k=0}^n b_{k,n+1}\left(\mu\kappa R_{s_k}R_{r_k} - \delta\kappa R_{r_k}\right).$$

$$R_{sfl_{n+1}}^P = R_{sfl_0} + \frac{h^\alpha}{\Gamma(\alpha+1)}\Sigma_{k=0}^n b_{k,n+1}\left(\rho\kappa R_{i_k}R_{s_k} + \gamma\kappa R_{s_k}\right).$$

Here, we have

$$b_{k,n+1} = (n-k+1)^\alpha - (n-k)^\alpha,$$

and

$$a_{k,n+1} = \begin{cases} n^{\alpha+1} - (n-\alpha)(n+1)^\alpha, & k=0, \\ (n-k+2)^{\alpha+1} + (n-k)^{\alpha+1} - 2(n-k+1)^{\alpha+1}, & 1 \le k \le n, \\ 1, & k=n+1. \end{cases}$$

With an analogy to the work of Wang *et al.* [58], we have assumed the initial conditions for the four group of populations as $R_{i_0} = \frac{N-1}{N}, R_{s_0} = \frac{1}{N}, R_{r_0} = 0, R_{sfl_0} = 0$. We have considered the values of the parameters as $\beta = 0.8$, $\rho = 0.1$, $\mu = 0.5$, $\gamma = 0.5$, $\delta = 0.5$, $\kappa = 6$, $N = 10^6$. Considering the values of fractional derivative α as $\alpha = 1, 0.9, 0.8, 0.7$, the evolution of densities of ignorants, spreaders, recallers and stiflers are observed to project the effect of fractional derivatives on them.

In Fig. (**2**), we have presented the development of the densities of ignorants, spreaders, recallers, and stiflers and compared them with the progress that happens in the absence of the recaller group in Fig. (**3**). For $\alpha = 1$, in Fig. (**2a**), we can observe a pattern of rumor spread. In the beginning, when a rumor starts there is only one spreader and no recallers and stiflers. Then spreader started increasing rapidly and as a result, the ignorant population started declining rapidly too and finally converges to zero. When the spreader population reaches the peak, the stifler population starts rising, whereas rumors are being recalled by the recaller again. In the meantime, with the continuous spread of the rumors, the stifler population keeps rising and when recaller and stifler converge to a certain value, rumors stop spreading with spreader converging to zero. When the fractional derivatives are introduced, in Fig. (**2b, c** and **d**), we see a significant difference in the nature of the rumor spread. The shape of the almost bell-shaped curves for spreaders is wider with decreasing fractional derivative. That means the rumor persists for a longer duration. But, even though, spreaders are spreading the rumor, it is being ignored and not being recalled as before. As a result ignorant population is increasing and the recaller population is decreasing with decreasing fractional derivatives. Similar kinds of observations can be found in Fig. (**3**). But in Fig. (**2**), the final density of ignorant is less than in Fig. (**3**). This is because of the absence of the recallers in Fig. (**3**), the rumors will not stay for a longer duration in networks. Stiflers are also less in Fig. (**3**) than in Fig. (**2**).

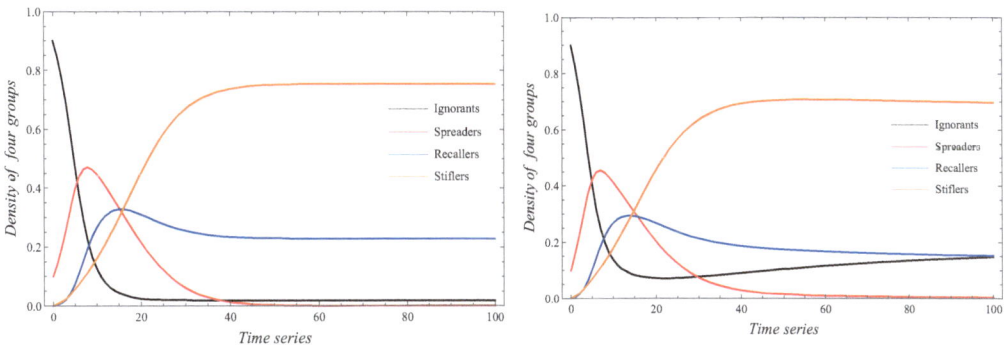

(Fig. 2) contd.....

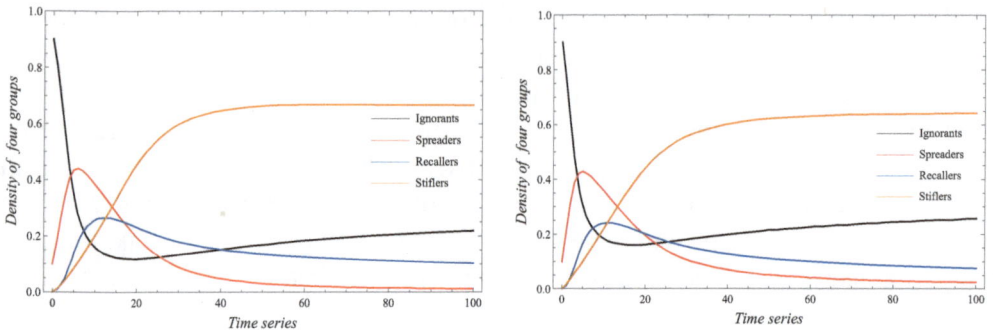

Fig. (2). The stability profile of density of four groups for (a) $\alpha = 1$, (b) $\alpha = 0.9$, (c)$\alpha = 0.8$ and (d)$\alpha = 0.7$.

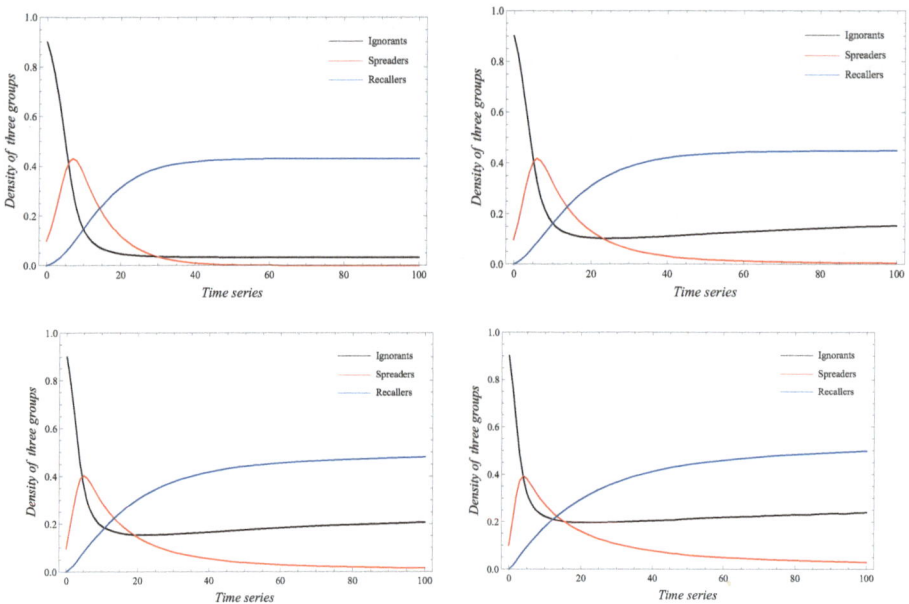

Fig. (3). The stability profile of density of three groups for (a) $\alpha = 1$, (b) $\alpha = 0.9$, (c) $\alpha = 0.8$ and (d) $\alpha = 0.7$.

In Fig. (**4**), we observe that the rate μ at which an initiating spreader becomes recaller and the rate δ at which a recaller becomes spreader is the driving force of this entire process of rumor spread. So, in Figs. (**4-6**), we have demonstrated, how the densities of spreaders change over time with different values of μ and δ under the influence of the fractional derivative. In Figs. (**4** and **5**), we observe that with the increasing μ and δ, the peak of the spreader density decreases. This implies that more spreaders have been converted to recallers. The peak value of the spreader density denotes the biggest fraction of spreaders in the rumor spreading process, which can be used to measure the maximum rumor influence. Whereas, Fig. (**6**) indicates that as the rate δ increases keeking μ fixed, the density of the spreader also increases *i.e.* more number of recallers become spreaders. The impact of the fractional derivatives can be observed in terms of delay in the termination of the rumor spreading. This again indicates that in the presence of fractional derivative, there is a higher scope for the rumor to persist for a long duration.

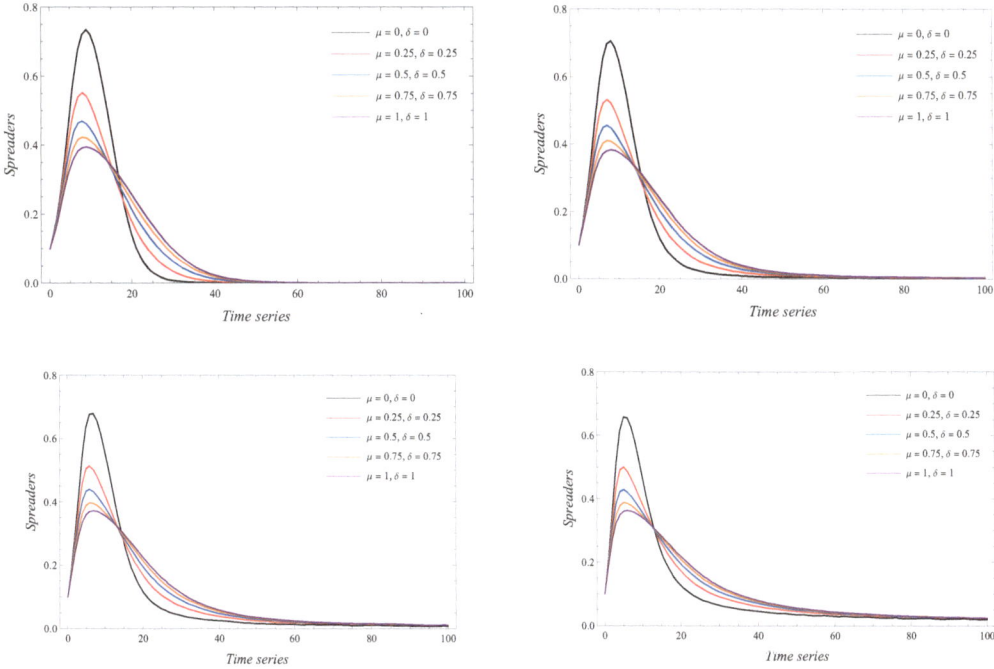

Fig. (4). Densities of the spreaders for parameters μ and δ for (a) $\alpha = 1$, (b) $\alpha = 0.9$, (c) $\alpha = 0.8$ and (d) $\alpha = 0.7$.

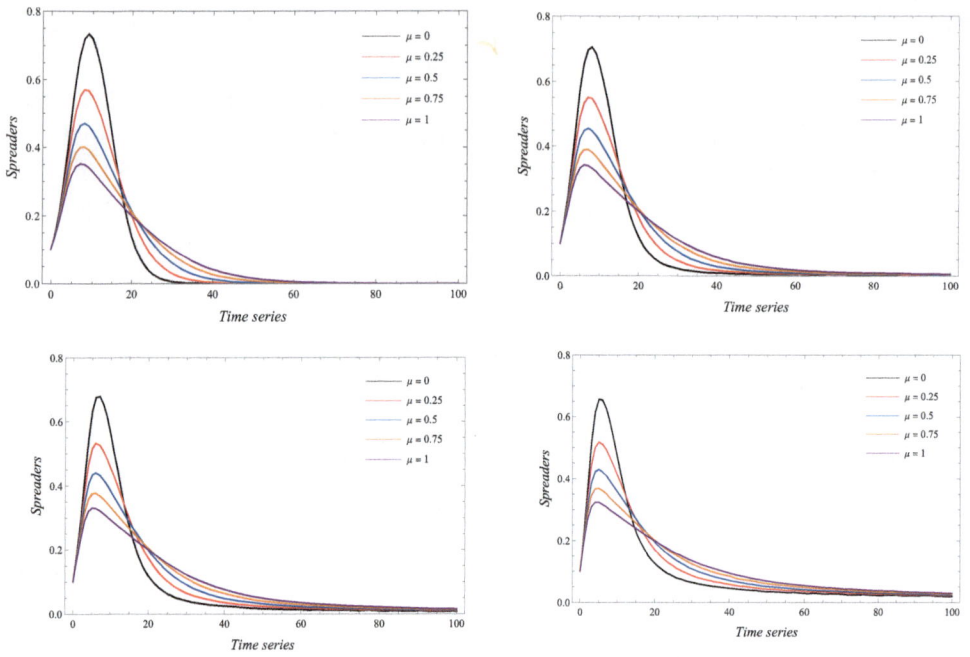

Fig. (5). Densities of the spreaders for parameters μ for (a) $\alpha = 1$, (b) $\alpha = 0.9$, (c) $\alpha = 0.8$ and (d) $\alpha = 0.7$.

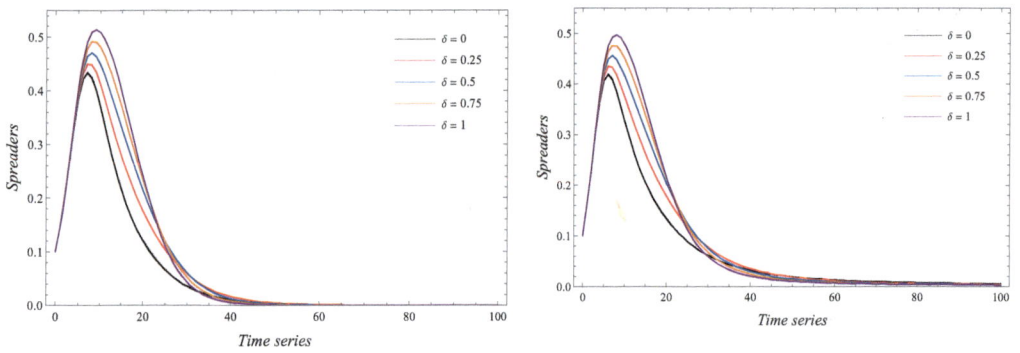

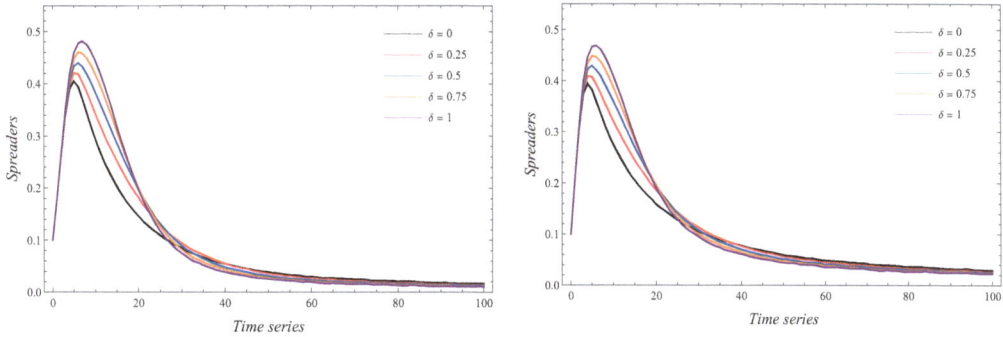

Fig. (6). Densities of the spreaders for parameters δ for (**a**) $\alpha = 1$, (**b**) $\alpha = 0.9$, (**c**) $\alpha = 0.8$ and (**d**) $\alpha = 0.7$.

Figs. (**7-9**) represent the changes in the densities of stiflers over time with changing rates of μ and δ for different values of α. By comparing the figures in Fig. (**7**), we observe that μ plays a major role in controlling the stifler densities. As more spreaders become recallers, the final stifler density decreases. The decrease in stifler density is prominent with the decrease in fractional derivative value. In Fig. (**8**), it is visible that as the rate of conversion of recaller to spreader increases, the stifler density increases. That is, the higher the value of δ, the lesser is the rumor spreading. On the other hand, Fig. (**9**) projects that when both μ and δ increase, then from $\mu = 0$ and $\delta = 0$ up to $\mu = 0.25$ and $\delta = 0.25$, we can observe a fall in stifler density convergence value. But it starts increasing thereafter and at $\mu = 1$ and $\delta = 1$ it reaches an intermediate convergence. In each of these cases, the presence of the fractional derivatives indicates the convergence of stifler density to a much lower value. That is, it influences delay in the termination of the rumor spreading again.

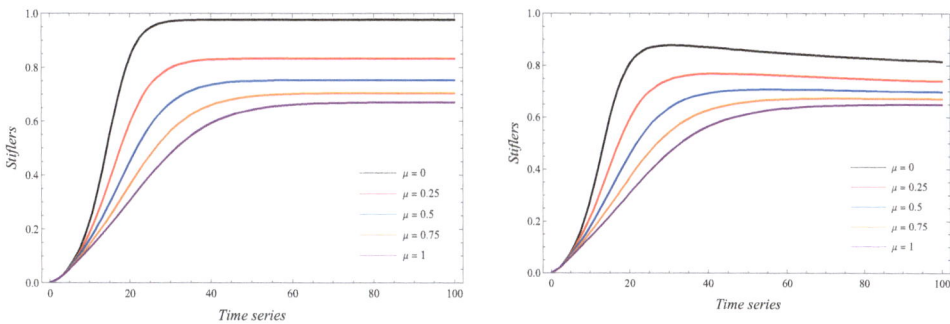

(Fig. 6) contd.....

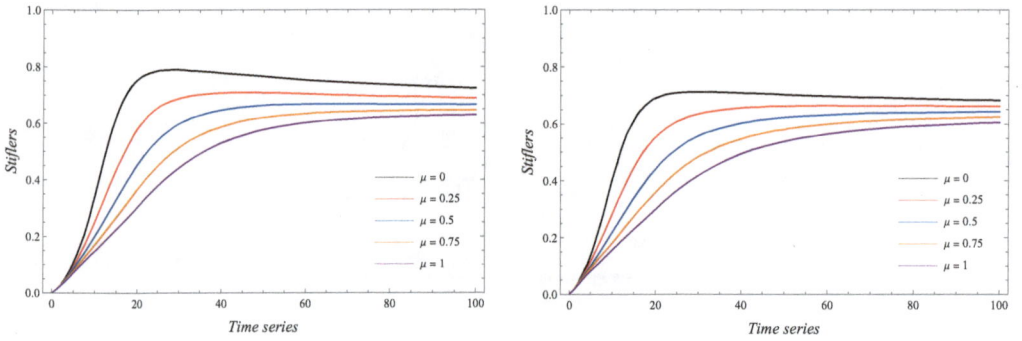

Fig. (7). Densities of the stiflers for parameters μ for (a) $\alpha = 1$, (b) $\alpha = 0.9$, (c) $\alpha = 0.8$ and (d) $\alpha = 0.7$.

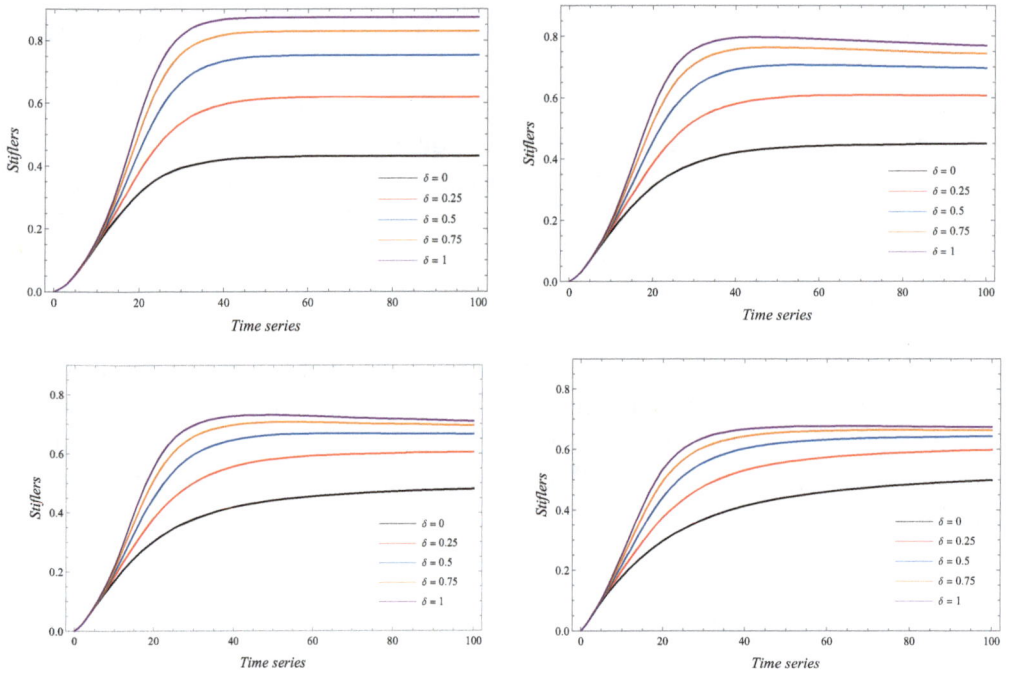

Fig. (8). Densities of the stiflers for parameters δ for (a) $\alpha = 1$, (b) $\alpha = 0.9$, (c) $\alpha = 0.8$ and (d) $\alpha = 0.7$.

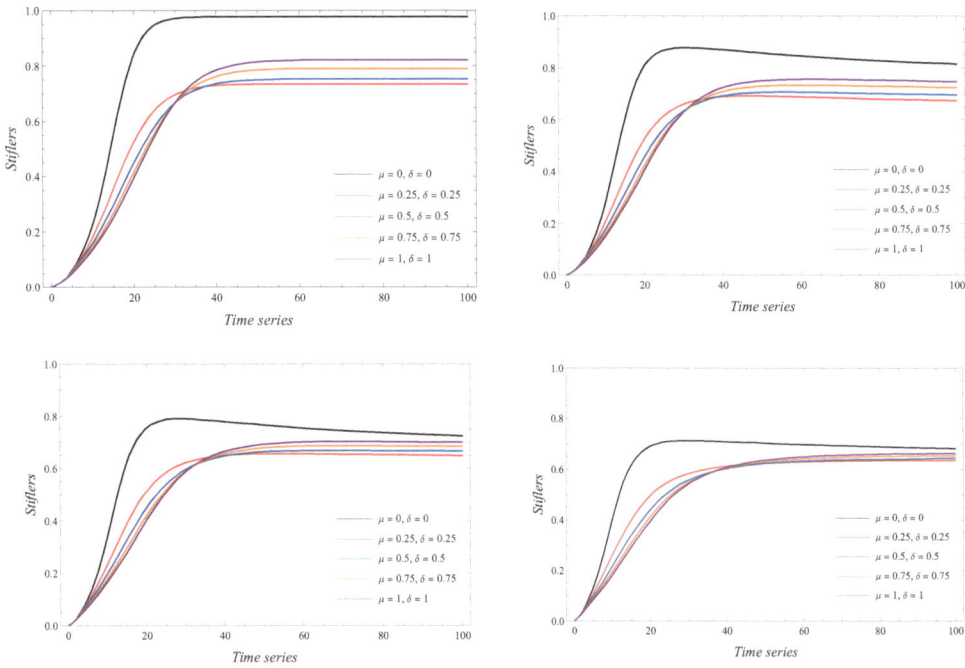

Fig. (9). Densities of the stiflers for parameters μ and δ for (a) $\alpha = 1$, (b) $\alpha = 0.9$, (c) $\alpha = 0.8$ and (d) $\alpha = 0.7$.

In Figs. (**10**, **11** and **12**), we have presented the observations relating to the changes in densities of spreaders, recallers, and stiflers respectively with the changing values of the rate of believing in rumors β, rate of not believing in rumors ρ, and fractional derivatives. It is observed that when β decreases from 0.8 and ρ increases from 0.1 simultaneously, both spreaders' and recallers' densities decrease. Due to this fall in the densities of spreaders and recallers, the rumor spread also decline. As a result, the stiflers' densities also project a declining trend. But, in the presence of fractional derivative even if the peak of these population densities shows a significant fall, the rumors seem to prevail for a longer duration in the networks.

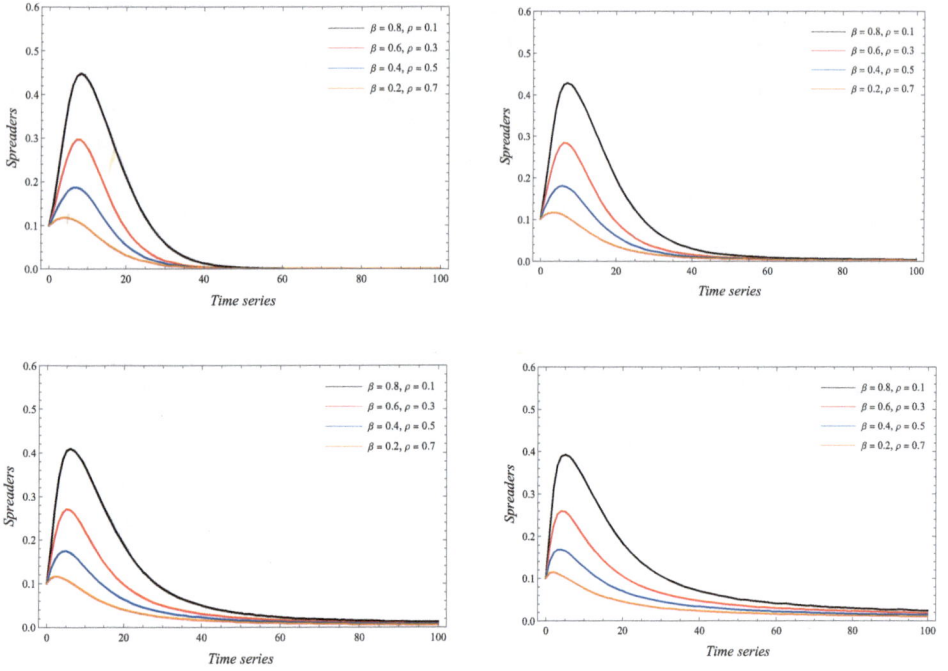

Fig. (10). Densities of the spreaders for parameters β and ρ for (a) $\alpha = 1$, (b) $\alpha = 0.9$, (c) $\alpha = 0.8$ and (d) $\alpha = 0.7$.

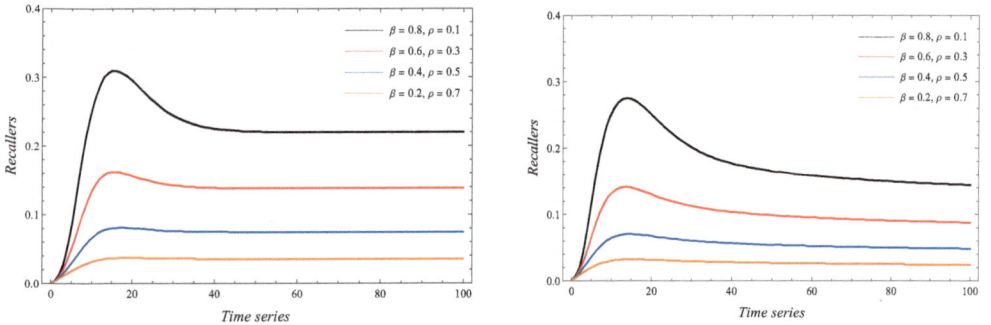

(Fig. 11) contd.....

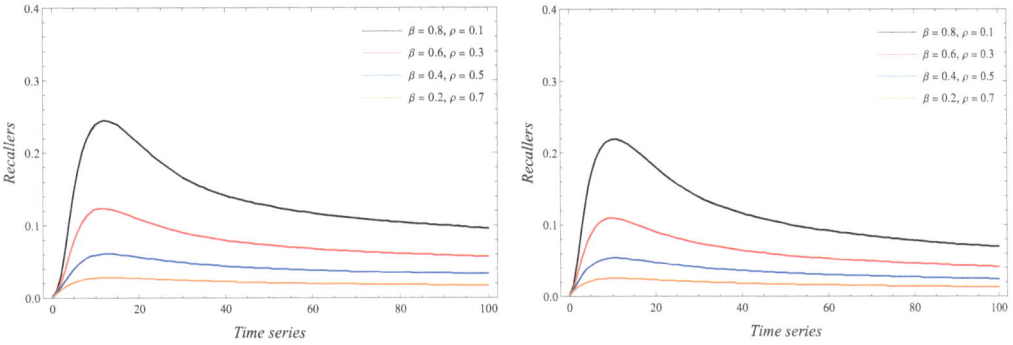

Fig. (11). Densities of the recallers for parameters β and ρ for (a) $\alpha = 1$, (b) $\alpha = 0.9$, (c) $\alpha = 0.8$ and (d) $\alpha = 0.7$.

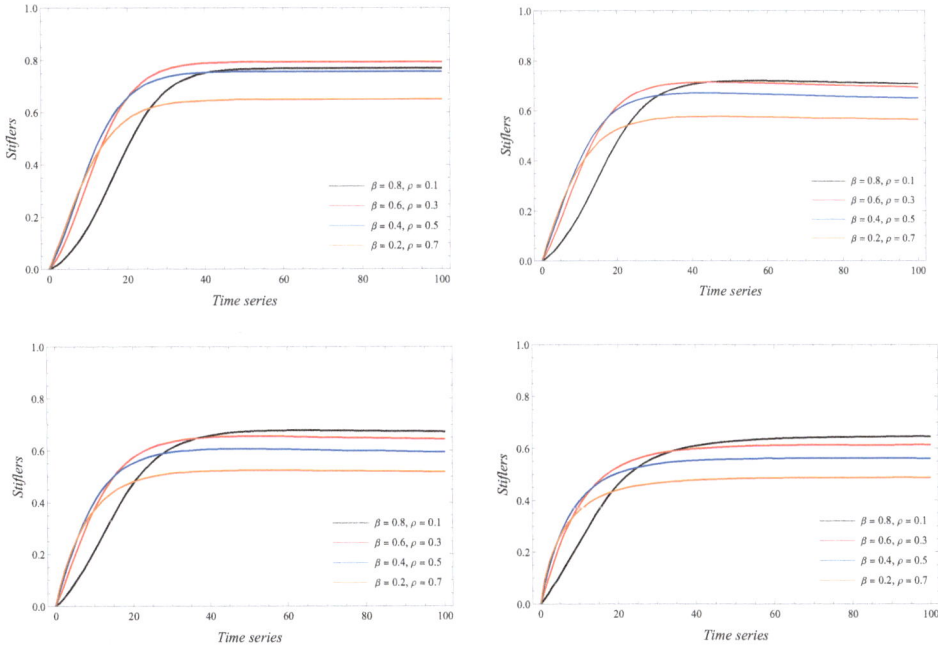

Fig. (12). Densities of the stiflers for parameters β and ρ for (a) $\alpha = 1$, (b) $\alpha = 0.9$, (c) $\alpha = 0.8$ and (d) $\alpha = 0.7$.

CONCLUSION

The current study emphasizes the relevance of utilizing a fractional mathematical model for tackling real-world situations like rumor spread. Here, the Caputo fractional-order derivative is used to examine the mathematical model of rumor transmission. We have investigated the theoretical aspects such as existence and uniqueness, and boundedness related to the solutions of the system (2). By using the computational mathematical tool Mathematica and predictor-corrector computational technique, we have numerically analyzed the projected model in the present work. It is observed that the rate at which spreaders become recallers, recallers become spreaders and ignorants believe in rumors, drive the entire process of rumor spreading. From the numerical stability profile, it is visible that rumor spreading reaches a peak and then gradually people started losing interest in the rumor and as a result, it terminates in the form of increasing stifler population. By varying the rate at which an initiating spreader becomes recaller and the rate at which a recaller becomes spreader, it is found that former one has more influence on the spreader than the later one. It is observed in the proof of the theorem related to the boundedness of the solutions that boundedness depends on the rate at which ignorant believe in the rumor. The same is visible in graphical representation of the effect of rate of believer and non-believer on spreader, recaller and stifler densities. Overall, the graphical representation demonstrated to showcase the influence of various parameters, establishes that, fractional derivative delays the termination of rumor in the social or virtual networks. Therefore, if a rumor is believed to be of the better interest of the nation, then fractional derivative can be observed as an interesting tool to model the spread of such rumor.

CONSENT FOR PUBLICATION

Not applicable.

CONFLICT OF INTEREST

The authors declare no conflict of interest, financial or otherwise.

ACKNOWLEDGEMENT

Declared none.

REFERENCES

[1]   A. J. Kimmel, "Rumors and rumor control," Journal of Behavioral Finance, vol. 5, 2004.
[2]   R. M. Anderson and R. M. May, Infectious diseases of humans: dynamics and control. Oxford university press, 1992.
[3]   F. Jin, E. Dougherty, P. Saraf, Y. Cao, and N. Ramakrishnan, "Epidemiological modeling of news and rumors on twitter," in Proceedings of the 7th workshop on social network mining and analysis, pp. 1–9, 2013.
[4]   H. W. Hetcot, "The mathematical of infectious diseases," SIAM Rev, vol. 42, no. 4, pp. 399–653, 2000.
[5]   F. Jia and G. Lv, "Dynamic analysis of a stochastic rumor propagation model," Physica A: Statistical Mechanics and Its Applications, vol. 490, pp. 613–623, 2018.

[6]   D. J. Daley and J. Gani, Epidemic modelling: an introduction. No. 15, Cambridge University Press, 2001.

[7]   D. P. Maki and M. Thompson, "Mathematical models and applications: with emphasis on the social life, and management sciences," Prentice-Hall, Englewood Cliffs, 1973.

[8]   C. Lefevre and P. Picard, "Distribution of the final extent of a rumour process," Journal of Applied Probability, pp. 244–249, 1994.

[9]   J. R. Piqueira, M. Zilbovicius, and C. M. Batistela, "Daley–Kendal models in fake-news scenario," Physica A: Statistical Mechanics and its Applications, vol. 548, 2020.

[10]   J. Jia and W. Wu, "A rumor transmission model with incubation in social networks," Physica A: Statistical Mechanics and Its Applications, vol. 491, pp. 453–462, 2018.

[11]   L. Zhao, Q. Wang, J. Cheng, Y. Chen, J. Wang, and W. Huang, "Rumor spreading model with consideration of forgetting mechanism: A case of online blogging LiveJournal," Physica A: Statistical Mechanics and its Applications, vol. 390, no. 13, pp. 2619–2625, 2011.

[12]   M. Kosfeld, "Rumours and markets," Journal of Mathematical Economics, vol. 41, no. 6, pp. 646–664, 2005.

[13]   S. Galam, "Modelling rumors: the no plane Pentagon French hoax case," Physica A: Statistical Mechanics and Its Applications, vol. 320, pp. 571–580, 2003.

[14]   X. Zhao and J. Wang, "Dynamical model about rumor spreading with medium," Discrete Dynamics in Nature and Society, vol. 2013, 2013.

[15]   J. Dhar, A. Jain, and V. K. Gupta, "A mathematical model of news propagation on online social network and a control strategy for rumor spreading," Social Network Analysis and Mining, vol. 6, no. 1, pp. 1–9, 2016.

[16]   Z. He, Z. Cai, and X. Wang, "Modeling propagation dynamics and developing optimized countermeasures for rumor spreading in online social networks," in 2015 IEEE 35Th international conference on distributed computing systems, pp. 205–214, IEEE, 2015.

[17]   J. Singh, "A new analysis for fractional rumor spreading dynamical model in a social network with Mittag-Leffler law," Chaos: An Interdisciplinary Journal of Nonlinear Science, vol. 29, no. 1, 2019.

[18]   K. Afassinou, "Analysis of the impact of education rate on the rumor spreading mechanism," Physica A: Statistical Mechanics and Its Applications, vol. 414, pp. 43–52, 2014.

[19]   G. Chen, "ILSCR rumor spreading model to discuss the control of rumor spreading in emergency," Physica A: Statistical Mechanics and its Applications, vol. 522, pp. 88–97, 2019.

[20]   C. Wang, Z. X. Tan, Y. Ye, L. Wang, K. H. Cheong, and N.-g. Xie, "A rumor spreading model based on information entropy," Scientific Reports, vol. 7, no. 1, pp. 1–14, 2017.

[21]   J. Wang, L. Zhao, and R. Huang, "2SI2R rumor spreading model in homogeneous networks," Physica A: Statistical Mechanics and its Applications, vol. 413, pp. 153–161, 2014.

[22]   Y. Hu, Q. Pan, W. Hou, and M. He, "Rumor spreading model with the different attitudes towards rumors," Physica A: Statistical Mechanics and its Applications, vol. 502, pp. 331–344, 2018.

[23]   K. S. Miller and B. Ross, An introduction to the fractional calculus and fractional differential equations, A Wiley-Interscience Publication, 1993.

[24]   I. Podlubny, Fractional differential equations: An introduction to fractional derivatives, fractional differential equations, to methods of their solution and some of their applications, Elsevier, 1998.

[25]   M. Caputo and M. Fabrizio, "A new definition of fractional derivative without singular kernel," Journal of Engineering Mechanics, vol. 1, pp. 73–85, 2015.

[26]   A. Atangana and D. Baleanu, "New fractional derivatives with nonlocal and non-singular kernel: Theory and application to heat transfer model," Thermal Science, vol. 20, no. 2, pp. 763–769, 2016.

[27]   M. A. Dokuyucu, E. Celik, H. Bulut, and H. M. Baskonus, "Cancer treatment model with the Caputo-Fabrizio fractional derivative," The European Physical Journal Plus, vol. 133, no. 3, pp. 1–6, 2018.

[28]   K. M. Owolabi, "Behavioural study of symbiosis dynamics via the Caputo and Atangana–Baleanu fractional derivatives," Chaos, Solitons & Fractals, vol. 122, pp. 89–101, 2019.

[29]   C. Baishya, "Dynamics of fractional stage structured predator prey model with prey refuge," Indian Journal of Ecology, vol. 47, no. 4, pp. 1118–1124, 2020.

[30]   D. Baleanu, S. Etemad, and S. Rezapour, "A hybrid Caputo fractional modeling for thermostat with hybrid boundary value conditions," Boundary Value Problems, vol. 2020, no. 1, pp. 1–16, 2020.

[31]   C. Baishya, S. J. Achar, P. Veeresha, and D. G. Prakasha, "Dynamics of a fractional epidemiological model with disease infection in both the populations," Chaos: An Interdisciplinary Journal of Nonlinear Science, vol. 31, no. 4, 2021.

[32]   C. Baishya, "Dynamics of fractional holling type-ii predator-prey model with prey refuge and additional food to predator," Journal of Applied Nonlinear Dynamics, vol. 10, no. 2, pp. 315–328, 2020.

[33]   I. Owusu-Mensah, L. Akinyemi, B. Oduro, and O. S. Iyiola, "A fractional order approach to modeling and simulations of the novel COVID-19," Advances in Difference Equations, no. 1, 2020.

[34]   L. Akinyemi, "A fractional analysis of Noyes–Field model for the nonlinear Belousov–Zhabotinsky reaction," Comp. Appl. Math., vol. 39, no. 3, 2020.

[35]   K. Logeswari and C. Ravichandran, "A new exploration on existence of fractional neutral integro-differential equations in the concept of Atangana–Baleanu derivative," Physica A: Statistical Mechanics and its Applications, vol. 544, 2020.

[36]   C. Ravichandran, K. Logeswari, and F. Jarad, "New results on existence in the framework of Atangana–Baleanu derivative for fractional integro-differential equations," Chaos, Solitons & Fractals, vol. 125, pp. 194–200, 2019.

[37]   S. K. Panda, T. Abdeljawad, and C. Ravichandran, "Novel fixed point approach to Atangana-Baleanu fractional and Lp-Fredholm integral equations," Alexandria Engineering Journal, vol. 59, no. 4, pp. 1959–1970, 2020.

[38]   C. Ravichandran, K. Logeswari, S. K. Panda, and K. S. Nisar, "On new approach of fractional derivative by Mittag-Leffler kernel to neutral integro-differential systems with impulsive conditions," Chaos, Solitons & Fractals, vol. 139, 2020.

[39]   S. K. Panda, C. Ravichandran, and B. Hazarika, "Results on system of Atangana–Baleanu fractional order Willis aneurysm and nonlinear singularly perturbed boundary value problems," Chaos, Solitons & Fractals, vol. 142, 2021.

[40]   M. Mirzazadeh, L. Akinyemi, M. Şenol, and K. Hosseini, "A variety of solitons to the sixth-order dispersive (3+1)-dimensional nonlinear time-fractional Schrödinger equation with cubic-quintic-septic nonlinearities," Optik, vol. 241, 2021.

[41]   K. S. Nisar, K. Jothimani, K. Kaliraj, and C. Ravichandran, "An analysis of controllability results for nonlinear Hilfer neutral fractional derivatives with non-dense domain," Chaos, Solitons & Fractals, vol. 146, 2021.

[42]   P. Veeresha, E. Ilhan, and H. M. Baskonus, "Fractional approach for analysis of the model describing wind-influenced projectile motion," Physica Scripta 96 (7), 2021.

[43]   M. Yavuz, and N. Sene, "Stability Analysis and Numerical Computation of the Fractional Predator–Prey Model with the Harvesting Rate," Fractal and Fractional, vol. 4 (3), 2020.

[44]   S. W. Yao, E. Ilhan, P. Veeresha, and H. M. Baskonus, "A powerful iterative approach for quintic complex Ginzburg-Landau equation within the frame of fractional operator," Fractals vol. 29 (5), 2021, DOI: 10.1142/S0218348X21400235.

[45]   M. Yavuz, and T. Abdeljawad, "On a new integral transformation applied to fractional derivative with Mittag-Leffler nonsingular kernel," Electronic Research Archive, vol. 28 (1), pp. 481-495, 2020.

[46]   P. Veeresha, and D. Baleanu, A unifying computational framework for fractional Gross–Pitaevskii equations, Physica Scripta, 2021, DOI: 10.1088/1402-4896/ac28c9.

[47]   M. Yavuz, and N. Özdemir, "Comparing the new fractional derivative operators involving exponential and Mittag-Leffler kernel, Discrete & Continuous Dynamical Systems-S," vol. 13 (3), pp. 995-1006, 2020.

[48]   P. Veeresha, and D. G. Prakasha, "An efficient technique for two-dimensional fractional order biological population model," International Journal of Modeling, Simulation, and Scientific Computing," vol. 11 (1) 2020, DOI: 10.1142/S1793962320500051.

[49]   P. A. Naik, M. Yavuz, S. Qureshi, J. Zu, and S. Stuart, "Modeling and analysis of COVID-19 epidemics with treatment in fractional derivatives using real data from Pakistan," The European Physical Journal Plus, vol. 135, 2020, DOI: 10.1140/epjp/s13360-020-00819-5.

[50]   E. Ilhan, P. Veeresha, and H. M. Baskonus, Fractional approach for a mathematical model of atmospheric dynamics of CO2 gas with an efficient method, Chaos, Solitons and Fractals, vol. 152, 2021.

[51]   P. A. Naik, K. M. Owolabi, M. Yavuz, and J. Zu, "Chaotic dynamics of a fractional order HIV-1 model involving AIDS-related cancer cells," Chaos, Solitons & Fractals, vol. 140, 2020, DOI: 10.1016/j.chaos.2020.110272.

[52]   M. Yavuz, and N. Özdemir, "Analysis of an epidemic spreading model with exponential decay law," Mathematical Sciences and Applications E-Notes, vol. 8, pp.142-154, 2020.

[53]   D. G. Prakasha, P. Veeresha, and H. M. Baskonus, "Analysis of the dynamics of hepatitis E virus using the Atangana-Baleanu fractional derivative," The European Physical Journal Plus, vol. 134 (5), pp. 1-11, 2019.

[54]   P. A. Naik, M. Yavuz, and J. Zu, The role of prostitution on HIV transmission with memory: A modeling approach, Alexandria Engineering Journal, vol. 59 (4,) pp. 2513-2531, 2020.

[55]   O. Iyiola, B. Oduro, and L. Akinyemi, "Analysis and solutions of generalized Chagas vectors re-infestation model of fractional order type," Chaos, Solitons & Fractals, vol. 145, 2021.

[56]   Y. Li, Y. Chen, and I. Podlubny, Stability of fractional order nonlinear dynamic systems: Lyapunov direct method and generalized mittag-leffler stability," *Computers & Mathematics with Applications*, vol. 59, pp. 1810-1821, 2010.

[57]   *H.L., Li, L., Zhang*, C. Hu, Y.L. Jiang, and Z. Teng, "Dynamical analysis of a fractional-order predator-prey model incorporating a prey refuge," Journal of Applied Mathematics and Computing, vol. 54, no. 1-2, pp. 435–449, 2017.

[58]   J. Wang, L. Zhao, R. Huang, and Y. Chen, "Rumor spreading model on social networks with consideration of remembering mechanism," IET International conference on smart and sustainable city, 2013.

[59]   K. Diethelm, and F. NJ, "A predictor-corrector approach for numerical solution of fractional differential euations," Nonlinear Dynamics, vol. 29, pp. 3–22, 2002.

[60]   P. Veeresha, A Numerical Approach to the Coupled Atmospheric Ocean Model Using a Fractional Operator. Mathematical Modelling and Numerical Simulation with Applications, vol. 1 (1), pp. 1-10, 2021.

[61]   A. Yokuş, Construction of Different Types of Traveling Wave Solutions of the Relativistic Wave Equation Associated with the Schrödinger Equation. Mathematical Modelling and Numerical Simulation with Applications, vol. 1 (1), pp. 24-31, 2021.

[62]   Z. Hammouch, M. Yavuz, and N. Özdemir, Numerical Solutions and Synchronization of a Variable-Order Fractional Chaotic System. Mathematical Modelling and Numerical Simulation with Applications, vol. 1 (1), pp. 11-23, 2021.

A Unified Approach for the Fractional System of Equations Arising in the Biochemical Reaction without Singular Kernel

P. Veeresha[1*], M.S. Kiran[2], L. Akinyemi[3] and **Mehmet Yavuz[4]**

[1]*Center for Mathematical Needs, Department of Mathematics, CHRIST (Deemed to be University), Bengaluru 560029, India*

[2]*Research Centre, Department of Chemistry, GM Institute of Technology, Davangere 577006, India*

[3]*Department of Mathematics, Lafayette College, Easton, Pennsylvania, USA*

[4]*Department of Mathematics and Computer Sciences, Necmettin Erbakan University, 42090 Konya, Turkey*

Abstract: The pivotal aim of the present work is to find the solution for the fractional system of equations arising in the biochemical reaction using q-homotopy analysis transform method (q-HATM). The hired scheme technique unification of Laplace transform with q-homotopy analysis method, and fractional derivative defined with Caputo-Fabrizio (CF) operator. To validate and illustrate the competence of the future method, we examined the model in terms of fractional order. The fixed-point theorem hired to demonstrates the existence and uniqueness. Moreover, the physical nature of achieved solutions has been captured in terms of plots for different order. The obtained results elucidate that the considered algorithm is easy to implement, highly methodical, and very effective as well as accurate to analyse the nature of nonlinear differential equations of fractional order arising in the connected areas of science and engineering.

Keywords: Biochemical reaction, Caputo-Fabrizio derivative, Enzyme kinetics, Mathematical model, Homotopy analysis method, Laplace transform.

INTRODUCTION

The concept of Fractional calculus (FC) was initiated in Newton's time. Nevertheless, it fascinated the attention of many authors recently. The concept of classical or integer-order is associated with power law, and it has a wide range of

*Corresponding author P. Veeresha: Center for Mathematical Needs, Department of Mathematics, Christ (Deemed to be University), Bengaluru 560029, India; E-mail: pundikala.veeresha@christuniversity.in

Mehmet Yavuz & Necati Özdemir (Eds.)

applications and admits numerous properties. However, mankind always looks for the invocation and modification for betterment to lead life systematically and happily. Recently, scientists and mathematicians notice that, the classical concept is not able to capture memory and hereditary-based consequences, and then they suggested with some essential results that the concept of FC is suitable for these types of studies. Moreover, the classical concept is a subset of FC, and we can be able to capture the behaviour of the corresponding system for classical values as a particular case. The main reason for the researchers attracted towards the study of FC is the open problems (including physical and geometrical interpretation of the concept), basic rules admitted by classical concept to extend the results and many others. Many researchers present their own notions and viewpoints with novel concepts in different forms [1-6].

Studying the biochemical models with the mathematical system is always a venue for innovation and development to understand and predict the corresponding complex behaviour of phenomena. Particularly, in biochemical systems, enzyme kinetics have been effectively exemplified with the aid of a system of ordinary differential equations [7-9]. The system was constructed uniquely on reactions that were deprived of spatial dependency of the several concentrations. Here, we consider the system studied by authors in the study [10, 11], and the corresponding enzyme reaction model is presented with enzyme E, substrate S, and product P as [12]

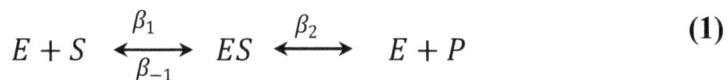

$$E + S \underset{\beta_{-1}}{\overset{\beta_1}{\longleftrightarrow}} ES \overset{\beta_2}{\longleftrightarrow} E + P \qquad (1)$$

where β_1, β_{-1} and β_2 are positive rate constants for each reaction. Here, ES the enzyme-substrate intermediate complex. Now, the reactants concentration is presented as follows for Eq. (1) $s = [S]$, $c = [C], p = [P]$, $e = [E]$. Then employing the law of mass action, we have [10-12] system with associated conditions

$$\frac{ds}{dt} = -\beta_1 es + \beta_{-1} c, \quad s(0) = s_0,$$

$$\frac{de}{dt} = -\beta_1 es + (k_{-1} + k_2)c, \quad e(0) = e_0, \qquad (2)$$

$$\frac{dc}{dt} = k_1 es - (k_{-1} + k_2)c, \quad c(0) = c_0,$$

$$\frac{dp}{dt} = k_2 c, \quad p(0) = p_0.$$

By the assist of Eq. (2), authors in [12] derived the following system

$$\frac{d\mathcal{P}(t)}{dt} = -\alpha\mathcal{P} + \alpha(\mathcal{P} + \beta - \varepsilon)\mathcal{Q},$$

$$\frac{d\mathcal{Q}(t)}{dt} = \mathcal{P} - (\mathcal{P} + \beta)\mathcal{Q}, \tag{3}$$

$$\frac{d\mathcal{S}(t)}{dt} = \varepsilon\mathcal{Q}.$$

The projected system is analyzed by many researchers to illustrate their viewpoints using numerical schemes. For instance, variational iteration scheme [10], multistage homotopy analysis algorithm, Adomian decomposition [13] and multistage homotopy-perturbation techniques [14] and others.

The fractional-order derivatives are familiarized by Leibnitz soon after the classical concept. As compared to classical calculus, it was soon discovered that fractional calculus (FC) is more suitable capturing complex phenomena [15-25]. The FC considered is the essential apparatus to illustrate the chemical and biological phenomena. Most of the mathematical models demonstrate the non-local distributed effects, hereditary properties and system memory. These properties are necessary to describe the above-cited phenomena. The pivotal aim of generalizing the integer to fractional order is to capture consequences related to non-locality, long-range memory and time-based properties and also anomalous diffusion aspects [26-31]. Most familiarly hired operators to analyze many models are Riemann, Liouville, Caputo, Fabrizio and others [1-6, 32, 33]. In this connection, Caputo and Fabrizio in 2015 overcome the many limitations raised by many mathematicians to generalize complex models, and then many scholars hired to present simulating consequences. It has been proved by many researchers that, the CF fractional operator has great results compared to other fractional operators.

In the present work, we consider the fractional-order system in order to include all the above-described consequences into the system cited in Eq. (3) and which as follows

$$\,^{CF}_{0}D^{\mu}_{t}\mathcal{P}(t) = -\alpha\mathcal{P} + \alpha(\mathcal{P} + \beta - \varepsilon)\mathcal{Q},$$

$$\,^{CF}_{0}D^{\mu}_{t}\mathcal{Q}(t) = \mathcal{P} - (\mathcal{P} + \beta)\mathcal{Q}, \qquad\qquad 0 < \mu \le 1 \qquad (4)$$

$$\,^{CF}_{0}D^{\mu}_{t}\mathcal{S}(t) = \varepsilon\mathcal{Q},$$

where $\,^{CF}_{0}D^{\mu}_{t}$ represents the Caputo-Fabrizio fractional derivative with respect to time (t) of order μ.

In the present work, we find the solution and examine behaviours of the system of four equations, demonstrating the chemical reaction evolved in the helium burning network by using q-HATM. The hired method is familiarized by Singh et al. [34] using Laplace transform (LT) with the concept of HAM [35]. Recently, due to its consistency and efficacy, many researchers extremely employed to find the solution for numerous types of nonlinear models [36-45]. The hired scheme can lessen the time and computational work while maintaining great accuracy as weigh against the other established technique. Moreover, the concept of fractional calculus with suitable numerical and analytical schemes offers us enormous freedom to the deliberate type of equation [46-57].

The present work is distributed as follows; in the subsequent section, preliminaries are recalled, playing a vital role in the present framework. The fundamental rule of the hired algorithm is discussed in Section 3 with fractional order and by the aid of this algorithm, we find the solution for the projected system with the newly proposed fractional operator in Section 4. The existence and uniqueness of the hired fractional operator with the projected system are illustrated in Section 5 and further. The behaviour of the achieved results is presented in terms of plots and discussed their corresponding consequences in Sections 6 and 7.

PRELIMINARIES

Here, we recalled some primary results [1-6, 32, 33].

Definition 1. The CF fractional derivative for $f \in H^{1}(a, b)$ is presented as [35]

$$D^{\mu}_{t}\big(f(t)\big) = \frac{\mathcal{B}(\mu)}{1 - \mu} \int_{a}^{t} f'(t) \exp\left[-\mu \frac{t - \vartheta}{1 - \mu}\right] d\vartheta, b > a,$$

where $\mathcal{B}(\mu)$ is a normalization function and admits $\mathcal{B}(0) = \mathcal{B}(1) = 1$. Further, if $f \notin H^{1}(a, b)$ then we have,

$$D_t^\mu(f(t)) = \frac{\mu B(\mu)}{1-\mu} \int_{-\infty}^{t} (f(t) - f(\vartheta)) \exp\left[-\mu \frac{t-\vartheta}{1-\mu}\right] d\vartheta.$$

Definition 2. The CF fractional integral for $f \in H^1(a,b)$ is presented as [54]

$$I_t^\mu(f(t)) = \frac{2(1-\mu)}{(2-\mu)B(\mu)} f(t) + \frac{2\mu}{(2-\mu)B(\mu)} \int_0^t f(\vartheta) d\vartheta,$$
$$0 < \mu < 1, t \geq 0.$$

Note: According to [54], the following must hold

$$\frac{2(1-\mu)}{(2-\mu)B(\mu)} + \frac{2\alpha}{(2-\mu)B(\mu)} = 1, 0 < \alpha < 1,$$

which gives $B(\mu) = \frac{2}{2-\mu}$. By the assist of above equation researchers in [54] proposed a novel Caputo derivative as follows

$$D_t^\mu(f(t)) = \frac{1}{1-\mu} \int_0^t f'(t) \exp\left[\mu \frac{t-\vartheta}{1-\mu}\right] d\vartheta, 0 < \mu < 1.$$

Definition 3. The *LT* for a CF derivative ${}_0^{CF}D_t^\mu f(t)$ is presented as [32] below

$$\mathcal{L}\left[{}_0^{CF}D_t^{(\mu+n)} f(t)\right]$$
$$= \frac{s^{n+1}\mathcal{L}[f(t)] - s^n f(0) - s^{n-1}f'(0) - \cdots - f^{(n)}(0)}{s + (1-s)\mu}.$$

FUNDAMENTAL IDEA OF THE CONSIDERED SCHEME

In this section, we hired the differential equation to present the basic procedure of the projected scheme with initial conditions

$${}_0^{CF}D_t^\mu v(x,t) + \mathcal{R}\, v(x,t) + \mathcal{N}v(x,t) = f(x,t), \quad 0 < \mu \leq 1, \tag{5}$$

and

$$v(x, 0) = g(x).$$

We obtained by applying LT on Eq. (5)

$$\mathcal{L}[v(x,t)] - \frac{g(x)}{s} + \frac{s+(1-s)\mu}{s}\{\mathcal{L}[\mathcal{R}v(x,t)] + \mathcal{L}[\mathcal{N} \ v(x,t)] - \mathcal{L}[f(x,t)]\} = 0.$$

For $\varphi(x, t; q)$, $\mathcal{N}$ is contracted as follows

$$\mathcal{N}[\varphi(x,t;q)] = \mathcal{L}[\varphi(x,t;q)] - \frac{g(x)}{s}$$

$$+ \frac{s+(1-s)\mu}{s}\{\mathcal{L}[\mathcal{R} \ \varphi(x,t;q)] + L[\mathcal{N}\varphi(x,t;q)] - L[f(x,t)]\},$$

where $q \in \left[0, \frac{1}{n}\right]$. Then, the homotopy is defined by results in [36]

$$(1 - nq)\mathcal{L}[\varphi(x,t;q) - v_0(x,t)] = \hbar q\mathcal{N}[\varphi(x,t;q)], \tag{6}$$

where L is signifying LT. For $q = 0$ and $q = \frac{1}{n}$, the following conditions satisfies

$$\varphi(x,t;0) = v_0(x,t), \ \varphi\left(x,t;\frac{1}{n}\right) = v(x,t).$$

By using Taylor theorem, we get

$$\varphi(x,t;q) = v_0(x,t) + \sum_{m=1}^{\infty} v_m(x,t)q^m,$$

where

$$v_m(x,t) = \frac{1}{m!}\frac{\partial^m \varphi(x,t;q)}{\partial q^m}\Big|_{q=0}.$$

For the proper chaise of $v_0(x,t)$, n and $\hbar$ the series (15) converges at $q = \frac{1}{n}$. Then

$$v(x,t) = v_0(x,t) + \sum_{m=1}^{\infty} v_m(x,t)\left(\frac{1}{n}\right)^m.$$

After differentiating Eq. (6) m-times with q and multiplying by $\frac{1}{m!}$ and substituting $q = 0$, one can get

$$\mathcal{L}[v_m(x,t) - k_m v_{m-1}(x,t)] = \hbar \mathfrak{R}_m(\vec{v}_{m-1}), \tag{7}$$

where the vectors are defined as

$$\vec{v}_m = \{v_0(x,t), v_1(x,t), \dots, v_m(x,t)\}.$$

Eq. (7) reduces after employing inverse LT to

$$v_m(x,t) = k_m v_{m-1}(x,t) + \hbar \mathcal{L}^{-1}[\mathfrak{R}_m(\vec{v}_{m-1})], \tag{8}$$

where

$$\mathfrak{R}_m(\vec{v}_{m-1}) = L[v_{m-1}(x,t)] - \left(1 - \frac{k_m}{n}\right)\left(\frac{g(x)}{s}\right) + \frac{s+(1-s)\mu}{s} L[f(x,t)]$$

$$+ \frac{s+(1-s)\mu}{s} L[Rv_{m-1} + \mathcal{H}_{m-1}], \tag{9}$$

and

$$k_m = \begin{cases} 0, & m \le 1, \\ n, & m > 1. \end{cases}$$

Here, $\mathcal{H}_m$ is homotopy polynomial and presented as

$$\mathcal{H}_m = \frac{1}{m!}\left[\frac{\partial^m \varphi(x,t;q)}{\partial q^m}\right]_{q=0} \text{ and } \varphi(x,t;q) = \varphi_0 + q\varphi_1 + q^2\varphi_2 + \cdots.$$

Using Eqs. (8) and (9), we found

$$v_m(x,t) = (k_m + \hbar)v_{m-1}(x,t) - \left(1 - \frac{k_m}{n}\right)\mathcal{L}^{-1}\left(\frac{g(x)}{s}\right)$$

$$+ \frac{s+(1-s)\mu}{s} L[f(x,t)] + \hbar \mathcal{L}^{-1}\left\{\frac{s+(1-s)\mu}{s} L[Rv_{m-1} + \mathcal{H}_{m-1}]\right\}. \tag{10}$$

By the help of q-HATM, the series solution is

$$v(x,t) = v_0(x,t) + \sum_{m=1}^{\infty} v_m(x,t)\left(\frac{1}{n}\right)^m. \tag{11}$$

IMPLEMENTATION OF THE q-HOMOTOPY ANALYSIS TRANSFORM METHOD

Consider the system of equation cited in Eq. (4) in CF fractional derivative

$$
{}^{CF}_0 D^\mu_t \mathcal{P}(t) + \alpha \mathcal{P} - \alpha(\mathcal{P} + \beta - \varepsilon)\mathcal{Q} = 0,
$$

$$
{}^{CF}_0 D^\mu_t \mathcal{Q}(t) - \mathcal{P} + (\mathcal{P} + \beta)\mathcal{Q} = 0, \tag{12}
$$

$$
{}^{CF}_0 D^\mu_t \mathcal{S}(t) - \varepsilon \mathcal{Q} = 0,
$$

associated to

$$
A(0) = \mathcal{P}_0, \quad \mathcal{Q}(0) = \mathcal{Q}_0 \text{ and } \mathcal{S}(0) = \mathcal{S}_0.
$$

Applying Laplace transform on Eq. (12) and then with the help of Eq. (11), we get

$$
L[\mathcal{P}(t)] - \frac{1}{s}(\mathcal{P}_0) + \frac{s+(1-s)\mu}{s} L\{\alpha\mathcal{P} - \alpha(\mathcal{P} + \beta - \varepsilon)\mathcal{Q}\} = 0,
$$

$$
L[\mathcal{Q}(t)] - \frac{1}{s}(\mathcal{Q}_0) + \frac{s+(1-s)\mu}{s} L\{-\mathcal{P} + (\mathcal{P} + \beta)\mathcal{Q}\} = 0,
$$

$$
L[\mathcal{S}(t)] - \frac{1}{s}(\mathcal{S}_0) - \frac{s+(1-s)\mu}{s} L\{\varepsilon\mathcal{Q}\} = 0.
$$

The non-linear operator $N[\varphi_1(t;q), \varphi_2(t;q), \varphi_3(t;q)]$ defined as

$$
N^1 = L[\varphi_1(t;q)] - \frac{1}{s}(\mathcal{P}_0)
$$

$$
+ \frac{s+(1-s)\mu}{s} L\{\alpha\varphi_1(t;q) - \alpha(\varphi_1(t;q) + \beta - \varepsilon)\varphi_2(t;q)\},
$$

$$
N^2 = L[\varphi_2(t;q)] - \frac{1}{s}(\mathcal{Q}_0)
$$

$$
+ \frac{s+(1-s)\mu}{s} L\{-\varphi_1(t;q) + (\varphi_1(t;q) + \beta)\varphi_2(t;q)\},
$$

$$
N^3 = L[\varphi_3(t;q)] - \frac{1}{s}(\mathcal{S}_0) - \frac{s+(1-s)\mu}{s} L\{\varepsilon\varphi_2(t;q)\}.
$$

The m-th order deformation equation by the projected scheme at $\mathcal{H}(t) = 1$ is given by

$$L[\mathcal{P}_m(t) - k_m \mathcal{P}_{m-1}(t)] = \hbar L^{-1}\{\Re_{1,m}[\vec{\mathcal{P}}_{m-1}, \vec{\mathcal{Q}}_{m-1}, \vec{\mathcal{S}}_{m-1}]\},$$

$$L[\mathcal{Q}_m(t) - k_m \mathcal{Q}_{m-1}(t)] = \hbar L^{-1}\{\Re_{2,m}[\vec{\mathcal{P}}_{m-1}, \vec{\mathcal{Q}}_{m-1}, \vec{\mathcal{S}}_{m-1}]\}, \qquad (13)$$

$$L[\mathcal{S}_m(t) - k_m \mathcal{S}_{m-1}(t)] = \hbar L^{-1}\{\Re_{3,m}[\vec{\mathcal{P}}_{m-1}, \vec{\mathcal{Q}}_{m-1}, \vec{\mathcal{S}}_{m-1}]\},$$

where

$$\Re_{1,m}[\vec{\mathcal{P}}_{m-1}, \vec{\mathcal{Q}}_{m-1}, \vec{\mathcal{S}}_{m-1}] = L[\mathcal{P}_{m-1}(t)] - \left(1 - \frac{k_m}{n}\right)\frac{1}{s}(\mathcal{P}_0)$$

$$+ \frac{s + (1-s)\mu}{s} L\left\{\alpha \mathcal{P}_{m-1} - \alpha\left(\sum_{i=0}^{m-1} \mathcal{P}_i \mathcal{Q}_{m-1-i} + (\beta - \varepsilon)\mathcal{Q}_{m-1}\right)\right\},$$

$$\Re_{2,m}[\vec{\mathcal{P}}_{m-1}, \vec{\mathcal{Q}}_{m-1}, \vec{\mathcal{S}}_{m-1}] = L[\mathcal{Q}_{m-1}(t)] - \left(1 - \frac{k_m}{n}\right)\frac{1}{s}(\mathcal{Q}_0)$$

$$+ \frac{s + (1-s)\mu}{s} L\left\{-\mathcal{P}_{m-1} + \sum_{i=0}^{m-1} \mathcal{P}_i \mathcal{Q}_{m-1-i} + \beta \mathcal{Q}_{m-1}\right\},$$

$$\Re_{3,m}[\vec{\mathcal{P}}_{m-1}, \vec{\mathcal{Q}}_{m-1}, \vec{\mathcal{S}}_{m-1}] = L[\mathcal{S}_{m-1}(t)] - \left(1 - \frac{k_m}{n}\right)\frac{1}{s}(\mathcal{S}_0)$$

$$- \frac{s + (1-s)\mu}{s} L\{\varepsilon \mathcal{Q}_{m-1}\}.$$

On employing inverse LT on Eq. (13), it simplifies to

$$\mathcal{P}_m(t) = k_m \mathcal{P}_{m-1}(t) + \hbar L^{-1}\{\Re_{1,m}[\vec{\mathcal{P}}_{m-1}, \vec{\mathcal{Q}}_{m-1}, \vec{\mathcal{S}}_{m-1}]\},$$

$$\mathcal{Q}_m(t) = k_m \mathcal{Q}_{m-1}(t) + \hbar L^{-1}\{\Re_{2,m}[\vec{\mathcal{P}}_{m-1}, \vec{\mathcal{Q}}_{m-1}, \vec{\mathcal{S}}_{m-1}]\},$$

$$\mathcal{S}_m(t) = k_m \mathcal{S}_{m-1}(t) + \hbar L^{-1}\{\Re_{3,m}[\vec{\mathcal{P}}_{m-1}, \vec{\mathcal{Q}}_{m-1}, \vec{\mathcal{S}}_{m-1}]\}.$$

By using $\mathcal{P}_0(t)$, $\mathcal{Q}_0(t)$ and $\mathcal{S}_0(t)$ and then solving the forgoing equations, we can obtain the terms of

$$\mathcal{P}(t) = \mathcal{P}_0(t) + \sum_{m=1}^{\infty} \mathcal{P}_m(t) \left(\frac{1}{n}\right)^m,$$

$$Q(t) = Q_0(t) + \sum_{m=1}^{\infty} Q_m(t) \left(\frac{1}{n}\right)^m,$$

$$S(t) = S_0(t) + \sum_{m=1}^{\infty} S_m(t) \left(\frac{1}{n}\right)^m.$$

EXISTENCE AND UNIQUENESS OF SOLUTIONS

In this section, the existence and uniqueness are illustrated for the considered system with the assist of fixed-point theory. We consider Eq. (12) as follows

$$\begin{cases} {}^{CF}_0 D_t^\mu [\mathcal{P}(t)] = \mathcal{G}_1(t,\mathcal{P}), \\ {}^{CF}_0 D_t^\mu [Q(t)] = \mathcal{G}_2(t,Q), \\ {}^{CF}_0 D_t^\mu [S(t)] = \mathcal{G}_3(t,S). \end{cases}$$

Now, using Eq. (10) and results derived in [46], we obtained

$$\begin{cases} \mathcal{P}(t) - \mathcal{P}(0) = {}^{CF}_0 I_t^\mu \{-\alpha\mathcal{P} + \alpha(\mathcal{P} + \beta - \varepsilon)Q\}, \\ Q(t) - Q(0) = {}^{CF}_0 I_t^\mu \{\mathcal{P} - (\mathcal{P} + \beta)Q\}, \\ S(t) - S(0) = {}^{CF}_0 I_t^\mu \{\varepsilon Q\}. \end{cases}$$

Then we have from [54] as follows

$$\begin{cases} \mathcal{P}(t) - \mathcal{P}(0) = \dfrac{2(1-\mu)}{B(\mu)} \mathcal{G}_1(t,\mathcal{P}) + \dfrac{2\alpha}{(2-\mu)B(\mu)} \displaystyle\int_0^t \mathcal{G}_1(\zeta,\mathcal{P})d\zeta, \\ Q(t) - Q(0) = \dfrac{2(1-\mu)}{B(\mu)} \mathcal{G}_2(t,Q) + \dfrac{2\alpha\mu}{(2-\alpha)B(\mu)} \displaystyle\int_0^t \mathcal{G}_2(\zeta,Q)d\zeta, \qquad (14) \\ S(t) - S(0) = \dfrac{2(1-\mu)}{B(\mu)} \mathcal{G}_3(t,S) + \dfrac{2\mu}{(2-\mu)B(\mu)} \displaystyle\int_0^t \mathcal{G}_3(\zeta,S)d\zeta. \end{cases}$$

Theorem 1. The kernel $\mathcal{G}_1$ admits the Lipschitz condition and contraction if $0 \leq \left(\alpha(1 - (1 + \beta - \varepsilon)\Lambda_2)\right) < 1$ satisfies.

Proof. Let us consider the two functions A and A_1 to prove the theorem, then

$$\|\mathcal{G}_1(t,\mathcal{P}) - \mathcal{G}_1(t,\mathcal{P}_1)\| = \|\alpha\big(\mathcal{P}(t) - \mathcal{P}(t_1)\big) - \alpha(Q\big(\mathcal{P}(t) - \mathcal{P}(t_1)\big) - (\beta - \varepsilon)Q)\|$$

$$= \|\alpha(1 - (1 + \beta - \varepsilon)Q)\big(\mathcal{P}(t) - \mathcal{P}(t_1)\big)\|$$

$$\leq \|\alpha(1 - (1 + \beta - \varepsilon)Q)\|\|\mathcal{P}(t) - \mathcal{P}(t_1)\|$$

$$\leq \left(\alpha(1 - (1 + \beta - \varepsilon)\Lambda_2)\right)\|\mathcal{P}(t) - \mathcal{P}(t_1)\|,$$

where $\|Q\| \leq \Lambda_2$ is a bounded function. Putting $\eta_1 = \alpha(1 - (1 + \beta - \varepsilon)\Lambda_2)$ in the above inequality, then we have

$$\|\mathcal{G}_1(t, \mathcal{P}) - \mathcal{G}_1(t, \mathcal{P}_1)\| \leq \eta_1 \|\mathcal{P}(t) - \mathcal{P}(t_1)\|. \tag{15}$$

Eq. (15) provides the Lipschitz condition for $\mathcal{G}_1$. Similarly, we can see that if $0 \leq \alpha(1 - (1 + \beta - \varepsilon)\Lambda_2) < 1$, then it implies the contraction. Similarly, we can prove

$$\|\mathcal{G}_2(t, Q) - \mathcal{G}_2(t, Q_1)\| \leq \eta_2 \|Q(t) - Q(t_1)\|,$$

$$\|\mathcal{G}_3(t, \mathcal{S}) - \mathcal{G}_3(t, \mathcal{S}_1)\| \leq \eta_3 \|\mathcal{S}(t) - \mathcal{S}(t_1)\|.$$

By the assist of the above equations, Eq. (14) simplifies to

$$\begin{cases} \mathcal{P}(t) = \mathcal{P}(0) + \dfrac{2(1 - \mu)}{(2 - \mu)B(\mu)}\mathcal{G}_1(t, \mathcal{P}) + \dfrac{2\mu}{(2 - \mu)B(\mu)}\displaystyle\int_0^t \mathcal{G}_1(\zeta, \mathcal{P})d\zeta, \\[3mm] Q(t) = Q(0) + \dfrac{2(1 - \mu)}{(2 - \mu)B(\mu)}\mathcal{G}_2(t, Q) + \dfrac{2\mu}{(2 - \mu)B(\mu)}\displaystyle\int_0^t \mathcal{G}_2(\zeta, Q)d\zeta, \\[3mm] \mathcal{S}(t) = \mathcal{S}(0) + \dfrac{2(1 - \mu)}{(2 - \mu)B(\mu)}\mathcal{G}_3(t, \mathcal{S}) + \dfrac{2\mu}{(2 - \mu)B(\mu)}\displaystyle\int_0^t \mathcal{G}_3(\zeta, \mathcal{S})d\zeta. \end{cases}$$

Then we get the recursive form as follows

$$\begin{cases} \mathcal{P}_n(t) = \dfrac{2(1 - \mu)}{(2 - \alpha)B(\mu)}\mathcal{G}_1(t, \mathcal{P}_{n-1}) + \dfrac{2\alpha}{(2 - \alpha)B(\mu)}\displaystyle\int_0^t \mathcal{G}_1(\zeta, \mathcal{P}_{n-1})d\zeta, \\[3mm] Q_n(t) = \dfrac{2(1 - \mu)}{(2 - \alpha)B(\mu)}\mathcal{G}_2(t, Q_{n-1}) + \dfrac{2\alpha}{(2 - \alpha)B(\mu)}\displaystyle\int_0^t \mathcal{G}_2(\zeta, Q_{n-1})d\zeta, \\[3mm] \mathcal{S}_n(t) = \dfrac{2(1 - \mu)}{(2 - \alpha)B(\mu)}\mathcal{G}_3(t, \mathcal{S}_{n-1}) + \dfrac{2\alpha}{(2 - \alpha)B(\mu)}\displaystyle\int_0^t \mathcal{G}_3(\zeta, \mathcal{S}_{n-1})d\zeta. \end{cases}$$

The associated initial conditions are

$$\mathcal{P}(0) = \mathcal{P}_0(t), \quad Q(0) = Q_0(t) \text{ and } \mathcal{S}(0) = \mathcal{S}_0(t).$$

Now, between the terms the successive difference is defined as

$$\phi_{1n}(t) = \mathcal{P}_n(t) - \mathcal{P}_{n-1}(t) = \frac{2(1-\mu)}{(2-\mu)B(\mu)}\left(\mathcal{G}_1(t,\mathcal{P}_{n-1}) - \mathcal{G}_1(t,\mathcal{P}_{n-2})\right)$$

$$+\frac{2\mu}{(2-\mu)B(\mu)}\int_0^t\left(\mathcal{G}_1(t,\mathcal{P}_{n-1}) - \mathcal{G}_1(t,\mathcal{P}_{n-2})\right)d\zeta,$$

$$\phi_{2n}(t) = \mathcal{Q}_n(t) - \mathcal{Q}_{n-1}(t) = \frac{2(1-\mu)}{(2-\mu)B(\mu)}\left(\mathcal{G}_2(t,\mathcal{Q}_{n-1}) - \mathcal{G}_2(t,\mathcal{Q}_{n-2})\right)$$

$$+\frac{2\mu}{(2-\mu)B(\mu)}\int_0^t\left(\mathcal{G}_2(t,\mathcal{Q}_{n-1}) - \mathcal{G}_1(t,\mathcal{Q}_{n-2})\right)d\zeta, \tag{16}$$

$$\phi_{3n}(t) = \mathcal{S}_n(t) - \mathcal{S}_{n-1}(t) = \frac{2(1-\mu)}{(2-\mu)B(\mu)}\left(\mathcal{G}_3(t,\mathcal{S}_{n-1}) - \mathcal{G}_3(t,\mathcal{S}_{n-2})\right)$$

$$+\frac{2\mu}{(2-\mu)B(\mu)}\int_0^t\left(\mathcal{G}_3(t,\mathcal{S}_{n-1}) - \mathcal{G}_3(t,\mathcal{S}_{n-2})\right)d\zeta.$$

Notice that

$$\begin{cases} \mathcal{P}_n(t) = \sum_{i=1}^n \phi_{1i}(t), \\ \mathcal{Q}_n(t) = \sum_{i=1}^n \phi_{2i}(t), \\ \mathcal{S}_n(t) = \sum_{i=1}^n \phi_{3i}(t). \end{cases}$$

Then we have

$$\|\phi_{1n}(t)\| = \|\mathcal{P}_n(t) - \mathcal{P}_{n-1}(t)\| = \left\|\frac{2(1-\mu)}{(2-\mu)B(\mu)}\left(\mathcal{G}_1(t,\mathcal{P}_{n-1}) - \mathcal{G}_1(t,\mathcal{P}_{n-2})\right)\right.$$

$$\left. +\frac{2\mu}{(2-\mu)B(\mu)}\int_0^t\left(\mathcal{G}_1(t,\mathcal{P}_{n-1}) - \mathcal{G}_1(t,\mathcal{P}_{n-2})\right)d\zeta\right\|.$$

Application of the triangular inequality, Eq. (16) reduces to

$$\|\phi_{1n}(t)\| = \|\mathcal{P}_n(t) - \mathcal{P}_{n-1}(t)\| = \frac{2(1-\mu)}{(2-\mu)B(\mu)}\left\|\left(\mathcal{G}_1(t,\mathcal{P}_{n-1}) - \mathcal{G}_1(t,\mathcal{P}_{n-2})\right)\right\|$$

$$+\frac{2\mu}{(2-\mu)B(\mu)}\left\|\int_0^t\left(\mathcal{G}_1(t,\mathcal{P}_{n-1}) - \mathcal{G}_1(t,\mathcal{P}_{n-2})\right)d\zeta\right\|.$$

The Lipschitz condition satisfied by the kernel T_1, so, we have

$$\|\phi_{1n}(t)\| = \|\mathcal{P}_n(t) - \mathcal{P}_{n-1}(t)\| \le \frac{2(1-\mu)}{(2-\mu)\mathcal{B}(\mu)}\eta_1\|\phi_{1(n-1)}(t)\|$$

$$+\frac{2\mu}{(2-\mu)\mathcal{B}(\mu)}\eta_1\int_0^t \|\phi_{1(n-1)}(t)\|d\zeta. \tag{17}$$

Similarly, we have

$$\|\phi_{2n}(t)\| \le \frac{2(1-\mu)}{(2-\mu)\mathcal{B}(\mu)}\eta_2\|\phi_{2(n-1)}(t)\| + \frac{2\mu}{(2-\mu)\mathcal{B}(\mu)}\eta_2\int_0^t\|\phi_{2(n-1)}(\zeta)\|d\zeta,$$

$$\|\phi_{3n}(t)\| \le \frac{2(1-\mu)}{(2-\mu)\mathcal{B}(\mu)}\eta_3\|\phi_{3(n-1)}(t)\| + \frac{2\mu}{(2-\mu)\mathcal{B}(\mu)}\eta_3\int_0^t\|\phi_{3(n-1)}(\zeta)\|d\zeta. \tag{18}$$

By the help of the above result, we state the following theorem:

Theorem 2. If we have specific t_0, then the solution for Eq. (12) will exist and unique. Further, we have

$$\frac{2(1-\mu)}{(2-\mu)\mathcal{B}(\mu)}\eta_i + \frac{2\mu}{(2-\mu)\mathcal{B}(\mu)}\eta_i t_0 < 1,$$

for $i = 1, 2$ and 3.

Proof. Let $\mathcal{P}(t)$, $\mathcal{Q}(t)$ and $\mathcal{S}(t)$ be the bounded functions admitting the Lipschitz condition. Then, we get by Eqs. (17) and (18)

$$\|\phi_{1i}(t)\| \le \|\mathcal{P}_n(0)\| \left[\frac{2(1-\mu)}{(2-\mu)\mathcal{B}(\mu)}\eta_1 + \frac{2\mu}{(2-\mu)\mathcal{B}(\mu)}\eta_1 t\right]^n,$$

$$\|\phi_{2i}(t)\| \le \|\mathcal{Q}_n(0)\| \left[\frac{2(1-\mu)}{(2-\mu)\mathcal{B}(\mu)}\eta_2 + \frac{2\mu}{(2-\mu)\mathcal{B}(\mu)}\eta_2 t\right]^n, \tag{19}$$

$$\|\phi_{3i}(t)\| \le \|\mathcal{S}_n(0)\| \left[\frac{2(1-\mu)}{(2-\mu)\mathcal{B}(\mu)}\eta_3 + \frac{2\mu}{(2-\mu)\mathcal{B}(\mu)}\eta_3 t\right]^n.$$

Therefore, for the obtained solutions, continuity and existence are verified. Now, to prove that Eq. (19) is a solution for Eq. (12), we consider

$$\mathcal{P}(t) - \mathcal{P}(0) = \mathcal{P}_n(t) - \mathcal{K}_{1n}(t),$$

$$Q(t) - Q(0) = Q_n(t) - \mathcal{K}_{2n}(t),$$

$$S(t) - S(0) = S_n(t) - \mathcal{K}_{3n}(t).$$

Let us consider

$$\|\mathcal{K}_{1n}(t)\| = \| \tfrac{2(1-\mu)}{(2-\mu)\mathcal{B}(\mu)} \big(\mathcal{G}_1(t,\mathcal{P}) - \mathcal{G}_1(t,\mathcal{P}_{n-1}) \big)$$

$$+ \tfrac{2\mu}{(2-\mu)\mathcal{B}(\mu)} \int_0^t \big(\mathcal{G}_1(\zeta,\mathcal{P}) - \mathcal{G}_1(\zeta,\mathcal{P}_{n-1}) \big) d\zeta \|$$

$$\leq \tfrac{2(1-\mu)}{(2-\mu)\mathcal{B}(\mu)} \big\| \big(\mathcal{G}_1(t,\mathcal{P}) - \mathcal{G}_1(t,\mathcal{P}_{n-1}) \big) \big\|$$

$$+ \tfrac{2\mu}{(2-\mu)\mathcal{B}(\mu)} \int_0^t \big\| \big(\mathcal{G}_1(\zeta,\mathcal{P}) - \mathcal{G}_1(\zeta,\mathcal{P}_{n-1}) \big) \big\| d\zeta$$

$$\leq \tfrac{2(1-\mu)}{(2-\mu)\mathcal{B}(\mu)} \eta_1 \|\mathcal{P} - \mathcal{P}_{n-1}\| + \tfrac{2\mu}{(2-\mu)\mathcal{B}(\mu)} \eta_1 \|\mathcal{P} - \mathcal{P}_{n-1}\| t.$$

This process gives

$$\|\mathcal{K}_{1n}(t)\| \leq \left(\frac{2(1-\mu)}{(2-\mu)\mathcal{B}(\mu)} + \frac{2\mu}{(2-\mu)\mathcal{B}(\mu)} t \right)^{n+1} \eta_1^{n+1} M.$$

Similarly, at t_0 we can obtain

$$\|\mathcal{K}_{1n}(t)\| \leq \left(\frac{2(1-\mu)}{(2-\mu)\mathcal{B}(\mu)} + \frac{2\mu}{(2-\mu)\mathcal{B}(\mu)} t_0 \right)^{n+1} \eta_1^{n+1} M. \tag{20}$$

As $n \to \infty$ and from Eq. (20), $\|\mathcal{K}_{1n}(t)\| \to 0$. Similarly, we can verify for $\|\mathcal{K}_{2n}(t)\|$ and $\|\mathcal{K}_{3n}(t)\|$.

Next, for the solution of the projected model, we prove the uniqueness. Suppose $\mathcal{P}^*(t), Q^*(t)$ and $S^*(t)$ be the set of other solutions, then

$$\mathcal{P}(t) - \mathcal{P}^*(t) = \tfrac{2(1-\mu)}{(2-\mu)\mathcal{B}(\mu)} \big(\mathcal{G}_1(t,\mathcal{P}) - \mathcal{G}_1(t,\mathcal{P}^*) \big)$$

$$+ \tfrac{2\mu}{(2-\mu)\mathcal{B}(\mu)} \int_0^t \big(\mathcal{G}_1(\zeta,\mathcal{P}) - \mathcal{G}_1(\zeta,\mathcal{P}^*) \big) d\zeta.$$

Now, employing the norm on above equation we get

$$\|\mathcal{P}(t) - \mathcal{P}^*(t)\| = \left\| \frac{2(1-\mu)}{(2-\mu)\mathcal{B}(\mu)} \big(\mathcal{G}_1(t,\mathcal{P}) - \mathcal{G}_1(t,\mathcal{P}^*)\big) \right.$$

$$\left. + \frac{2\mu}{(2-\mu)\mathcal{B}(\mu)} \int_0^t \big(\mathcal{G}_1(\zeta,\mathcal{P}) - \mathcal{G}_1(\zeta,\mathcal{P}^*)\big)d\zeta \right\|$$

$$\leq \frac{2(1-\mu)}{(2-\mu)\mathcal{B}(\mu)} \eta_1 \|\mathcal{P}(t) - \mathcal{P}^*(t)\| + \frac{2\mu}{(2-\mu)\mathcal{B}(\mu)} \eta_1 t \|\mathcal{P}(t) - \mathcal{P}^*(t)\|.$$

On simplification

$$\|\mathcal{P}(t) - \mathcal{P}^*(t)\| \left(1 - \frac{2(1-\mu)}{(2-\mu)\mathcal{B}(\mu)}\eta_1 - \frac{2\mu}{(2-\mu)\mathcal{B}(\mu)}\eta_1 t\right) \leq 0.$$

From the above condition, it is clear that $\mathcal{P}(t) = \mathcal{P}^*(t)$, if

$$\left(1 - \frac{2(1-\mu)}{(2-\mu)\mathcal{B}(\mu)}\eta_1 - \frac{2\mu}{(2-\mu)\mathcal{B}(\mu)}\eta_1 t\right) \geq 0. \tag{21}$$

Hence, Eq. (21) proves our required result.

RESULTS AND DISCUSSION

As important as formulating the nature of chemical reaction and finding the solution for the corresponding system, it is vital to capture the response of the attained results with different order and parameters associated with the method. Here, we present the nature of the considered model with two different cases

In Figures 1 and 2, we presented the nature of two different cases with the obtained solution. From these, we can notice the change in corresponding behaviour for distinct parameters and confirm that the change in the system order helps us capture more consequence of the system and aid us to understand and predict nature. In Figure 3, we presented $\hbar$-curves and these can offer the convergence providence of the results attained.

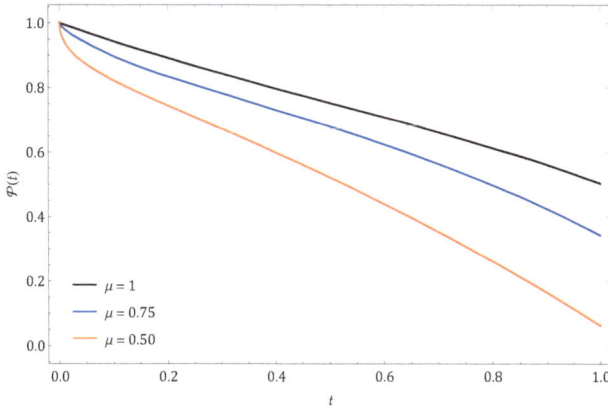

$$(\boldsymbol{a})$$

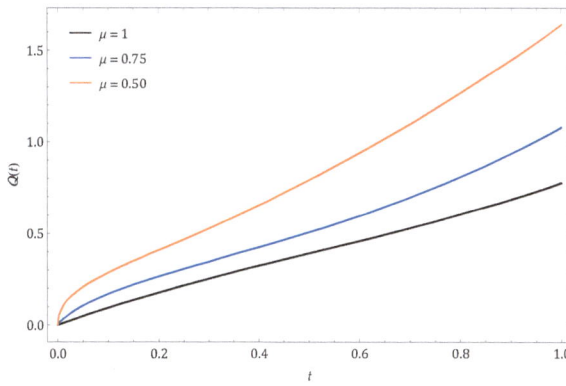

$$(\boldsymbol{b})$$

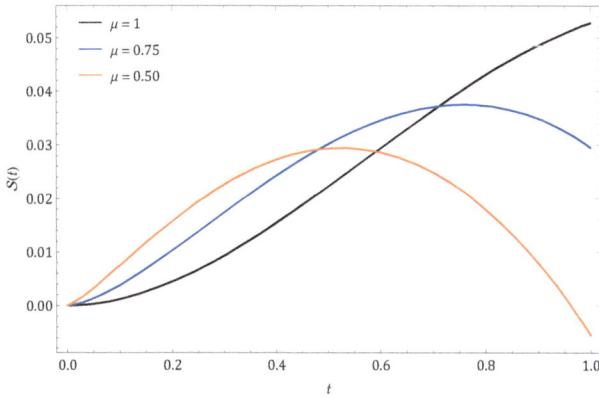

$$(\boldsymbol{c})$$

Fig. (1). Nature of obtained solution for $(\boldsymbol{a})$ $\mathcal{P}(t)$, $(\boldsymbol{b})$ $\mathcal{Q}(t)$, $(\boldsymbol{c})$ $\mathcal{S}(t)$ with the change in time (t) for diverse μ at $n = 1, \beta = 1, \alpha = 0.6, \varepsilon = 0.5$ and $\hbar = -1$.

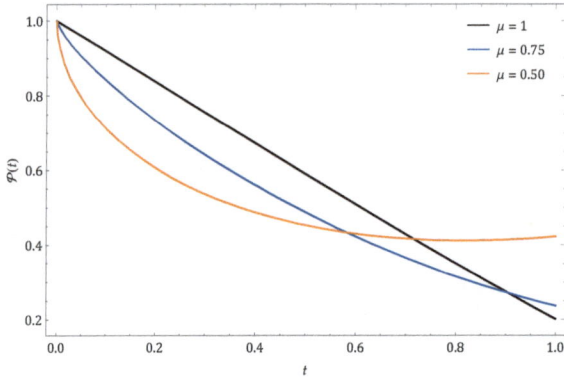

$$(a)$$

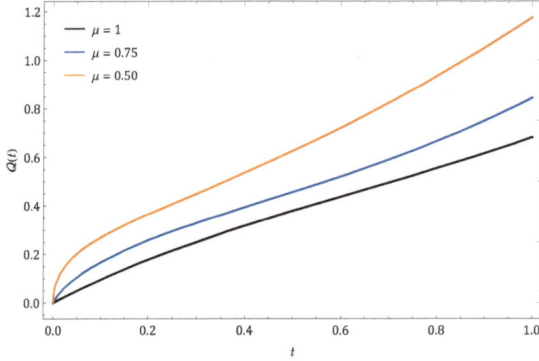

$$(b)$$

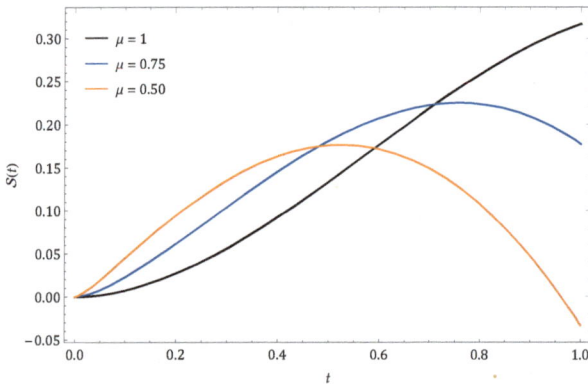

$$(c)$$

Fig. (2). Response of obtained solution for (**a**) $\mathcal{P}(t)$, (**b**) $\mathcal{Q}(t)$, (**c**) $\mathcal{S}(t)$ with the change in time (t) for diverse μ at $n = 1, \beta = 0.8, \alpha = 0.8, \varepsilon = 3$ and $\hbar = -1$.

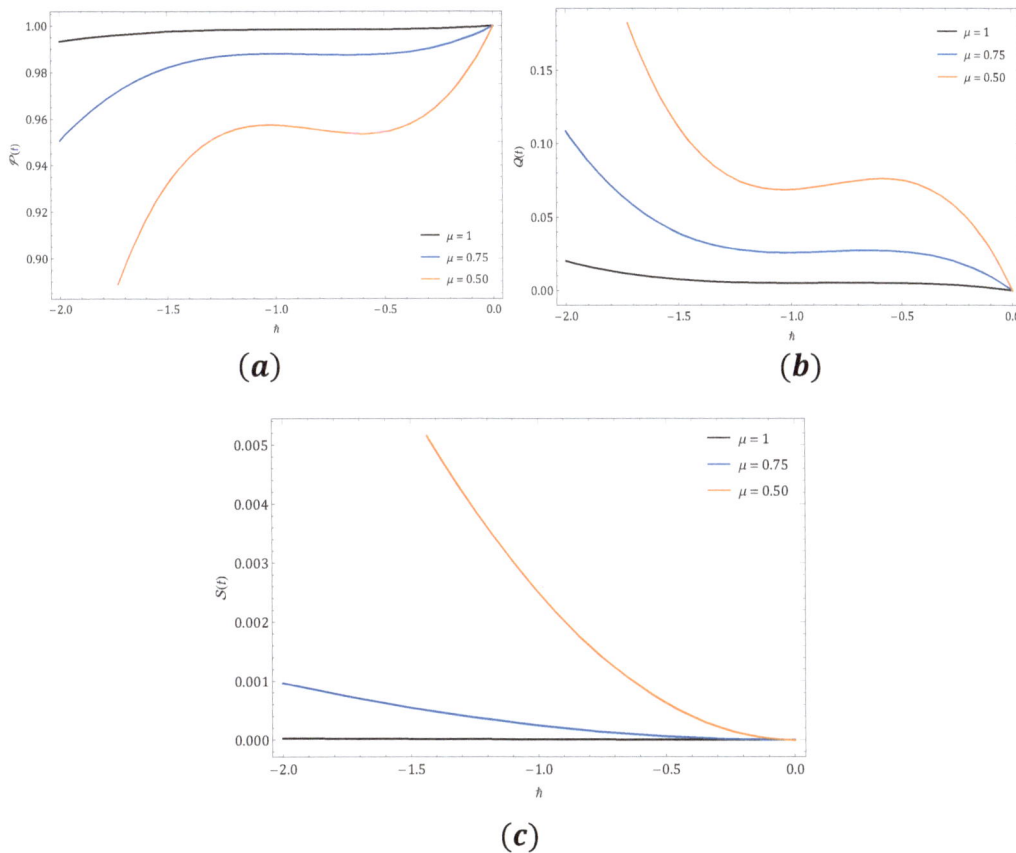

Fig. (3). $\hbar$-curves drown for q-HATM solution of (**a**) $\mathcal{P}(t)$, (**b**) $\mathcal{Q}(t)$, (**c**) $\mathcal{S}(t)$ for distinct μ $n = 1, \beta = 1, \alpha = 0.6, \varepsilon = 0.6$ and $t = 0.005$.

CONCLUSION

In this paper, we find the approximated analytical solution for the system of nonlinear ordinary differential equations exemplifying the arising in the biochemical reaction without a singular kernel using an efficient unifying computational method. The considered scheme is the unification of two algorithms associated with the topological concept, and the projected operator is capable of capturing more essential consequences of the system. The uniqueness and convergence are illustrated for the hired system with a fixed-point theorem. For different order and parameters, related methods are captured in terms of figures to demonstrate the physical behaviour of the solution. The present investigation shows the reliability of the scheme and the effect of fractional order, which helps the reader investigate and predict the nature of the nonlinear models.

CONSENT FOR PUBLICATION

Not applicable.

CONFLICT OF INTEREST

The authors declare no conflict of interest, financial or otherwise.

ACKNOWLEDGEMENT

Declared none.

REFERENCES

[1]   G. F. B. Riemann, Versuch Einer Allgemeinen Auffassung der Integration und Differentiation, Gesammelte Mathematische Werke, Leipzig, 1896.

[2]   M. Caputo, Elasticita e Dissipazione, Zanichelli, Bologna, 1969.

[3]   K. S. Miller, and B. Ross, An introduction to fractional calculus and fractional differential equations, A Wiley, New York, 1993.

[4]   I. Podlubny, Fractional Differential Equations, Academic Press, New York, 1999.

[5]   A. A. Kilbas, H. M. Srivastava, and J. J. Trujillo, Theory and applications of fractional differential equations, Elsevier, Amsterdam, 2006.

[6]   D. Baleanu, Z.B. Guvenc, J.A. T. Machado, New trends in nanotechnology and fractional calculus applications, Springer Dordrecht Heidelberg, London New York, 2010.

[7]   S. Schnell, C. Mendoza, Closed form solution for time-dependent enzyme kinetics, Journal of Theoretical Biology, vol. 187, pp. 207–212, 1997.

[8]   B. O. Palsson, and E. N. Lightfoot, Mathematical modelling of dynamics and control in metabolic networks. I. On michaelis–menten kinetics, Journal of Theoretical Biology, vol. 111 (2), pp. 273–302, 1984.

[9]   A. Meena, A. Eswari, and L. Rajendran, Mathematical modelling of enzyme kinetics reaction mechanisms and analytical solutions of non-linear reaction equations, Journal of Mathematical Chemistry, vol. 48, pp. 179–186, 2010.

[10]   S. M. Goha, and M. S. M. Noorani, and I. Hashim, Introducing variational iteration method to a biochemical reaction model, Nonlinear Analysis: Real World Applications, vol. 11, pp. 2264–2272, 2010.

[11]   L. Michaelis, and M. L. Menten, Die kinetik der invertinwirkung. Biochemische Zeitschrift, vol. 49, pp. 333-369, 1913.

[12]   M. Zurigat, S. Momani, and A. Alawneh, The multistage homotopy analysis method: application to a biochemical reaction model of fractional order, International Journal of Computer Mathematics, vol. 91 (5), pp. 1030-1040, 2013.

[13]   A. K. Sen, An application of the Adomian decomposition method to the transient behavior of a model biochemical reaction, Journal of Mathematical Analysis and Applications, vol. 131, pp. 232–245, 1988.

[14]   I. Hashim, M. S. H. Chowdhury, and S. Mawa, On multistage homotopy-perturbation method applied to nonlinear biochemical reaction model, Chaos, Solitons and Fractals, vol. 36 (4), pp. 823–827, 2008.

[15]   V. P. Dubey, S. Dubey, D. Kumar, and J. Singh, A computational study of fractional model of atmospheric dynamics of carbon dioxide gas, Chaos, Solitons and Fractals, 2020, DOI: 10.1016/j.chaos.2020.110375.

[16] M. S. Kiran, *et al.*, A mathematical analysis of ongoing outbreak COVID-19 in India through nonsingular derivative, Numerical Methods for Partial Differential Equations, vol 37 (2), pp. 1282-1298. 2021.

[17] M. Yavuz, and N. Özdemir, Comparing the New Fractional Derivative Operators Involving Exponential and Mittag-Leffler Kernel, Discrete and Continuous Dynamical Systems – Series S, vol. 13 (3), pp. 995-1006, 2020.

[18] P. Veeresha, and D. G. Prakasha, Solution for fractional Kuramoto–Sivashinsky equation using novel computational technique, International Journal of Applied and Computational Mathematics, vol. 7 (33), 2021, DOI: 10.1007/s40819-021-00956-0.

[19] W. Gao, *et al.*, Iterative method applied to the fractional nonlinear systems arising in thermoelasticity with Mittag-Leffler kernel, Fractals, vol. 28 (8), 2020, DOI: 10.1142/S0218348X2040040X.

[20] W. Gao, H. M. Baskonus, and L. Shi, New investigation of bats-hosts-reservoir-people coronavirus model and application to 2019-nCoV system, Advances in Difference Equations, vol. 391, 2020, DOI: 10.1186/s13662-020-02831-6.

[21] E. K. Akgül, A. Akgül, and M. Yavuz, New Illustrative Applications of Integral Transforms to Financial Models with Different Fractional Derivatives, Chaos, Solitons and Fractals, vol. 146, 2021, 110877.

[22] P. Veeresha, D. G. Prakasha, *et al.*, An efficient approach for fractional nonlinear chaotic model with Mittag-Leffler law, Journal of King Saud University-Science, vol. 33 (2), 2021, DOI: 10.1016/j.jksus.2021.101347.

[23] L. Akinyemi, P. Veeresha, and M. Senol, Numerical solutions for coupled nonlinear schrodinger-Korteweg-de Vries and Mac-cari's systems of equations, Modern Physics Letters B, vol. 35(20), 2021, DOI: 10.1142/s0217984921503395.

[24] M. Yavuz, *et al.*, The Schrödinger-KdV Equation of Fractional Order with Mittag-Leffler Non singular Kernel, Alexandria Engineering Journal, vol. 60 (2), pp. 2715-2724, 2021.

[25] P. Veeresha, and D. G. Prakasha, Novel approach for modified forms of Camassa-Holm and Degasperis-Procesi equations using fractional operator, Communications in Theoretical Physics, vol. 72 (10), 2021.

[26] M. Yavuz, and N. Sene, Approximate solutions of the model describing fluid flow using generalized ρ-Laplace transform method and heat balance integral method, Axioms, vol. 9(4), 2020, DOI: 10.3390/axioms9040123.

[27] P. Veeresha, and D. G. Prakasha, A reliable analytical technique for fractional Caudrey-Dodd-Gibbon equation with Mittag-Leffler kernel, Nonlinear Engineering, vol. 9 (1), 319–328, 2020.

[28] L. Akinyemi, and O. S. Iyiola, Analytical Study of (3+1)-Dimensional Fractional-Reaction Diffusion Trimolecular Models, International Journal of Applied and Computational Mathematics, vol. 7 (92), 2021, DOI: 10.1007/s40819-021-01039-w.

[29] S. K. Panda, C. Ravichandran, and B. Hazarika, Results on system of Atangana–Baleanu fractional order Willis aneurysm and nonlinear singularly perturbed boundary value problems, Chaos, Solitons and Fractals, vol. 142, (2021).

[30] P. Veeresha, D. G. Prakasha, and D. Kumar, Fractional SIR epidemic model of childhood disease with Mittag-Leffler memory, Fractional Calculus in Medical and Health Science, CRC Press, pp. 229-248, 2020.

[31] S. K. Panda, T. Abdeljawad, and C. Ravichandran, A complex valued approach to the solutions of Riemann-Liouville integral, Atangana-Baleanu integral operator and non-linear Telegraph equation via fixed point method, Chaos, Solitons and Fractals vol. 130, 109439, 2021.

[32] M. Caputo, and M. Fabrizio, A new definition of fractional derivative without singular kernel, Progress in Fractional Differentiation and Applications, vol. 1 (2), pp. 73-85, 2015.

[33] A. Atangana, and D. Baleanu, New fractional derivatives with nonlocal and non-singular kernel: theory and application to heat transfer model, Thermal Science, vol. 20 (2), pp. 763-769, 2016.

[34] J. Singh, D. Kumar, and R. Swroop, Numerical solution of time- and space-fractional coupled Burgers' equations via homotopy algorithm, Alexandria Engineering Journal, vol. 55 (2) (2016) pp. 1753-1763.

[35]   S.J. Liao, Homotopy analysis method: a new analytic method for nonlinear problems, Applied Mathematics and Mechanics, vol. 19, pp. 957-962, 1998.

[36]   P. Veeresha, D. G. Prakasha, and Z. Hammouch, An efficient approach for the model of thrombin receptor activation mechanism with Mittag-Leffler function, Nonlinear Analysis: Problems, Applications and Computational Methods, pp. 44-60, 2020, DOI: 10.1007/978-3-030-62299-2_4.

[37]   H. Bulut, D. Kumar, J. Singh, R. Swroop, and H. M. Baskonus, Analytic study for a fractional model of HIV infection of CD4+T lymphocyte cells, Mathematics in Natural Science, vol. 2 (1), pp. 33–43, 2018.

[38]   P. Veeresha, D.G. Prakasha, and D. Baleanu, Analysis of fractional Swift-Hohenberg equation using a novel computational technique, Mathematical Methods in the Applied Sciences, vol. 43 (4), pp. 1970-1987, 2019.

[39]   H. M. Srivastava, D. Kumar, and J. Singh, An efficient analytical technique for fractional model of vibration equation, Applied Mathematical Modelling, vol. 45, pp. 192–204, 2017.

[40]   P. Veeresha, D. G. Prakasha, and J. Singh, A novel approach for nonlinear equations occurs in ion acoustic waves in plasma with Mittag-Leffler law, Engineering with Computers, vol. 37 (6), pp. 1865-1897, 2020.

[41]   L. Akinyemi, and S. N. Huseen, A powerful approach to study the new modified coupled Korteweg-de Vries system, Mathematics and Computers in Simulation, vol. 177, pp. 556-567, 2021.

[42]   P. Veeresha, D. G. Prakasha, J. Singh, D. Kumar, and D. Baleanu, Fractional Klein-Gordon-Schrödinger equations with Mittag-Leffler memory, Chinese Journal of Physics, vol. 68, pp. 65-78, 2020.

[43]   L. Akinyemi, M. Şenol, and S. N. Huseen, Modified homotopy methods for generalized fractional perturbed Zakharov–Kuznetsov equation in dusty plasma, Advances in Difference Equations, vol. 45, 2021, DOI: 10.1186/s13662-020-03208-5.

[44]   D. Kumar, R. P. Agarwal, and J. Singh, A modified numerical scheme and convergence analysis for fractional model of Lienard's equation, Journal of Computational and Applied Mathematics, vol. 399, pp. 405–413, 2018.

[45]   D. G. Prakasha, N. S. Malagi, and P. Veeresha, New approach for fractional Schrödinger–Boussinesq equations with Mittag-Leffler kernel, Mathematical Methods in the Applied Sciences, vol. 43, pp. 9654-9670, 2020.

[46]   M. Şenol, O. S. Iyiola, H. D. Kasmaei, and L. Akinyemi, Efficient analytical techniques for solving time-fractional nonlinear coupled Jaulent–Miodek system with energy-dependent Schrödinger potential, Advances in Difference Equations, vol. 462, 2019, DOI: 10.1186/s13662-019-2397-5.

[47]   M. Yavuz, and E. Bonyah, New approaches to the fractional dynamics of schistosomiasis disease model, Physica A, vol. 525, 373-393, 2019.

[48]   W Gao, *et al.*, New approach for the model describing the deathly disease in pregnant women using Mittag-Leffler function, Chaos, Solitons and Fractals, vol. 134, DOI: 10.1016/j.chaos.2020.109696, 2020.

[49]   M. Yavuz, T. A. Sulaiman, F. Usta, and H. Bulut, Analysis and numerical computations of the fractional regularized long-wave equation with damping term, Mathematical Methods in the Applied Sciences, vol. 44 (9), pp. 7538-7555, 2021.

[50]   M. Senol, L. Akinyemi, A. Ata, and O. S. Iyiola, Approximate and generalized solutions of conformable type Coudrey–Dodd–Gibbon–Sawada–Kotera equation, International Journal of Modern Physics B, vol. 35 (2), 2021, DOI: 10.1142/S0217979221500211.

[51]   M. Yavuz, and T. Abdeljawad, Nonlinear regularized long-wave models with a new integral transformation applied to the fractional derivative with power and Mittag-Leffler kernel, Advances in Difference Equations, vol. 367, 2020, DOI: 10.1186/s13662-020-02828-1.

[52]   W Gao, *et al.*, New numerical results for the time-fractional Phi-four equation using a novel analytical approach, Symmetry, vol. 12 (3), 2020, DOI: 10.3390/sym12030478.

[53]   M. Yavuz, and A. Yokus, Analytical and numerical approaches to nerve impulse model of fractional-order, Numerical Methods for Partial Differential Equations, vol. 36, pp. 1348-1368, 2020.

[54]   J. Losada, and J. J. Nieto, Properties of the new fractional derivative without singular Kernel, Progress in Fractional Differentiation and Applications, vol. 1 (2015), pp.87-92.

[55]   P. Veeresha, A Numerical Approach to the Coupled Atmospheric Ocean Model Using a Fractional Operator. Mathematical Modelling and Numerical Simulation with Applications (MMNSA), vol. 1 (1), pp. 1-10, 2021.

[56]   A. Yokuş, Construction of Different Types of Traveling Wave Solutions of the Relativistic Wave Equation Associated with the Schrödinger Equation. Mathematical Modelling and Numerical Simulation with Applications (MMNSA), vol. 1 (1), pp. 24-31, 2021.

[57]   Z. Hammouch, M. Yavuz, and N. Özdemir, Numerical Solutions and Synchronization of a Variable-Order Fractional Chaotic System. Mathematical Modelling and Numerical Simulation with Applications (MMNSA), vol. 1 (1), pp. 11-23, 2021.

CHAPTER 10

Floating Object Induced Hydro-morphological Effects in Approach Channel

Onur Bora[1,*], M. Sedat Kabdaşlı[2], Nuray Gedik[1], Emel İrtem[3]

[1]*Department of Civil Engineering, Balıkesir University, Balıkesir, Turkey*

[2]*Faculty of Civil Engineering, Istanbul Technical University, Istanbul, Turkey*

[3]*Department of Civil Engineering, Doğuş University, Istanbul, Turkey*

Abstract: Transversal and diverging waves, return flows, propeller induced jet flows, and other hydrodynamic effects induced by a floating object may cause significant movement and/or suspension of bottom and bank sediments in the marine environment, especially in approach channels. Using the CFD (Computational Fluid Dynamics) process, the hydro-morphodynamic effects induced by a non-powered floating object navigating in an approach channel are investigated in this study. The approach channel dimensions depth, width, and channel slope are determined according to PIANC (2014) [1]. The floating object locations and velocities are used in nine different scenarios. In these cases, the floating object is 0.90, 1.10, and 1.30 meters from the bottom of the approach channel, respectively. According to the findings, when the floating object is located nearest to the bottom and its speed is fastest, there is a significant amount of sediment suspension and sediment movement in the channel slope, which is mostly attributed to super-critical return flows. When the floating object is farthest from the channel bottom and the floating object speed is lowest, however, there is a noticeable reduction in the acceleration and suspension of the sediment. As a result, the velocity and location of the floating object, channel slope, the kinematics of ship-generated waves, and particularly the return flows are found to have a significant impact on sediment movement and suspension.

Keywords: CFD, Floating object, Hydrodynamic, Morphodynamic, Sediment suspension, Sediment transport.

INTRODUCTION

Waterborne commerce has increased continuously over the last decades and this situation has led to an increase in ship dimensions and ship numbers. Because of

*Corresponding author Onur Bora:** Department of Civil Engineering, Balıkesir University, Balıkesir, Turkey; E-mail: bora.onur@gmail.com

Mehmet Yavuz & Necati Özdemir (Eds.)

the increasing ship sizes and the number of ships, safe navigation and economical requirements have gained great importance. In order to ensure safe navigation in the approach channel, the channel must be wide and deep enough for vessel traffic, but it is not so deep or large as to require excessive dredging. Therefore, vessel or floating object-induced sediment transport must be well studied in the design phase of the approach channel.

Floating objects navigating in a natural or an artificial approach channel, in rivers or inland waterways cause several hydrodynamic disturbances in the form of waves and currents. Floating objects generate two main types of waves, namely primary (drawdown) and secondary waves [2]. The primary wave system consists of significant water level depression along the hull of the floating object and return flow. The primary wave system is dominant where floating object induced the cross-sectional blockage is significant [3]. Secondary waves are gravity waves generated by pressure peaks along the floating objects and these waves are short waves. The secondary waves are dominant in canals for ocean-going floating object and on most rivers, where the blockage factors of the floating object are usually very low [4].

In previous, many researchers have worked on ship-generated waves [5-7]. These studies show that there is a relationship between wave height and ship type, draught, speed, and distance to the banks. Ship-generated waves cause intensive sediment resuspension and sediment transport in the approach channel. These actions are so important that Houser (2011) found that the vessel-generated wakes (including drawdown and surge waves) have much more effects on sediment resuspension than wind waves and suspended sediment concentration (SSC) increases with the increment of turbulent kinetic energy (TKE) of the supercritical pilot-boat wakes [8].

Floating object-induced hydrodynamic effects such as water level drawdown, transversal, and diverging waves, return flows, propeller jet flows, etc., lead to bank erosion, sediment resuspension, and environmental impact on plankton, fish, plants, etc. Rapaglia *et al.*, (2011) measured water velocity, water depth, and sediment concentration on the shoals alongside the shipping channel after the passage of forty vessels. They found that higher return velocities and ten vessel-induced wakes led to SSC concentration above 400 mg/L, which is 30 times higher than the average background concentration [9]. Ji *et al.*, (2014) investigated that ship induced suspended particulate matter (SPM) concentration navigating in an approach channel with and without a propeller [10]. Schroevers *et al.* (2015) carried out a 1/1 scale physical experiment using a heavy loaded barge to observe the canal bottom stability in the 36 km long Juliana Canal in the south of the Netherlands. During each passage of the barge, the flow velocities under the ship and the bed change were measured. They found that the amount of erosion in the middle of the channel reached 1 cm at each 10 passages of the barge and 6 cm erosion value in total at the end of the experiment (60 passages) [11]. McConchie and Toleman (2003) investigated boat wakes-induced riverbank erosion. They measured wake wave characteristics and suspended sediment concentration at several sites, and they found that boat wakes 2-80 times larger than background wind-generated waves. So, boat-generated waves are more erosive than wind-generated waves in riverine environments, particularly where fetch lengths are restricted [12].

Also, several studies have been carried out on various parameters such as water depth, turbulence energy, ship type, ship velocity, eddies, etc., that may affect sediment resuspension. Smaoui *et al.*, (2011) investigated the quantitatively and relatively accurate relationship between sediment transport and boat traffic via a one-dimensional vertical model [13].

Some researchers have studied ship-induced current with the help of physical experiments [14 - 17]. Maynord (2000) investigated the physical forces under the ship to determine ship-induced sediment transport and sediment suspension. Lenselink (2011) studied loaded barges and investigated the velocity profile under the barge and its effect on the seabed [14]. In this study, the hydro-morphodynamic effects caused by a non-powered floating object navigating in an approach channel are investigated using a 3-D numerical model.

MATERIALS AND METHODS

Numerical Model

FLOW-3D software was used in this analysis to build a 3D hydro-morphodynamic numerical model. Flow Science, Inc. created FLOW-3D, a commercial software kit. Flow-3D uses a finite volume approach to solve the continuity equation (Eq. (1)) and the unsteady Reynolds-averaged Navier-Stokes equations governing fluid motion (Eq. (2)) [18].

$$\frac{\partial}{\partial x_i} U_i A_i = 0, \tag{1}$$

$$\frac{\partial U_i}{\partial t} + \frac{1}{V_f}\left(U_j A_j \frac{\partial u}{\partial x_j}\right) = -\frac{1}{\rho}\frac{\partial P}{\partial x_i} + G_i + f_i, \tag{2}$$

$$\rho V_f f_i = \tau_{b,i} - \left[\frac{\partial}{\partial x_j}(A_j S_{ij})\right] ; S_{ii} = -2\mu_{tot}\left[\frac{\partial U_i}{\partial x_i}\right] ; S_{ij} = -\mu_{tot}\left[\frac{\partial U_i}{\partial x_j} + \frac{\partial U_j}{\partial x_i}\right],$$

where P indicates pressure, U_i is the mean velocity, time by t, A_i fractional open area open to flow in the i direction, V_f fractional volume open to flow, G_i represents the body accelerations, f_i indicates the viscous accelerations, S_{ij} strain rate tensor, $\tau_{b,i}$ denotes wall shear stress, ρ density of water, μ_{tot} total dynamic viscosity, which includes the effects of turbulence ($\mu_{tot} = \mu + \mu_T$), μ dynamic viscosity, and μ_T eddy viscosity.

Sediment transport modeling requires accurate estimates of the near-wall shear stresses, so a good turbulence model should be selected for turbulence flows. In FLOW-3D, there are six turbulence models available: the Prandtl mixing length model, the one-equation, the two-equation k-ε, RNG, and k-ω models, and a large eddy simulation, LES, model. Renormalization Group (RNG) turbulence model [19] is used in this study. The RNG model uses equations similar to the equations for the k - ε model. However, equation constants that are found empirically in the standard k - ε model are derived explicitly in the RNG model. In particular, the RNG model is known to describe low intensity turbulence flows and flows having strong shear regions more accurately. RNG turbulence model equations are shown below.

$$\frac{\partial(\rho k)}{\partial t} + \frac{\partial(\rho k u_i)}{\partial x_i} = \frac{\partial}{\partial x_j}\left[\left(\mu + \frac{\mu_t}{\sigma_k}\right)\frac{\partial k}{\partial x_j}\right] + P_k - \rho\varepsilon$$

$$\frac{\partial(\rho\varepsilon)}{\partial t} + \frac{\partial(\rho\varepsilon u_i)}{\partial x_i} = \frac{\partial}{\partial x_j}\left[\left(\mu + \frac{\mu_t}{\sigma_\varepsilon}\right)\frac{\partial(\varepsilon)}{\partial x_j}\right] + C_{1\varepsilon}\frac{\varepsilon}{k}P_k - C_{2\varepsilon}\rho\frac{\varepsilon^2}{k}$$

where k is turbulence kinetic energy (henceforth, TKE), ε is dissipation rate, ρ is density, t is time, x_i is coordinate in the (i) axis, μ is dynamic viscosity, μ_t is turbulent dynamic viscosity, and P_k is the production of TKE. The default values of the $C_{1\varepsilon}$, $C_{2\varepsilon}$ and σk=σe are 1,42, 1,68 and 1.39, respectively. The turbulent viscosity is computed using the parameter Cμ=0.085 using Equation (3).

$$\mu_T = \rho C_\mu \frac{K^2}{\epsilon} \tag{3}$$

Sediment is entrained by the picking up and re-suspension due to shearing and small eddies at the packed sediment interface. Because it is not possible to compute the flow dynamics about each individual grain of sediment, an empirical model must be used. The model used here is based on Mastbergen and Van den Berg. The first step to computing the critical Shields parameter is calculating the dimensionless parameter d∗,i:

$$d_{*,i} = d_i \left[\frac{\rho_f(\rho_i - \rho_f)\|g\|}{\mu_f^2} \right]^{1/3}$$

where ρ_i is the density of the sediment species I, ρ_f is the fluid density, d_i is the diameter, μ_f is the dynamic viscosity of fluid, $\|g\|$ is the magnitude if the acceleration of gravity g.

From this, the dimensionless critical Shields parameter is computed using the Soulsby-Whitehouse equation:

$$\theta_{cr,i} = \frac{0.3}{1 + 1.2d_{*,i}} + 0.055[1 - \exp(-0.02d_{*,i})]$$

The critical Shields parameter can be modified for sloping surfaces to include the angle of repose. The modification further alters $\theta_{cr,i}$

$$\theta'_{cr,i} = \theta_{cr,i} \frac{\cos\psi \sin\beta + \sqrt{\cos^2\beta \tan^2\varphi_i - \sin^2\psi \sin^2\beta}}{\tan\varphi_i}$$

where β is the angle of slope of bed, φ is the user-defined angle of repose for sediment species i (default is 32), and ψ is the angle between the flow and the upslope direction [18].

The local Shields parameter is computed based on the local bed shear stress, τ:

$$\theta_i = \frac{\tau}{\|g\|d_i(\rho_i - \rho_f)}$$

where τ is calculated using the law of the wall and quadratic law of bottom shear stress for 3D turbulent flow shallow water turbulent flow, respectively, with consideration of bed surface roughness. The settling velocity equation proposed by Soulsby (Soulsby, 1997) is used:

$$u_{settling,i} = [(10.36^2 + 1.049d_*^3)^{0.5} - 10.36]$$

Bed-load transport is the mode of sediment transport due to rolling or bouncing over the surface of the packed bed of sediment. Volumetric transport rate of sediment per width of bed (van Rijn, 1984):

$$\Phi_i = \beta_{VT,i} d_{*,i}^{-0.3} \left(\frac{\theta_i}{\theta'_{cr,i}} - 1.0\right)^{2.1} c_{b,i}$$

where $\beta_{VT,i}$ coefficient is 0.053. $c_{b,i}$ is the volume fraction of species i in the bed material.

For each species, the suspended sediment concentration is calculated by solving its own transport equation,

$$\frac{\partial C_{s,i}}{\partial t} + \nabla \cdot \left(u_{s,i} C_{s,i}\right) = \nabla \cdot \nabla (D C_{s,i})$$

Here $C_{s,i}$ is the suspended sediment mass concentration of species i, which is defined as the sediment mass per volume of fluid-sediment mixture; D is the diffusivity; $u_{s,i}$ is the suspended sediment velocity [18].

The FAVOR (Fractional Area/Volume Obstacle Representation) method for the description of the geometry and the Volume-of-Fluid (VOF) method as originally described by Hirt and Nichols (1981) for tracking fluid interfaces are used in the model [20]. Typically, the fractional volume is represented by the quantity. In interior regions of liquid, the value of F would be 1.0, while outside of the liquid, in regions of gas (air for example), the value of F is zero. The location of a free surface is where F changes from 0.0 to 1.0. Thus, any element having an F value lying between 0.0 and 1.0 must contain a surface. In addition, the normal to the surface can be calculated from the direction in which F changes most rapidly applying boundary conditions to the surface.

$$\frac{\partial F}{\partial t} + \frac{\partial F}{\partial t} + \frac{1}{V_f}\left[A_X u \frac{\partial F}{\partial X} + A_y v \frac{\partial F}{\partial Y} + A_z w \frac{\partial F}{\partial z}\right] = 0$$

In the FAVOR method, a geometric surface can cut through a rectangular mesh cell dividing it into blocked and open portions. The ratio of the open volume in a cell to its total volume is called the fractional volume. The intersections of the surface with the faces of the cell (six in three dimensions) are computed and stored as fractional areas, which are the ratios of the open area to the total area at the respective cell face.

In Flow-3D software, the determination of free surfaces with the water-air interface is carried out by the VOF method depending on whether the cells are completely filled with water, empty, or partially filled. In this method, the cell represents value

1 if it is completely filled with fluid, 0 if it is completely empty, and therefore the free surface when it is between 0 and 1 (Flow-3D Hydro Manual, 2021).

Design of Approach Channel

The following formula, proposed by PIANC (2014), is used for designing the width of the approach channel.

$$W = W_{BM} + \Sigma W_i + 2W_B$$

where W_{BM} is width of basic maneuvering lane as a multiple of the design ship's beam B; ΣW_i = additional widths to allow for the effects of wind, current, etc.; W_B is Bank clearance on the "red" and "green" sides of the channel.

The width of the channel is determined by the maneuverability of the design vessel. Since the floating object does not have an engine in this study, it does not have any maneuverability on its own, but this maneuverability will depend on the vessel which is pulling the floating object.

In this study, it was assumed that there was no meteorological phenomenon in the study area, therefore, the design criteria depend on the meteorological phenomena (wind and wave) in PIANC (2014) were not taken into consideration in the channel width design. The slope of the approach channel is taken as 1/3. Since the velocity of the floating object is 4 knots (≈ 2 m/s), it is in the "slow" category. Also, it has any risk of heavy load. In the light of these information, the approach channel was designed to be 10 m twice as wide as the width of the floating object (5 m).

Design of Approach Depth

The following factors have been taken into account in the design of the approach channel:

•Maximum loaded draft of the maximum design ship. The floating object draft is 1.5 m in this study.

•The squat, or the reduction of the under-keel clearance, is due to the suction effect induced by the higher current velocity between the sea bottom and the ship. This causes a reduction in the water level near the ship, and the ship, therefore, sinks bodily in the water. The squat can calculate the following formula according to PIANC (2014):

$$S = C_s \frac{\nabla}{L_{BP}^2} \frac{F_r^2}{\sqrt{(1-F_r^2)}}$$

where Fr is Froud number; L_{BP} is floating object length; ∇ is displacement volume; C_B is block coefficient (0,9); B is floating object width; D is floating object draft; C_S is Huuska/Guliev coefficient (2.4).

•*Wave amplitude*; change in water level due to the wave motion. It is thought that no meteorological condition has occurred in the region within the scope of this study.

•*Tidal effect;* change in water level due to tide. It is thought that no meteorological condition has occurred in the region within the scope of this study.

•The increase in the average draft due to the unbalanced loading; since there was no unbalanced loading in this study, this criterion was not considered in the design of the approach channel.

•*Siltation;* changes in the seabed due to sediment transport. Since this parameter is investigated in this study, it is not considered in the design phase.

•Safety factor allowance for seabed type; it is not considered in the design phase.

•Safety factor allowance for seabed level uncertainties, it is not considered in the design phase.

The approach channel depth was designed as 2.15 m as a result of the aforementioned factors. The length of the approach channel was chosen as 100 m depending on the floating object dimensions, floating object sailing velocity, and the model run times. The approach channel and the study area dimensions can be seen in the Fig. (**1**).

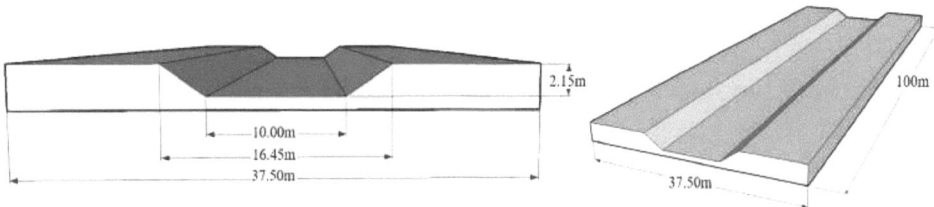

Fig. (1). Dimensions of the study area and approach channel.

Model Setup and Scenarios

Flow-3D allows users to import solids in STL (StereoLithography) format represent complex geometries. In this study, the floating object and seabed bathymetry was prepared in STL format. The seabed bathymetry with the approach channel and the floating object can be seen in the Fig. (**2**).

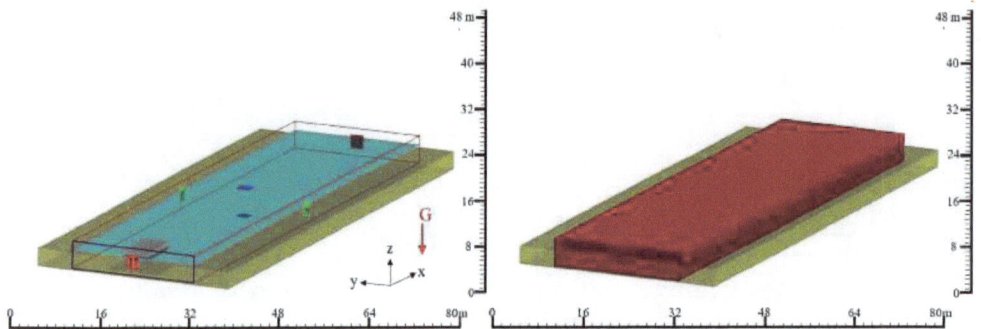

Fig. (2). Boundary conditions can be seen on the left side and computational domain; grid properties, and seabed bathymetry can be seen on the right side of the figure.

The study area was formed to be 24 m wide and 100 m long. The channel width reaches 10 m at the bottom of the approach channel and 16.45 m at the top of the trench. The floating object has 8 m length, 5 m width, and 2.5 m depth. The floating object is placed in the center of the approach channel along the Y-axis and 5 m after the beginning of the X-axis.

Because no external force acts on the studying area except for atmospheric pressure, boundary conditions for minimum and maximum X-axis were considered "outflow" condition. The boundary condition for minimum and maximum Y-axis were considered "symmetry", which indicates that there is no drag on this boundary also as there are zero velocity derivatives across the boundary, hence zero turbulence production. The boundary condition for maximum and minimum Z-axis was considered atmospheric pressure and wall, respectively. Fig. (**2**) represents boundary conditions on the X-Y-Z planes. There is erosional sediment packed at the bottom of the approach channel at a height of 0.85 m (above the wall boundary). It was determined to assume a uniform surface sediment composition with a D_{50} value of 0.0625 mm for the entire studying area. The angle of repose was taken as 40°. The specific weight of sediments was taken as 2650 kg/m³, bedload transport coefficient and entrainment coefficient were taken as 0.06 and 0.018, respectively in the numerical modeling. Critical Shields number was calculated at every time step using the Soulsby-Whitehouse formula. Van Rijn (1984) equation was used as the transport method in sediment transport modeling. Model parameters used in the modeling can be seen in the Table 1.

Table 1. Summary of model parameters used in the modeling.

Number of Sediment Species	1
Sediment Diameter (mm)	0.0625
Sediment Density (kg/m^3)	2650
Critical Shields Number	Every timestep from Soulsby-Whitehouse Equation
Entrainment Coefficient	0.018
Bed Load Coefficient	0.06
Angle of Repose (°)	40
Bed Load Transport Rate Equation	Van Rijn (1984)
Bed Roughness/d$_{50}$ Ratio	2.50
External Force (Such as wind, wave, current etc.)	No

A general moving object (GMO) is a rigid body with any kind of motion that is either user-prescribed or dynamically coupled with fluid flow. It can have six-degrees-of-freedom or motion constraints such as a fixed axis/point. Prescribed forces and torques can be applied to a GMO under coupled motion. The GMO model allows multiple rigid bodies under independent motion types as well as rigid body interactions including collisions and continuous contact. In this study, it is assumed that a floating object is pulled along the X-axis in the approach channel. The displacement of the floating object in the Y-axis direction (sway) and the rotation in the Z-axis (yaw) were restricted. In the other directions, it is possible to move freely. Pulling velocity of the floating object varies with time also it was examined in this study with different scenarios.

A total of nine scenarios were conducted in the modeling. In the first 3 scenarios, under-keel clearance (h_c) are 0.90 m, 1.1 m, and 1.3 m respectively (Fig. **3**). Also, the velocity of the floating object (V_o) is considered 2 m/s in these scenarios. In the next six scenarios, under-keel clearance was considered to be the same as the first 3 scenarios but the velocity of the floating object was reduced to 1.5 m/s in the second three scenarios and 1 m/s in the last three scenarios. Brief explanations of these scenarios are given in the Table **2**.

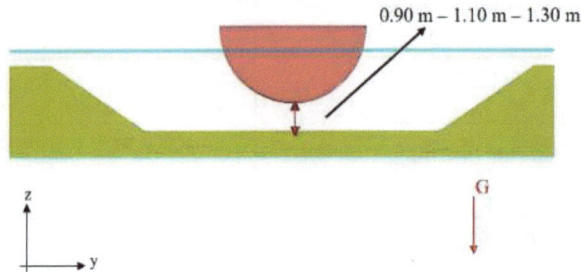

Fig. (3). Sketch of the under-keel clearance of the floating object used in the scenarios.

Table 2. Explanation of the scenarios used in the modeling.

Scenario	Under-keel Clearance (h_c, m)	Floating Object Velocity (V_o, m/s)
Scenario-1	0.90	
Scenario-2	1.10	2.0
Scenario-3	1.30	
Scenario-4	0.90	
Scenario-5	1.10	1.5
Scenario-6	1.30	
Scenario-7	0.90	
Scenario-8	1.10	1.0
Scenario-9	1.30	

Considering the time required for the floating object to move from the beginning to the end of the study area, the run time of the numerical model was taken as 55 s in the first three scenarios, 67 s in second three scenarios and 100 s in the last three scenarios.

RESULTS AND DISCUSSION

The importance of floating object location along the Z-axis to the sediment transport and sediment resuspension was examined in the first three scenarios. Fig. (4) depicts the suspended sediment concentrations with the general of the computational area for the first three scenarios. The legend is limited to 100 kg/m³ in all scenarios in order to observe the model results comfortably.

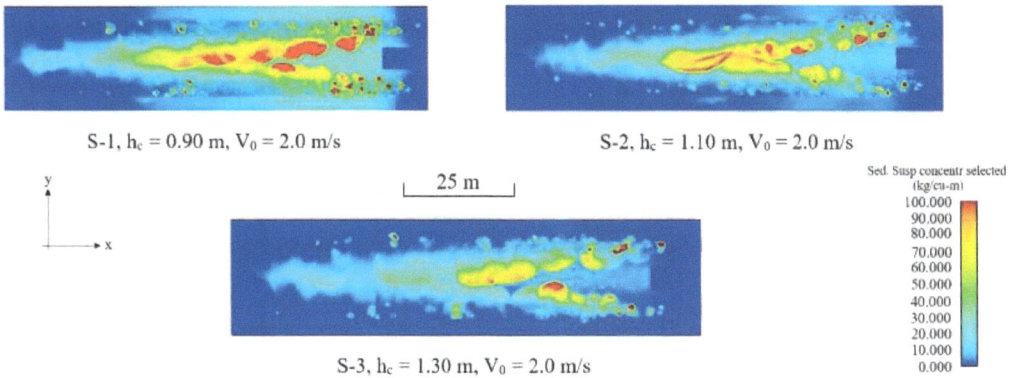

Fig. (4). Suspended sediment concentration according to the first three scenarios (plan view).

Sea-bottom morphology change at the last time step can be seen in the Fig. (5) according to the first three scenarios. White areas represent "No-change zones", red color show deposition areas, and blue color show erosion areas.

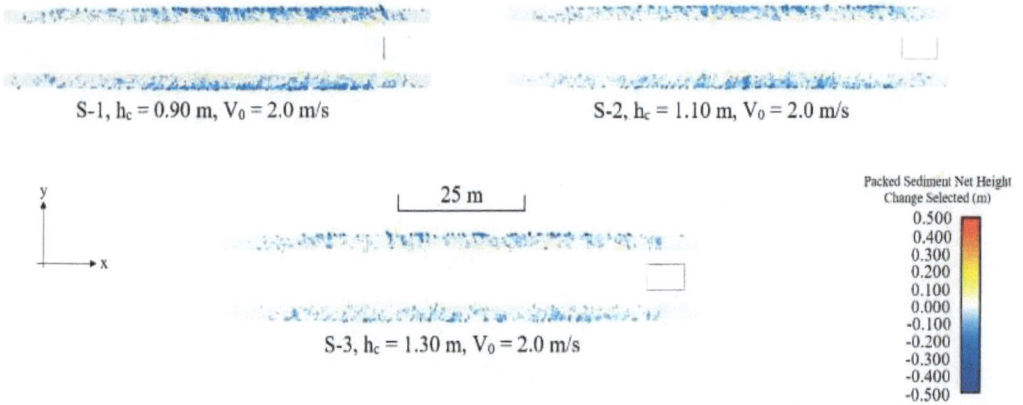

S-1, $h_c = 0.90$ m, $V_0 = 2.0$ m/s   S-2, $h_c = 1.10$ m, $V_0 = 2.0$ m/s

25 m

S-3, $h_c = 1.30$ m, $V_0 = 2.0$ m/s

Packed Sediment Net Height
Change Selected (m)

| 0.500 |
| 0.400 |
| 0.300 |
| 0.200 |
| 0.100 |
| 0.000 |
| -0.100 |
| -0.200 |
| -0.300 |
| -0.400 |
| -0.500 |

Fig. (5). Seabed morphology change according to the first three scenarios (plan view).

A floating object in the water causes a water rise as its volume displaced in the bow and stern part of the floating object. When the floating object starts to move, it pushed the water mass in the direction of motion. This causes an increase in the water level in front of the floating object, thus causing the positive pressure in front of the ship. Below the floating object, water displacing due to its volume creates negative pressure zones. These pressure differences cause a current from high pressure to low pressure. This current is the return currents seen on each side of the ship and under the ship. Furthermore, due to the moving pressure zones, floating object-induced waves are formed. The return currents and floating object-induced waves are shown in the Fig. (**6**).

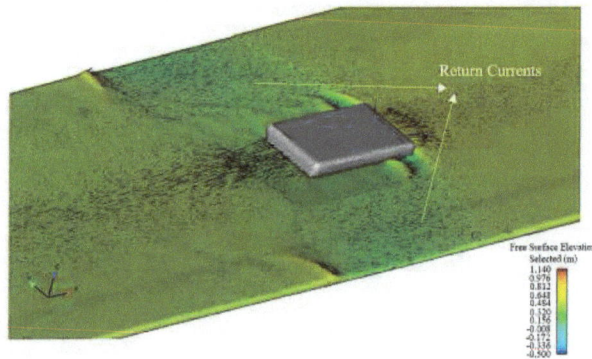

Fig. (6). Floating object-induced waves and return current according to the first scenario.

Due to the supercritical return currents and waves, shear-stresses increase at the channel slope and seabed, which causes the sediment resuspension and transport. The sediment motion is toward the bottom of the approach channel, due to the slope and gravitational acceleration. The suspended sediment is moving toward the center of the channel due to the wake region behind the floating object.

As the floating object approaches the bottom of the channel, the hydrodynamic effects described above increase so that the amount of sediment transport in the seabed and resuspension of sediment also increases. Fig. (**7**) shows the time variation of the total amount of suspended sediment and the amount of the eroded sediment, respectively.

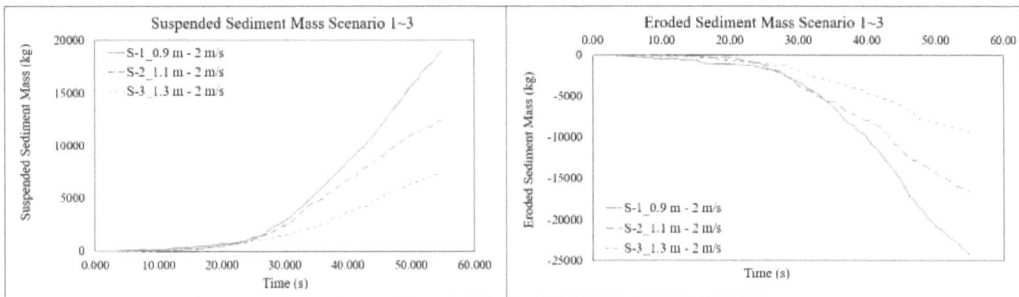

Fig. (7). Variation of the total amount of the sediment suspension and amount of the eroded sediment according to the first three scenarios.

When the under-keel clearance decreases, the amount of suspended sediment and also sediment transport at the seabed increase. As can be seen from the model results, particularly high-rate sediment transport occurs in the slope of the channel.

In the second three scenarios, it is observed that the sediment transport and suspension reduce due to the decrease in the velocity of the floating object. Fig. (**8**) shows the time variation of the total amount of sediment suspension and the amount of the eroded seabed sediment.

In the last three scenarios, sediment transport and suspension are reduced prominently according to the reduction of the floating object velocity. Suspended sediment concentration can be seen in the Fig. (**9**) according to the last three scenarios.

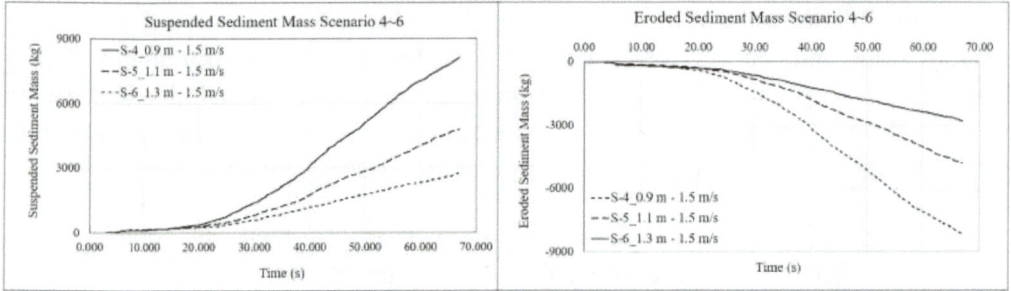

Fig. (8). Variation of the total amount of the sediment suspension and amount of the eroded sediment according to the fourth, fifth, and sixth scenarios.

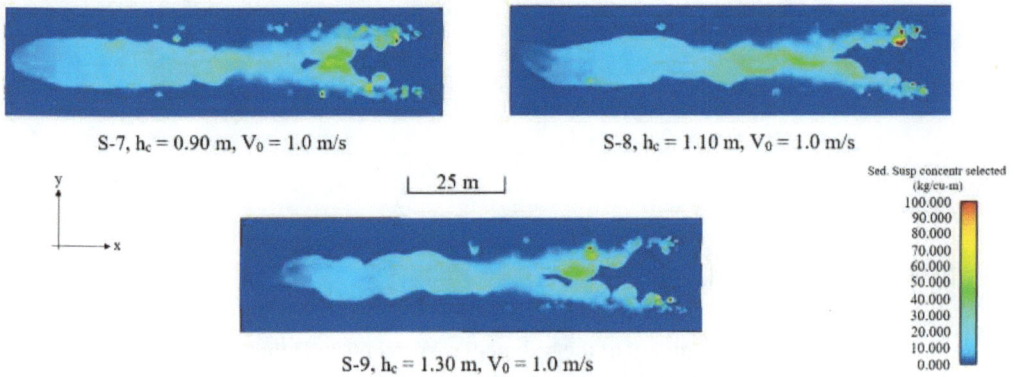

S-7, h_c = 0.90 m, V_0 = 1.0 m/s   S-8, h_c = 1.10 m, V_0 = 1.0 m/s

S-9, h_c = 1.30 m, V_0 = 1.0 m/s

Fig. (9). Suspended sediment concentration according to last three scenarios (plan view).

Also, sea-bottom morphology change at the last time step can be seen in the Fig. **(10)**.

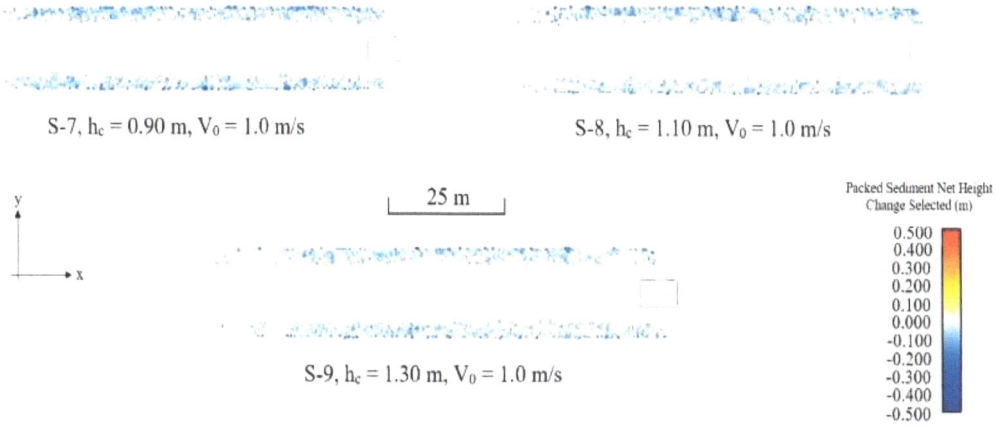

S-7, $h_c = 0.90$ m, $V_0 = 1.0$ m/s

S-8, $h_c = 1.10$ m, $V_0 = 1.0$ m/s

25 m

S-9, $h_c = 1.30$ m, $V_0 = 1.0$ m/s

Packed Sediment Net Height Change Selected (m)

Fig. (10). Seabed morphology change according to last three scenarios (plan view).

Fig. (**11**) shows time variation of the total amount of sediment suspension and the amount of the seabed sediment, respectively.

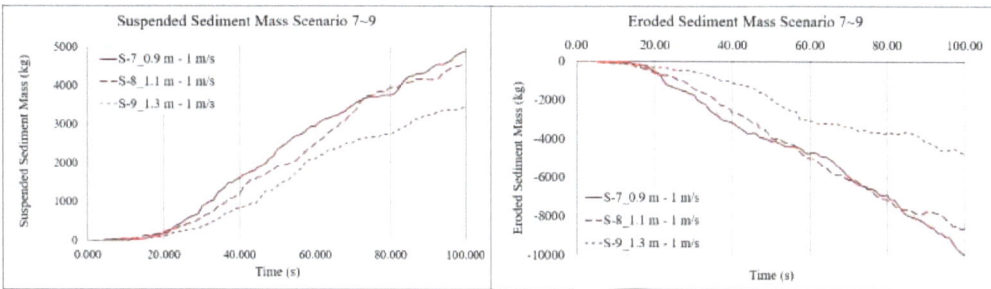

Fig. (11). Variation of the total amount of the sediment suspension and amount of the eroded sediment according to last three scenarios.

According to all scenarios, variation of the total amount of suspended sediment and amount of the eroded sediment can be seen in Fig. (**12**).

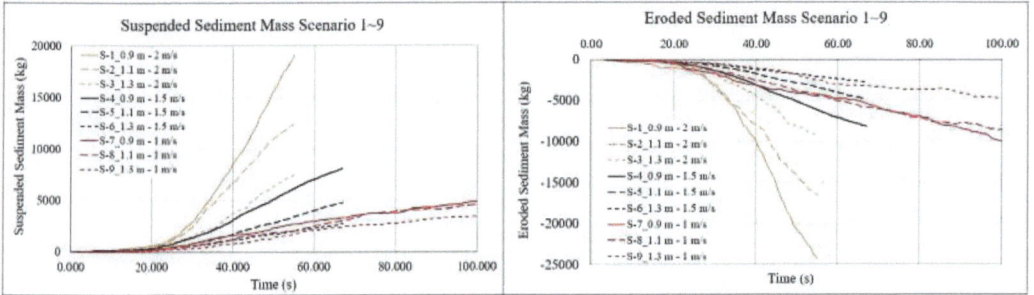

Fig. (12). Variation of the total amount of the sediment suspension and amount of the eroded sediment.

The amount of suspended sediment was decreased considerably with decreasing the floating object velocity. In the last three scenarios, the rate of sediment transport occurs in the slope of the approach channel, which is considerably reduced compared to the first three scenarios. In addition, the amount of the suspended sediment is approached with decrease in the velocity of the floating object. In other words, as the velocity of the floating object decreases, the importance of the vertical distance of the floating object to the seabed decreases in terms of suspended sediment amount.

The velocity of the floating object is 2 m/s in the first scenario, 1,5 m/s in the fourth scenario, 1 m/s in the seventh scenario and as well as depending on these velocities the total mass of suspended sediment is about 19000 kg in the first scenario, 8000 kg in the fourth scenario and 4800 kg in the seventh scenario. As it is understood from the values, the blockage volume of the front of the floating object increases with the velocity of the floating object, and also the return current velocity increases with increasing the blockage ratio.

CONCLUDING REMARKS

In the scope of this study, hydro-morphodynamic effects of the floating object navigating in the approach channel are investigated according to its distance from the bottom of the channel and the variation of its speed. As the floating object closes to the bottom of the channel, floating object-induced return currents increase, and accordingly, sediment transport and suspension are observed on a large scale on the slope of the approach channel because of increasing shear stresses and wake region behind the floating object. On the other hand, as the speed of the floating object is decreased, the amount of transported and suspended sediment, especially on the slopes, decreases, and also the total amount of suspended sediments is approached each other (as can be seen in Fig. **12**). In other words, as the speed of the floating object is decreased, the importance of the vertical distance of the floating object to the seabed is diminished in terms of the amount of suspended sediment. In the seventh scenario, (the under-keel clearance is 0,90 m and floating object velocity is 1 m/s), the amount of suspended sediment is decreased by about 75% compared to the first scenario (the under-keel clearance is 0,90 m and floating object velocity is 2 m/s).

This study is the first impression of an ongoing doctoral thesis, and the investigation of the subject is still continuing. The entire study will carry out on 96 different scenarios. The aim of future studies is to generalize a mathematical formula to represent the non-powered floating object-induced sediment suspension.

CONSENT FOR PUBLICATION

Not applicable.

CONFLICT OF INTEREST

The authors declare no conflict of interest, financial or otherwise.

ACKNOWLEDGEMENT

This work was supported by Balıkesir University Research Grant No: 2019/021.

REFERENCES

[1]   M. McBride, *et al.*, "Harbour approach channels—Design guidelines," PIANC Report No. 121. World Association for Waterborne Transport Infrastructure, Brussels, Belgium, 2014.

[2]   V. Bertram, *Practical Ship Hydrodynamics, Second Edition.* Oxford, UK: Butterworth-Heinemann, 2012.

[3]   G. J. Schiereck, *Introduction to Bed, Bank and Shore Protection*, Tylor & Fr. London: Spon Press, 2004.

[4]   J. Bouwmeester, "Recent studies on push-towing as a base for dimensioning waterways" in PIANC, XXIVth International Navigation Congress, Leningard, Sptember 6-14, 1977, pp 1-54.

[5]   P. D. Osborne and E. H. Boak, "Sediment suspension and morphological response under vessel-generated wave groups: Torpedo Bay, Auckland, New Zealand," *J. Coast. Res.*, vol. 15, no. 2, pp. 388–398, 1999.

[6]   G. C. Nanson, A. Von Krusenstierna, E. A. Bryant, and M. R. Renilson, "Experimental measurements of river-bank erosion caused by boat-generated waves on the gordon river, Tasmania," *Regul. Rivers Res. Manag.*, vol. 9, no. 1, pp. 1–14, 1994, doi:

10.1002/rrr.3450090102.

[7]   T. Aagaard and B. Greenwood, "Suspended sediment transport and the role of infragravity waves in a barred surf zone," *Mar. Geol.*, vol. 118, no. 1–2, pp. 23–48, 1994, doi: 10.1016/0025-3227(94)90111-2.

[8]   C. Houser, "Sediment resuspension by vessel-generated waves along the Savannah River, Georgia," *J. Waterw. Port, Coast. Ocean Eng.*, vol. 137, no. 5, pp. 246–257, 2011, doi: 10.1061/(ASCE)WW.1943-5460.0000088.

[9]   J. Rapaglia, L. Zaggia, K. Ricklefs, M. Gelinas, and H. Bokuniewicz, "Characteristics of ships' depression waves and associated sediment resuspension in Venice Lagoon, Italy," *J. Mar. Syst.*, vol. 85, no. 1–2, pp. 45–56, 2011, doi: 10.1016/j.jmarsys.2010.11.005.

[10]   S. Ji, A. Ouahsine, H. Smaoui, and P. Sergent, "3D numerical simulation of ship-induced waves and sediment transport in restricted waterways," *Houille Blanche*, vol. 2014-Decem, no. 6, pp. 68–73, 2014, doi: 10.1051/lhb/2014065.

[11]   M. Schroevers, K. D. Berends, T. Vermaas, and H. J. Verheij, "Measuring ship-induced currents in a canal," *2015 IEEE/OES 11th Curr. Waves Turbul. Meas. CWTM 2015*, pp. 2–7, 2015, doi: 10.1109/CWTM.2015.7098126.

[12]   J. A. McConchie and I. E. J. Toleman, "Boat wakes as a cause of riverbank erosion: A case study from the Waikato River, New Zealand," *J. Hydrol. New Zeal.*, vol. 42, no. 2, pp. 163–179, 2003.

[13]   H. Smaoui, A. Ouahsine, D. P. Van Bang, P. Sergent, and F. Hissel, "Numerical Modelling of the Sediment Re-Suspension Induced by Boat Traffic," in *Sediment Transp.*, S. S. Ginsberg, Ed., In Tech, 2011, pp. 55-70, 2014, 2011, doi: 10.5772/16053.

[14]   R. J. Lenselink, "Interaction between loaded barges and bed material," M. S. thesis, Dept. Hydro. Eng., Delft Univ., Netherlands, 2011.

[15]   S. T. Maynord, "Velocities induced by commercial navigation," Vicksburg, MS: USACE Waterways Experiment Station, USA, Tec. Rep. TR-90-15, 1990.

[16]   S. T. Maynord, "Physical forces near commercial tows," Vicksburg, MS: USACE, Env. Rep. 19, 2000.

[17]   R. L. Stockstill, S. K. Martin, and R. C. Berger, "Hydrodynamic model of vessel-generated currents," *Regul. Rivers Res. Manag.*, vol. 11, July 1994, pp. 211–225, 1995.

[18]   Flow Science Inc., *Flow-3D User Manual Release 11.2*, 2016.

[19]   V. Yakhot and L. M. Smith, "The renormalization group, the ε-expansion and derivation of turbulence models," *J. Sci. Comput.*, vol. 7, no. 1, pp. 35–61, 1992, doi: 10.1007/BF01060210.

[20]   C. W. Hirt and B. D. Nichols, "Volume of fluid (VOF) method for the dynamics of free boundaries," *J. Comput. Phys.*, vol. 39, pp. 201–225, 1981, doi: 10.1016/0021-9991(81)90145-5.

SUBJECT INDEX

A

Acceleration 232, 236, 245
  gravitational 245
Acquired immune deficiency syndrome 125
Activity 125, 126, 127, 186
  reverse transcriptase 125
Adams-Bashforth-Moulton method 186, 189, 194
Adomian decomposition method 129
Algebraic 162, 167, 178, 182
  equation system 162, 178, 182
  system 167
Algorithm 2, 210, 213, 227
  hired 213
Analysis 30, 63, 162
  epidemiological 30
  theoretical 63
  wave behavior 162
Analytical 3, 6, 31, 44, 49, 68, 108, 109, 121, 144, 163, 165
  solutions 3, 6, 31, 44, 49, 68, 108, 109, 121, 163, 165
  techniques 144
Applications 1, 4, 188
  illustrative 188
  innovative 4
  real-world 1
Applying Laplace transform 108
Asymptotic stability, local 15
Attractors 1, 3, 5, 19, 20
  hyperchaotic 5

B

Bed-load transport 236, 241
  rate equation 241
Behaviors 1, 5, 8, 11, 13, 17, 18, 19, 61, 62, 63, 75, 83, 110, 115, 128, 142, 180
  actual macroeconomic 75
  anomalous 110
  chaos's 11

complicated 61, 63, 83
  dynamic 17
  hyperchaotic 1
  mathematical 62
Bifurcation 1, 2, 3, 5, 12, 13, 14, 17, 18, 25, 26
  period-doubling 14
  maps 1, 2, 3, 5, 12, 13, 14, 17, 18
Boundary conditions 85, 87, 88, 99, 100, 106, 111, 120, 237, 240
  applying 237
  harmonic 120
  time-dependent 85, 87, 88, 99, 106

C

Calculus 30, 32, 63, 212
  classical 212
  integral 63
  traditional 30, 32
Caputo 2, 3, 4, 5, 8, 9, 10, 11, 12, 13, 16, 17, 23, 30, 31, 63, 67, 75, 85, 91, 186, 188, 212
  classical 91
  derivative of order 63
  sense 30, 31, 67, 186, 188
  type fractional operator 85
Caputo-Fabrizio 61, 68, 77, 78, 83, 85, 86, 90, 91, 93, 96, 98, 100, 103, 106
  derivatives 77, 78, 83, 93, 96, 106
  non-integer order 83
  operators 85, 86, 90, 91, 96, 98, 100, 103, 106
  sense 61, 68
  transform 90
Caputo-Fabrizio fractional 74, 75, 86, 105, 108, 213
  operator 86
  order 75
  technique 74
Caputo fractional 30, 190
  derivative operator (CFDO) 30

www.ingramcontent.com/pod-product-compliance
Lightning Source LLC
Chambersburg PA
CBHW050819220326
41598CB00006B/260